Autonomous Vehicles and Robotics

Autonomous Vehicles and Robotics

Guest Editors

Nan Ma
Taohong Zhang
Yang Yang

MDPI

Basel • Beijing • Wuhan • Barcelona • Belgrade • Novi Sad • Cluj • Manchester

Guest Editors

Nan Ma
Faculty of Information and Technology
Beijing University of Technology
Beijing
China

Taohong Zhang
School of Computer and Communication Engineering
University of Science and Technology Beijing
Beijing
China

Yang Yang
School of Electronic and Information Engineering
Xi'an Jiaotong University
Xi'an
China

Editorial Office
MDPI AG
Grosspeteranlage 5
4052 Basel, Switzerland

This is a reprint of the Special Issue, published open access by the journal *Applied Sciences* (ISSN 2076-3417), freely accessible at: https://www.mdpi.com/journal/applsci/special_issues/682FNCX21X.

For citation purposes, cite each article independently as indicated on the article page online and as indicated below:

Lastname, A.A.; Lastname, B.B. Article Title. *Journal Name* **Year**, *Volume Number*, Page Range.

ISBN 978-3-7258-7557-3 (Hbk)
ISBN 978-3-7258-7558-0 (PDF)
https://doi.org/10.3390/books978-3-7258-7558-0

Contents

Preface

This Special Issue addresses the ongoing transformation of the automobile industry, driven by the transition to electric engines, autonomous driving, and mobility as a service. Its scope covers autonomous vehicle software, robotic perception, machine vision, intelligent interaction, and related technologies. The Special Issue presents the latest research on software vulnerability assessment, risk and threat analysis, and advanced visual and interactive systems, highlighting both challenges and opportunities in the development of safe, efficient, and connected mobility. Motivated by the need for cleaner energy, safer transportation, and more effective use of urban infrastructure, this work provides a comprehensive resource for researchers, engineers, policymakers, and industry professionals interested in the future of automotive technologies and smart mobility solutions.

Nan Ma, Taohong Zhang, and Yang Yang
Guest Editors

applied sciences

MDPI

Article

Visual Odometry of a Low-Profile Pallet Robot Based on Ortho-Rectified Ground Plane Image from Fisheye Camera

Soon-Yong Park [1,*], Ung-Gyo Lee [2] and Seung-Hae Baek [3]

1 School of Electronic and Electrical Engineering, Kyungpook National University, 80 Daehak-ro, Puk-gu, Daegu 41566, Republic of Korea
2 HL Klemove, 224 Harmony-ro, Yeonsu-gu, Incheon 22011, Republic of Korea; dndry123@naver.com
3 KLA Corporation, 830 Dongtansunwhan-daero, Hwaseong-si 18468, Republic of Korea; eardrop77@gmail.com
* Correspondence: sypark@knu.ac.kr; Tel.: +82-53-950-7575

Abstract: This study presents a visual-only odometry technique of a low-profile pallet robot using image feature tracking in ground plane images generated from a fisheye camera. The fisheye camera is commonly used in many robot vision applications because it provides a larger field of view (FoV) around a robot. However, because of the large radial distortion, the fisheye image is generally converted to a pinhole image for visual feature tracking or matching. Although the radial distortion can be eliminated via image undistortion with the lens calibration parameters, it causes several side effects, such as degraded image resolution and a significant reduction in the FoV. In this paper, instead of using the pinhole model, we propose to generate a ground plane image (GPI) from the fisheye image. GPI is a virtual top-view image that only contains the ground plane at the front of the robot. First, the original fisheye image is projected to several virtual pinhole images to generate a cubemap. Second, the front and bottom faces of the cubemap are projected to a GPI. Third, the GPI is homographically transformed again to further reduce image distortion. As a result, an accurate ortho-rectified ground plane image is obtained from the virtual top-view camera. For visual odometry using the ortho-rectified GPI, a number of 2D motion vectors are obtained using feature extraction and tracking between the previous and current frames in the GPI. By calculating a scaled motion vector, which is the measurement of the virtual wheel encoder of the mobile robot, we estimate the velocity and steering angle of the virtual wheel using the motion vector. Finally, we estimate the pose of the mobile robot by applying a kinematic model to the mobile robot.

Keywords: fisheye lens; cubemap; ground plane image; pallet mobile robot

1. Introduction

Nowadays, almost all robots utilize visual information for their tasks such as grasping, random bin picking, autonomous navigation, and so on. In these robotic applications, visual information, similar to human vision, provides plenty of important and useful information to robots. For this reason, more and more computational or artificial vision techniques have been developed. In particular, in the field of autonomous vehicles or mobile robots, obtaining visual information with a large field of view (FoV) is advantageous because it can reduce the blind areas around the vehicles or robots. In this regard, using the fisheye camera is an efficient way for a robot to obtain images with a large field of view. In addition, a small number of fisheye cameras can be used to obtain around-view images from the robot rather than using normal FoV cameras. Therefore, the fisheye camera is commonly used nowadays in many robotic applications.

As one of the robotic applications, the vision-based or visual localization of a mobile robot is an essential technology in autonomous mapping and navigation. However, for certain types of robots, such as pallet or cleaning robots, it is inevitable that fisheye cameras are attached close to the ground level, as shown in Figure 1. Consequently, the ground area occupies half of the captured fisheye image. In this situation, vision-based localization is challenging mainly because of two reasons. First, half of image feature extraction and tracking is conducted in the ground area, while the texture pattern on the ground is generally uniform compared with other cluttered areas. Common feature extraction algorithms in computer vision are subject to extract a smaller number of features in the uniform ground area than the other surrounding areas. Thus, the ground area in a fisheye image is not cost effective in terms of feature extraction and tracking. Second, the undistortion of the fisheye image to a pinhole image causes inevitable image loss along the boundary of the original image. As a result, only the center area of the original image, where half of the center area is also the ground area, can be used in the undistorted pinhole image for feature extraction and tracking.

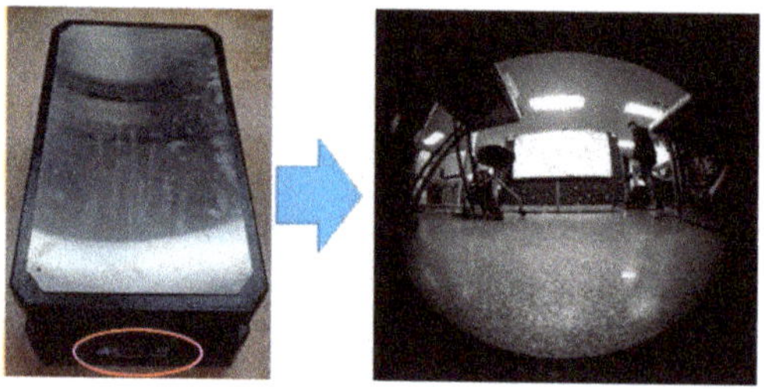

Figure 1. A fisheye image from a low-profile robot. (**Left**): A fisheye camera is mounted at the front of a pallet robot. (**Right**): A sample image from the fisheye camera.

Even though there are several disadvantages to using the fisheye camera in the low-profile robot, the ground information can be employed efficiently for visual navigation by assuming that the ground is a planar surface. Because almost all indoor mobile robots are assumed to move on the planar ground plane, the motion direction of a mobile robot can be considered parallel to the ground plane. This study is also motivated by the fact that the low-profile robot moves on a 2D planar space, which is parallel to the ground plane, and the fisheye camera captures the ground images at a very close distance to the ground.

In this study, we propose using a visual odometry method with a fisheye camera for low-profile pallet robots. By only using the ground plane images generated from the fisheye camera, the pose of the robot is estimated. The proposed method consists of several steps, as shown in Figure 2. First, a fisheye image is converted to a cubemap image [1]. Second, a ground plane image (GPI) is obtained from the cubemap image. GPI is a virtual top-view image generated by the bottom and front faces of the cubemap. Third, a ground plane image is again converted to an ortho-rectified GPI to compensate for any distortion error due to lens distortion parameters. Fourth, image feature points are extracted and tracked using the optical flow in the ortho-rectified GPI. The robot motion is then modeled as a 2D vector in the image space [2]. Finally, as shown in Figure 2, the odometry of the robot is estimated by applying the bicycle kinematic model with the motion vector between consecutive image frames [3]. The architecture of the proposed visual odometry is named Ground Plane Odometry (GPO). In several experiments using a low-profile pallet robot, we compare the localization performance with T265 stereo odometry, GPO, and the Extended Kalman Filtering (EKF) of GPO and T265 [4,5]. By applying EKF with GPO and T265, we achieve a 0.058 m mean localization error in meters based on three indoor odometry experiments.

The structure of this study is as follows. In Section 2, we briefly describe conventional studies related to this paper. In Section 3, we explain the details of each step of the proposed method. In Section 4, we show several experiments and evaluation results. Finally, in Section 5, we conclude this paper.

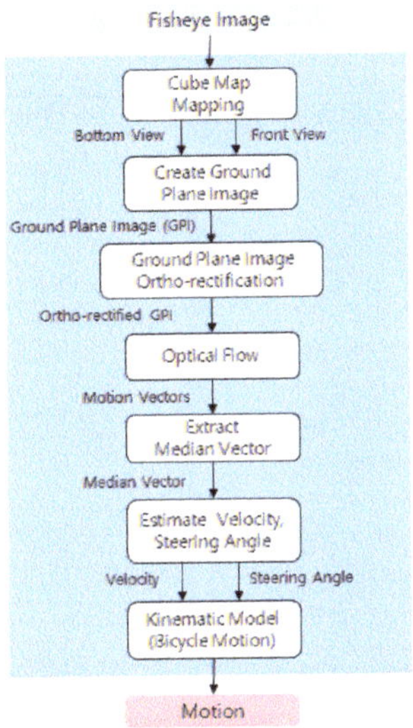

Figure 2. The flow diagram of the proposed visual odometry method.

2. Literature Review

In this section, we present the previous literature related to the visual odometry techniques of mobile robots, which employ image features on the surface of the ground plane with fisheye or wide-angle lens cameras.

M. Flores et al. [6] compare the visual localization performance of four well-known image features, which are extracted from fisheye images: Oriented FAST and Rotated BRIEF (ORB) [7], Speed-Up Robust Features (SURF) [8], Features from Accelerated Segment Test (FAST) [9], and KAZE [10]. They measure the translation and rotation errors of visual odometry computed from two different approaches: the standard visual odometry and the adaptive probability-oriented feature matching methods. A fisheye image dataset by Zhang et al. [11] is used for the experiment. The experimental results show that the performance of ORB and KAZE is relatively better than that of SURF and FAST.

W. Yahui et al. [1] propose a novel visual SLAM (Simultaneous Localization and Mapping) method using cubemap images. They employ ORB-SLAM [12] as the baseline of their proposed CubemapSLAM method. Cubemap images are consecutively generated from fisheye images, and they are used as the input of ORB-SLAM. The cubemap camera model can use all image information of the fisheye camera without affecting the performance of feature descriptors. In this study, CubemapSLAM is efficiently implemented, and it can run in real time. Despite the limited angular resolution of the sensor, CubemapSLAM shows better accuracy in localization compared with the pinhole camera model [13].

A. J. Swank [14] utilizes a single downward-pointing camera for the localization of a mobile robot. Image features are obtained from the camera, which is equipped with a wide-angle lens. The optical flow algorithm is employed to measure the location of the robot. In this study, the application of the proposed method was specifically targeted to robotic platforms in unseen areas. The author suggests applying the proposed method to any indoor or outdoor area when other localization sensors are not available.

J. Zhang et al. [15] present a robust localization method based on monocular visual odometry. Accurate position estimation is presented even when a mobile robot is operated in undulating terrain. Their method employs a steering model to separately estimate the rotation and translation of the robot. Moreover, they also address the localization problem in an undulating terrain. The undulating terrain is approximated by multiple patches of locally flat surface. In each local flat surface, the inclination of the surface is estimated at the same time as the robot's motion. The surface inclination accuracy is about 1% error.

J. Jordan et al. [16] present a visual odometry technique of a mobile robot that is working on a 2D plane by applying the kinematic model. A monocular camera facing downward is used to capture the images of the floor. The motion of the robot is estimated by the alignment between two consecutive images. The motion is then combined with the kinematic model of the robot to remove the outlier image regions that do not reflect the actual motion. To obtain the ortho-rectified floor images, they use a simple image homography method.

D. Zhou et al. [17] address a ground plane-based odometry for the purpose of autonomous vehicle driving. They use a single normal vision camera to capture the ground image in front of a road-driving vehicle. To estimate the vehicle motion in real metric space, they propose a divide-and-conquer motion scale estimation method based on the ground plane and camera height. Their investigation is a typical example that the ground plane-based odometry can be used in the autonomous vehicle driving technique.

J. Jordan et al. [18] introduce a ground plane-based visual odometry using a downward-facing RGBD camera. The RGBD camera captures the ground images and 3D point clouds frame by frame, and the images and points are transformed into a virtual orthogonal camera. Then, the projected data are split into multiple orthographic grid data blocks. Each image block is registered frame by frame using Efficient Second-order Minimization (ESM) [19]. Finally, the robot motion is estimated by the weighted average of all block motions. This approach uses an RGBD camera; thus, 3D ground information helps to find accurate robot motion.

M. Ouyang et al. [20] integrate the gyroscope and wheel encoder information into visual odometry to reduce the odometry drift problem during planar motion. They use the gyroscope information to compensate for the orientation from the wheel encoder. In addition, the ground plane from camera images is segmented to extract the image features on the ground. As the backend of the odometry measurement, the graph optimization is used to minimize the measurement residuals of wheel encoder, gyroscope, vision, and ground feature constraint, respectively.

X. Chen et al. [21] introduce a SLAM technique called StreetMap. For the localization and mapping from a mobile robot, they use a downward-facing camera to utilize the ground textures. They categorize the ground textures into two groups: feature based and line based. In each category, they proposed different localization mapping methods. They used the downward-facing camera to obtain the ground texture images. However, no image transformation, such as ortho-rectification, is used.

H. Lin et al. [22] address the self-localization of a mobile robot using a single catadioptric camera. The geometric property of the catadioptric camera projection model is utilized to obtain the intersection of vertical lines and the ground plane. From the catadioptric camera image, the ground plane is segmented and followed by the vertical line extraction. Using the geometry of the catadioptric camera projection model, the 3D line equation corresponding to the extracted vertical image line is calculated. Then, the robot motion between two consecutive frames is estimated using the relationship of the 3D vertical lines.

In our previous investigation, we design a pallet-type mobile robot and mount a fisheye camera at the front of the robot to calculate the visual odometry [23]. The fisheye image is first converted to a cubemap and next a ground plane image to track the image pattern on the ground. Using the motion of the feature tracking, the visual odometry of the robot is calculated.

3. Proposed Approach

3.1. Overview

The algorithm steps of the proposed method are shown in Figure 3, and it is summarized as follows. As the first step, a fisheye image of the front view of a pallet robot is converted to a cubemap image. From the cubemap image, we generate a virtual pinhole camera image, Ground Plane Image (GPI), which is introduced in our previous study [23]. GPI contains only the images of the ground plane at the front of the robot as if a virtual pinhole camera is facing perpendicular to the ground plane, as shown in Figure 3. Ideally, GPI must be the ortho-rectified image captured from the virtual top-view camera. However, due to the image transformation error both from the fisheye to the cubemap and from cubemap to GPI, it is not perfectly ortho-rectified. The transformation error is mainly due to errors in fisheye lens modeling and errors in homography computation between the cubemap and GPI.

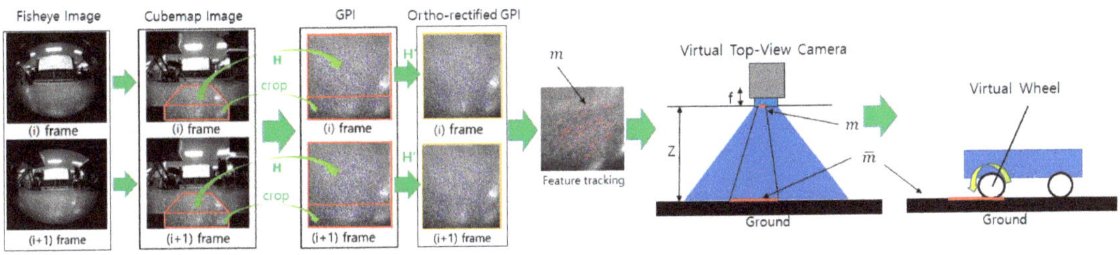

Figure 3. The algorithm steps of the proposed Ground Plane Odometry.

To perfectly generate an ortho-rectified GPI, we also propose to transform the ground plain image again to the ortho-rectified image (ortho-rectified GPI). In ortho-rectified GPI, the ground plane is perfectly rectified as the top-view camera captures perpendicular to the ground. As the next step, we track image features in ortho-rectified GPI using the Kanade–Lucas–Tomashi (KLT) optical flow algorithm to determine the motion vector of the robot [24]. The motion vector between two GPIs is then scaled to determine the robot's motion in the metric space. Then, from the scaled vector, the velocity and steering angle of a virtual front wheel of the robot are estimated. Because the fisheye camera is mounted at the front of the robot, we employ a kinematic bicycle model for the motion estimation of the virtual front wheel of the robot [3].

3.2. Calibration of Fisheye Camera Using EUCM

In normal visual odometry, using a fisheye image as the original form is not common because of the complexity of multi-view geometry. Thus, usually, the fisheye camera model is converted to the pinhole model to simplify the multi-view geometry. However, transforming the fisheye model to the pinhole model causes a serious distortion in the undistorted pinhole image. In addition, only the center part of the undistorted image can be used in the rasterization step, and the remaining image areas are removed.

Therefore, instead of the pinhole model, we use the Cubemap [1] model to utilize all image areas after undistortion. To transform the fisheye image to a cubemap image, all calibration parameters of the fisheye camera must be calibrated. As the calibration model, we use the enhanced unified camera model (EUCM) [25], which is a nonlinear camera model that is generally used in omnidirectional cameras such as catadioptric systems or fisheye cameras. The EUCM has six parameters, $[f_x, f_y, c_x, c_y, \alpha, \beta]$. The curvature of the fisheye lens is calculated using α and β. Therefore, by using these parameters, the EUCM can represent the fisheye camera distortion. Figure 4 shows the image projection model to the image plane through the fisheye lens. A 3D point $\boldsymbol{X}$ is projected on a curved plane P to a 3D point X_P, as shown in Equations (1)–(5).

$$\boldsymbol{X} = [x\ ,\ y\ ,z\ ], \tag{1}$$

$$\rho = \sqrt{\beta(x^2 + y^2) + z^2}, \tag{2}$$

$$\boldsymbol{X}_P = [x_P, y_P, z_P], \tag{3}$$

$$x_P = \frac{x}{\alpha\rho + (1-\alpha)z}, \tag{4}$$

$$y_P = \frac{y}{\alpha\rho + (1-\alpha)z} \tag{5}$$

Then, X_P is orthogonally projected into the M plane, which is the normal image plane, as shown in Equation (6).

$$\mathbf{X_M} = [x_P, y_P,\ 1] = [x_M, y_M,\ 1]. \tag{6}$$

Finally, as shown in Equation (7), we multiply the normal plane point $\mathbf{X}_M$ with the intrinsic matrix K to obtain the projection pixel $\mathbf{X}_I$ on the fisheye image.

$$\mathbf{X_I} = K\mathbf{X_M} = \begin{bmatrix} f_x & 0 & c_x \\ 0 & f_y & c_y \\ 0 & 0 & 1 \end{bmatrix} \mathbf{X_M} = \begin{bmatrix} u \\ v \\ 1 \end{bmatrix}. \tag{7}$$

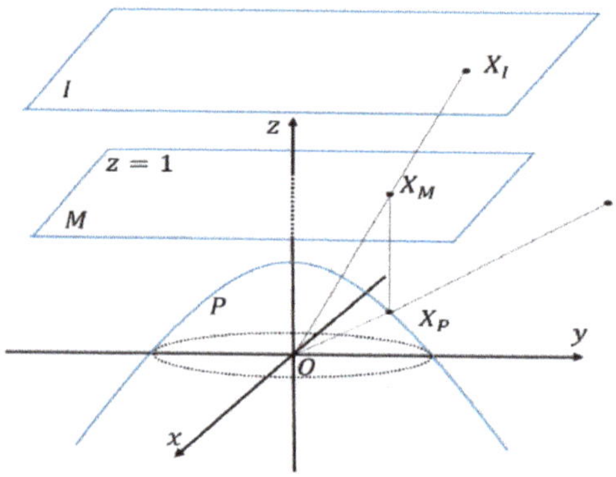

Figure 4. A graphical model of the enhanced unified camera model.

In this study, to calibrate the fisheye camera's EUCM, we use a calibration toolbox Kalibr [26–30], which is extremely useful for camera calibration. Kalibr offers a variety of camera models, including EUCM, pinhole model, omnidirectional model, and double sphere models.

3.3. Mapping Fisheye Image to Cubemap Image

After calibrating the fisheye camera, we use the calibration parameters to generate the cubemap image from the fisheye image. As shown in Figure 5a, the cubemap model consists of five image planes generated by virtual pinhole camera models. Each virtual pinhole model has the same intrinsic but different extrinsic parameters because of different viewing directions. Thus, the fisheye image is divided into five virtual pinhole cameras. Meanwhile, the intrinsic parameters of the virtual cubemap depend on the size of each cubemap face. We calibrate the intrinsic parameters $[f_{Face\_x}, f_{Face\_y}, c_{Face\_x}, c_{Face\_y}]$ of the virtual pinhole cameras so that the pinhole image resolution is half of the cubemap image resolution. Here, f is the focal length, c is the principal point, and the subscript $Face$ is the index of the five cubemap faces such that $Face = \{Front,\ Left,\ Right,\ Bottom,\ and\ Top\}$.

Figure 5b shows the relationship between the cubemap image plane and the cubemap model. To generate a cubemap image, we first define the transformation from the cubemap image coordinates to the cubemap model coordinates, as shown in Equation (14). Here, R_{Face} is a 3D rotation matrix of the index $Face$ cubemap plane.

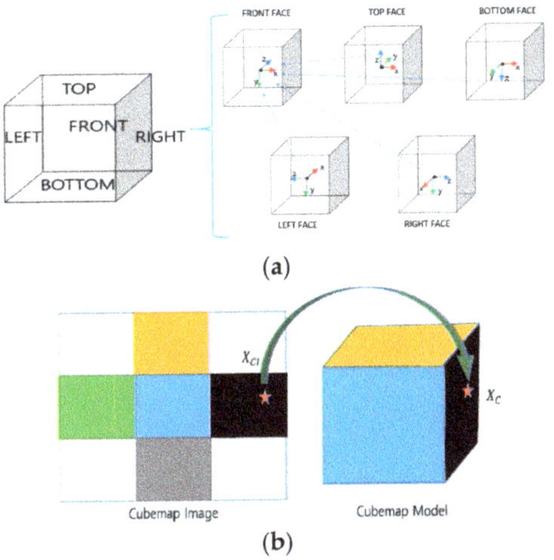

Figure 5. (**a**) Illustration of cubemap model. (**b**) Illustration of relation between cubemap image and cubemap model.

$$R_{Front} = \begin{bmatrix} 1 & 0 & 0 \\ 0 & 1 & 0 \\ 0 & 0 & 1 \end{bmatrix}, \tag{8}$$

$$R_{Left} = \begin{bmatrix} 0 & 0 & -1 \\ 0 & 1 & 0 \\ 1 & 0 & 0 \end{bmatrix}, \tag{9}$$

$$R_{Right} = \begin{bmatrix} 0 & 0 & 1 \\ 0 & 1 & 0 \\ -1 & 0 & 0 \end{bmatrix}, \tag{10}$$

$$R_{Bottom} = \begin{bmatrix} 1 & 0 & 0 \\ 0 & 0 & 1 \\ 0 & -1 & 0 \end{bmatrix}, \tag{11}$$

$$R_{Top} = \begin{bmatrix} 1 & 0 & 0 \\ 0 & 0 & -1 \\ 0 & 1 & 0 \end{bmatrix}, \tag{12}$$

$$\mathbf{X}_{CI} = [\mathrm{u},\ \mathrm{v},\ 1], \tag{13}$$

$$\mathbf{X}_C = R_{Face}\,\mathbf{X}_{CI}. \tag{14}$$

We transform the cubemap image coordinates $\mathbf{X}_{CI}$ into the model coordinates $\mathbf{X}_C$, which is on one of the cubemap faces, as shown in Figure 6. The concept of the EUCM-Cubemap projection model involves the use of the EUCM parameters to transform the point of the cubemap face to the pixel of the fisheye image or vice versa. First, we transform point $\mathbf{X}_C$ to another point $\mathbf{X}_P$ on the curve plane P of the EUCM, as shown from Equations (15) to (18). We replace $\mathbf{X}$ in Equation (1) with $\mathbf{X}_C$. Here, the back-projection z_P is not required in this step.

$$\mathbf{X}_C = [x_C, y_C, z_C], \tag{15}$$

$$\rho = \sqrt{\beta(x_C{}^2 + y_C{}^2) + z_C{}^2}, \tag{16}$$

$$x_P = \frac{x_C}{\alpha\rho + (1-\alpha)z_C}, \tag{17}$$

$$y_P = \frac{y_C}{\alpha\rho + (1-\alpha)z_C}. \quad (18)$$

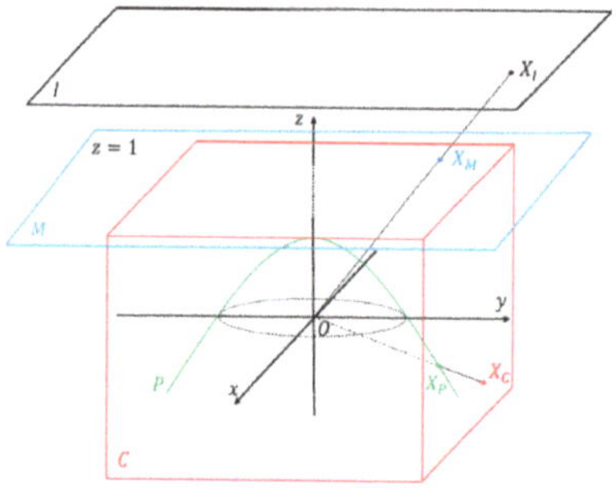

Figure 6. Illustration of enhanced unified camera model based cubemap projection model.

Next, we project $\boldsymbol{X_P}$ to $\boldsymbol{X_M}$ on the EUCM normal image plane M using orthogonal projection, as shown in Equation (6). Finally, $\boldsymbol{X_M}$ is projected to a pixel point $\boldsymbol{X_I}$ on the fisheye image by multiplying it with the intrinsic matrix K, as shown in Equation (7).

By following this process, pixel coordinates in the fisheye image are transformed into the cubemap image coordinates. Using the backward transformation from the cubemap image to the fisheye image, we can identify the correspondences between the cubemap and the fisheye images. Because this process is computationally expensive, the coordinate transformation was performed once when the proposed method starts. Then, the index table of the pixel coordinates mapping between images is stored and used. Figure 6 shows the EUCM–cubemap projection model, and Figure 7 shows the difference between the fisheye image and the cubemap image projections.

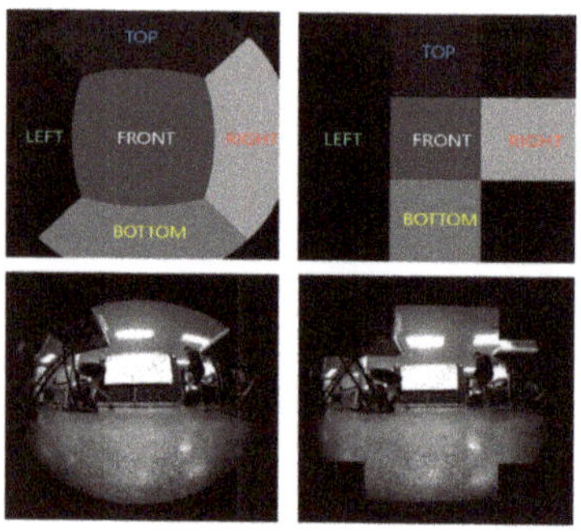

Figure 7. Illustration of relation between fisheye image and cubemap image projection.

3.4. Generating GPI from Cubemap Image

This section describes how to generate the proposed ground plane image (GPI) from the cubemap image. As shown in Figure 8, the front and the bottom faces in the cubemap model are used to generate GPI, which is an ortho-rectified image from a virtual top-view camera. To generate a GPI in the front face, as shown in Figure 8, the front-face image is warped to the ground plane using a homography matrix. We project a point $\boldsymbol{q}$ on the front-face image to a homogeneous point $\boldsymbol{Q}$ on the ground plane image using Equation (22).

$$Q = \begin{bmatrix} X \\ Y \\ 1 \end{bmatrix}, \quad (19)$$

$$q = \begin{bmatrix} u \\ v \\ 1 \end{bmatrix}, \quad (20)$$

$$H = \begin{bmatrix} h_{11} & h_{12} & h_{13} \\ h_{21} & h_{22} & h_{23} \\ h_{31} & h_{32} & 1 \end{bmatrix}, \quad (21)$$

$$Q \cong Hq. \quad (22)$$

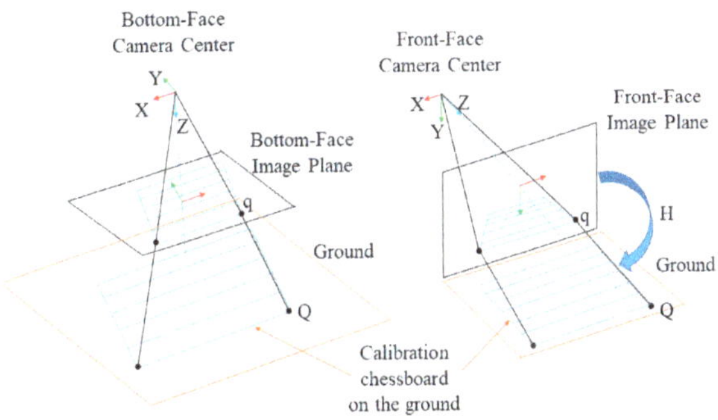

Figure 8. Illustration of the relationship between the ground plane and front-face and bottom-face images, respectively.

In Equation (22), the symbol ≅ means that the left and the right terms are homogeneously the same. The homography matrix H is calculated from the relationship between the four corner positions of a chessboard. The chessboard is placed on the ground, as shown in Figure 8, and the cubemap of the chessboard is generated. The four corners of the chessboard are selected from the cubemap image.

The bottom-face image is just cropped because the bottom-face is already obtained using a virtual top-view camera, as shown in Figure 5a. Figure 9 shows an example of GPI generation from a fisheye image. A trapezoid ground area in the front face of the cubemap image is warped to a rectangular area in a GPI, and a cropped bottom-face area is concatenated at the bottom of the GPI. In an ideal case, the generated ground plane image shows the ortho-rectified image from a virtual top-view camera.

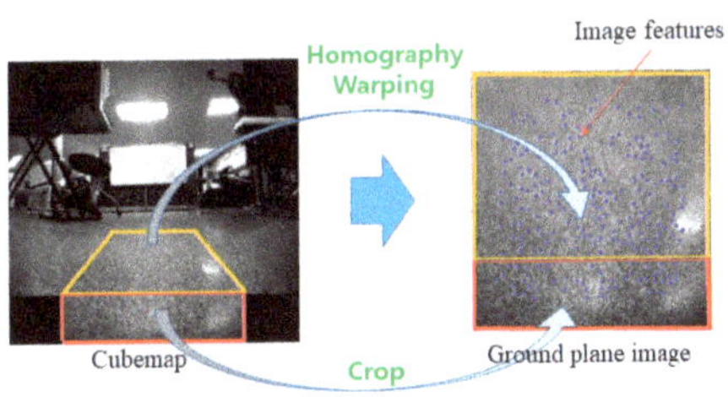

Figure 9. Generation of a ground plane image from a cubemap image.

3.5. Ortho-Rectification of GPI

Ideally, GPI is the ortho-rectified image obtained from the virtual top-view camera. Therefore, there should be no distortion in the image of the ground plane. However, due to the inherent errors in the EUCM calibration parameters and homography transformation from the front-view face of the cubemap model to the ground, there exists image distortion in GPI. An example of image distortion in GPI is shown in Figure 10. From the left in the figure, an original fisheye image is converted to a cubemap image and finally to a GPI. In this figure, we intentionally place the chessboard pattern in front of the robot to identify the distortion easily. In the generated GPI, we can notice that there is image distortion of the chessboard pattern. This kind of image distortion must be corrected. Otherwise, the visual odometry using the distorted GPI yields erroneous results.

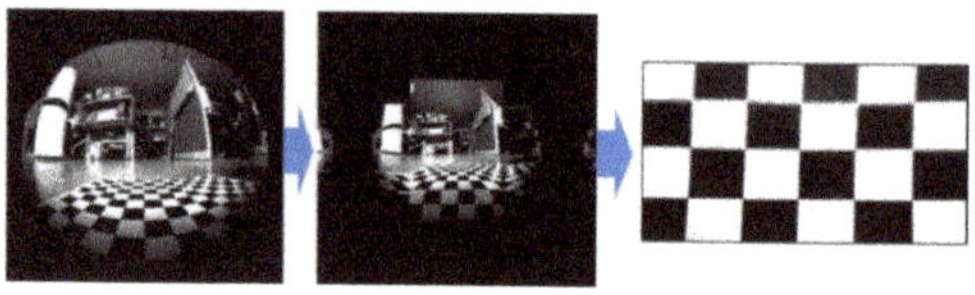

Figure 10. The distorted ground plane image of a chessboard placed at the front of the robot.

To correct the image distortion in GPI, we transform the GPI again to a new image, ortho-rectified GPI. The method of distortion correction is as follows. We place the chessboard at the front of the robot, obtain a fisheye image, and generate the GPI, as shown in Figure 11. Then, we select four corner corners in the image, as shown in the red-colored rectangle in the figure. Finally, a homography matrix is calculated so that the red-colored rectangle area is transformed into a real rectangle, and the area of the rectangle is the same as the real metric space of the 2×4 squares area. To convert the pixel to metric space unit, we use the pixel-to-meter scale factor, which is used in Section 3.6.

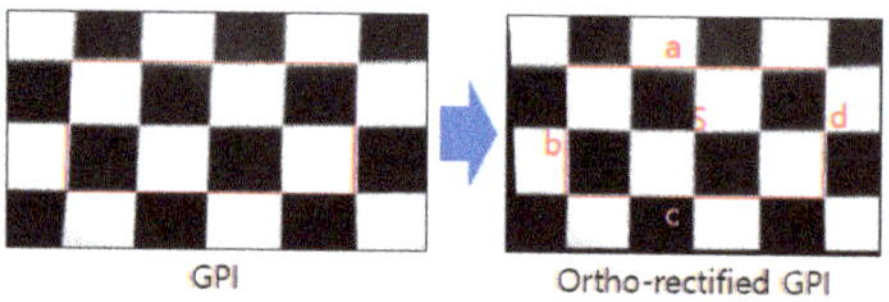

Figure 11. Ortho-rectification of GPI. After ortho-rectification, the chessboard image is almost perfective undistorted.

To compare the ortho-rectification accuracy of the GPI, we measure the side length and area of the 2×4 chessboard squares, as shown in Table 1. As shown in Figure 11, four sides of the red-colored rectangle, a, b, c, and d, are compared before and after ortho-rectification. The area of the rectangle S is also compared in Table 1. As shown in the table, the ortho-rectification corrects image distortion in GPI, and the dimension of the ortho-rectified GPI is almost close to the true values.

Table 1. Comparison of chessboard dimension in GPI and ortho-rectified GPI.

Measurement (Figure 11)	Chessboard Ground Truth	GPI (Error)	Ortho-Rectified GPI (Error)
(a + c)/2 (mm)	4×20.007	84.664 (4.636)	80.085 (0.058)
(b + d)/2 (mm)	2×20.050	36.660 (3.44)	40.036 (0.064)
Area (S) (mm^2)	3209.139	3127.102 (82.037)	3207.881 (1.258)

3.6. Feature Tracking in Orth-Rectified GPI

The proposed visual odometry method in this study is using the ground motion in ortho-rectified GPI. Because the ortho-rectified GPI is the top-view image from the virtual camera, the ground motion in GPI can be measured in real metric dimensions using the focal length and the height of the camera. To obtain the ground motion in GPI, we use a simple and robust KLT optical flow algorithm [24]. Moreover, to obtain a reliable motion vector that represents the motion of the robot, we obtain only the median vector between previous and current frames, as shown in Figure 10.

Usually, the height of a pallet robot is very low, and the camera mounted close to the ground captures the details of the ground or floor patterns. The ground pattern in the ortho-rectified GPI is random enough to extract and track the features in consecutive image frames. Therefore, many features are usually obtained in a single image frame, and they are tracked using KLT. In an ideal case, all motion vectors via KLT are the same because the ortho-rectified GPI is a perfectly orthogonal view of the ground. However, there can be tracking errors in KLT, thus using the median of the motion vectors is the best calculation for motion estimation.

3.7. Metric Motion in Ortho-Rectified GPI

As shown in Figure 12, the median motion vector m obtained from the ortho-rectified GPI matches with the robot motion with an unknown scale because vector m is obtained in the pixel space. Thus, we need the scale factor between the pixel space to the metric space to measure the actual robot motion. To find the scale of the motion vector, some assumptions are defined. Assuming that an ortho-rectified GPI is obtained using the virtual camera that is perfectly perpendicular to the flat ground, the scale of the motion vector is calculated using the height z from the ground to the camera's lens center and the focal length of the camera f. The scale factor is expressed in Equation (23).

$$\text{scale} = \frac{z}{f}, \tag{23}$$

$$\overline{m} = scale * m. \tag{24}$$

By using the scale factor, the robot's actual motion vector $\overline{m}$, as shown in Figure 13, is calculated by multiplying the scale factor by the pixel motion m.

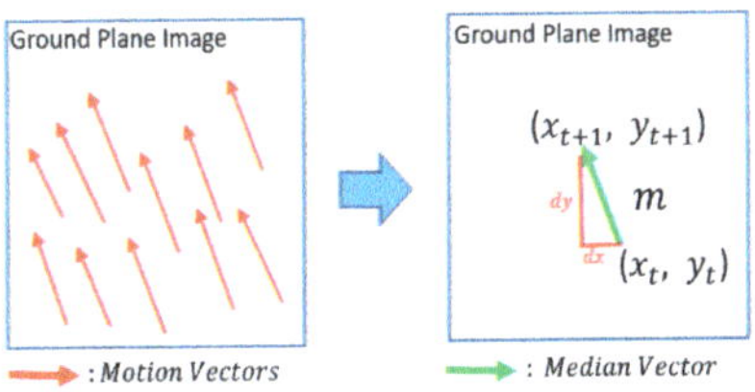

Figure 12. The motion vector on the ground plane image.

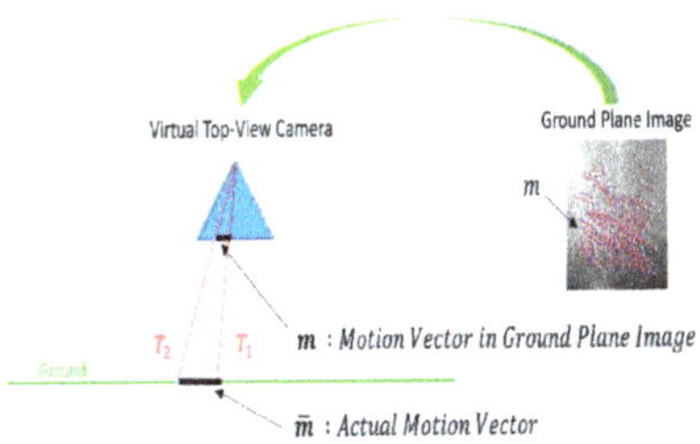

Figure 13. Relation of the motion vector in ground plane image and actual motion vector.

3.8. Robot Odometry Using a Kinematic Motion Model

As the motion of the camera mounted on the robot depends on the robot's movement, we apply the motion vector of the ground plane image to a kinematic model of the robot to estimate the robot's odometry. In this step, the motion vector of the ortho-rectified GPI is regarded as the measurement of a virtual front wheel. Thus, we employ the bicycle motion model and estimate the velocity V of the virtual front wheel and the steering angle δ using Equations (25) and (26).

$$V = \frac{\sqrt{\overline{dx}^2 + \overline{dy}^2}}{dt}, \tag{25}$$

$$\delta = tan^{-1}\left(\frac{\overline{dy}}{\overline{dx}}\right) - \frac{\pi}{2}, \tag{26}$$

where $\overline{dx}$ and $\overline{dy}$ are the components of the median motion vector, and dt is the interval time between image frames.

Figure 14 shows the bicycle motion model we employ in this study. The bicycle motion model rotates and moves around IC, the center of motion in a 2D plane. We use the median motion vector $\overline{m}$ of the GPI to measure the velocity V of the virtual front wheel and the steering angle δ. As a result, the 2D odometry of the robot is estimated, and it is named Ground Plane Odometry (GPO). The bicycle motion follows three assumptions:

- The wheel-based mobile robot moves in a 2D plane;
- The center of mass of the mobile robot is located in the center of the robot body;
- Non-slip, no longitudinal/lateral slip occurs on the wheels.

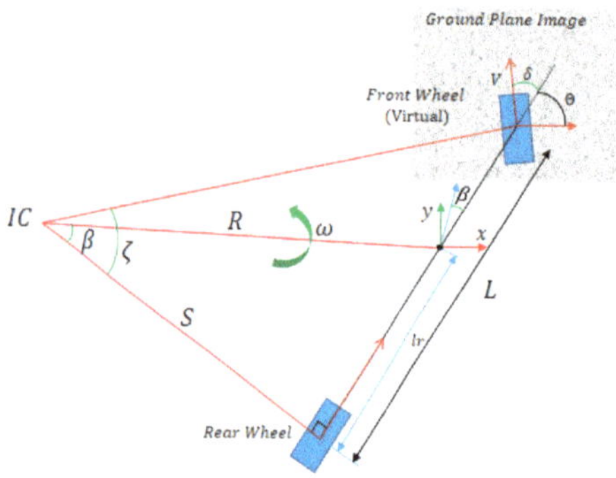

Figure 14. Motion vector on the ground plane image applying to the bicycle motion model.

Following these assumptions, the robot's pose is estimated by applying the velocity and the steering angle of the virtual wheel to the bicycle model of the robot as follows. All parameters of the bicycle model in Figure 14 are calculated using the following equations:

$$S = \frac{L}{\tan(\delta)}, \tag{27}$$

$$\beta = atan\left(\frac{lr}{S}\right) = atan\left(lr\frac{\tan(\delta)}{L}\right), \tag{28}$$

$$R = \frac{S}{\cos(\beta)}, \tag{29}$$

$$\dot{\theta} = \omega = \frac{V}{R} = V\frac{\tan(\delta)\cos(\beta)}{L}, \tag{30}$$

$$\theta = \theta + \dot{\theta}dt, \tag{31}$$

where lr is the distance between the rear wheel axis center and the mass center of the robot, L is the distance between the rear wheel axis center and the front wheel axis center, and ω

is the angular velocity of the robot that rotates around the IC. Using these parameters, the velocity of the robot is determined as shown in Equations (32) and (33).

$$\dot{x} = V\cos(\theta + \beta), \tag{32}$$

$$\dot{y} = V\sin(\theta + \beta). \tag{33}$$

Finally, the visual odometry of the mobile robot is estimated using Equation (34).

$$\begin{bmatrix} x(t+1) \\ y(t+1) \\ \theta(t+1) \end{bmatrix} = \begin{bmatrix} x(t) \\ y(t) \\ \theta(t) \end{bmatrix} + \begin{bmatrix} \dot{x} \\ \dot{y} \\ \dot{\theta} \end{bmatrix} dt. \tag{34}$$

4. Experiment and Evaluation

4.1. Experimental Setup

The robot used in this experiment is a pallet-type mobile robot, as shown in Figure 15, and it operates on ROS [31]. Table 2 lists the hardware specifications of the mobile robot. As shown in Figure 15b, an Intel tracking camera T265 is mounted at the front of the robot. T265 has stereo fisheye cameras, and the FoV is 163 ± 5 degrees. We use the left camera of T265 to obtain fisheye images for our experiment. Each fisheye image has a resolution of 848 × 800. T265 also has an IMU because it performs localization based on visual-inertial odometry (VIO).

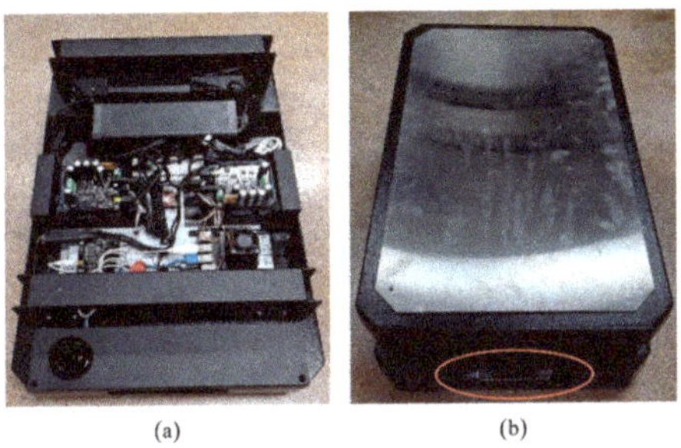

Figure 15. (**a**) A pallet-type robot; (**b**) An Intel T265 fisheye camera is attached in front of the pallet robot.

Table 2. Experimental pallet robot's hardware specification.

Component	Hardware
Embedded Board	NVIDIA Jetson-Nano
Wheel Encoder	PHIDGET Optical Rotary Encoder
Motor	PHIDGET NEMA-17 Bipolar Stepper
Motor Controller	PHIDGET Stepper Bipolar HC
LiDAR	SLAMTEC RPLIDAR-S1
Camera	Intel RealSense T265 Tracking Camera
Battery	NEXT-408PB-UPS (40,800 mAh)

The evaluation of the proposed method is carried out using an optical tracking system, as shown in Figure 16a. The ground truth position of the robot's odometry is measured using the OptiTrack V120 Trio pose-tracking device [32]. To track the robot, four optical markers are attached to the four corners of the pallet robot as shown in Figure 16b. In this study, we used the trajectory evaluation toolbox proposed in [33] to calculate the odometry translation error compared with GT.

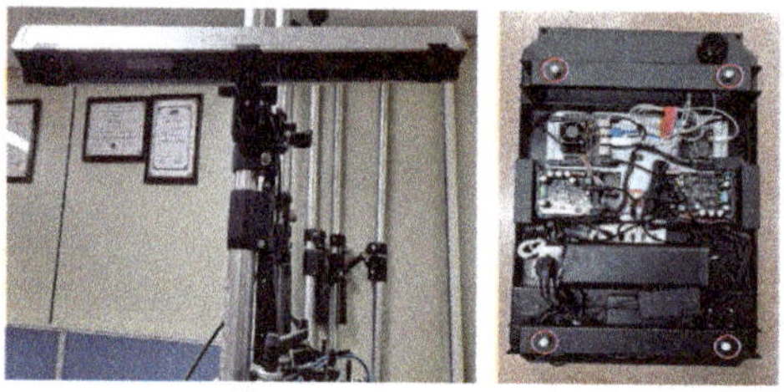

(a) (b)

Figure 16. (**a**) Optitrack's optical pose tracking device V120 Trio; (**b**) Four optical markers attached to the pallet robot.

4.2. Motion Vector Evaluation

In this study, we introduce the Ground Plane Odometry (GPO) of a low-profile pallet robot using ortho-rectified GPI, a virtual top-view image. GPO is performed on the assumption that a virtual top-view camera captures the flat ground plane. If this assumption is correct, ideally, all the motion vectors on an ortho-rectified GPI will have the same direction and length. Figure 17a shows an example of the ortho-rectified GPI and motion vectors via the KLT optical flow. Figure 17b shows the distribution of the motion vectors in the $(\overline{dx}, \overline{dy})$ space. As seen in the figure, the distribution of the motion vectors is almost uniformly distributed, and we take the median vector as the robot motion between frames.

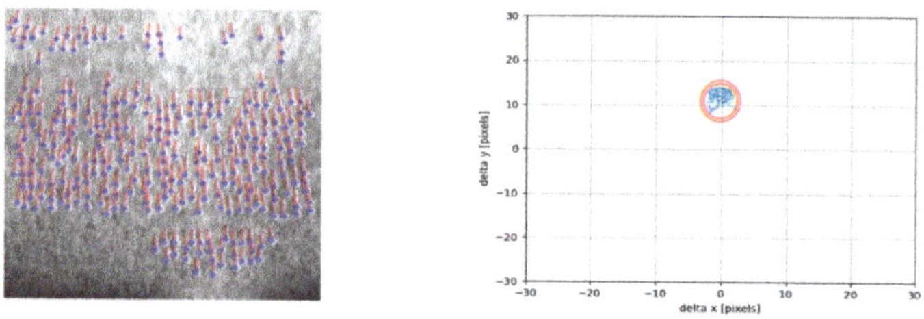

(**a**) (**b**)

Figure 17. (**a**) Unfiltered motion vectors obtained from the ground plane image (**b**) A $\left(\overline{dx}, \overline{dy}\right)$ plot of motion vectors.

In this study, we assume that the median vector is the motion of the robot. Using the median vector is more accurate than using the mean vector, which is the average of all the motion vectors. Because motion vectors can contain some outliers due to incorrect motion tracking, the mean vector is rather biased from the real robot motion. If the mean vector is erroneous due to the outlier, the final robot odometry is also affected by the erroneous motion vector. To compare the performance of GPO using the median and the mean vectors, we have three experiments using short trajectories. Table 3 shows a comparison of GPO results using median and mean vectors. The table shows the minimum, maximum, and average values of the relative translation error of GPO. Three experimental trajectories for comparison are shown in Figure 18. This experiment shows that using the median vector is more robust to noise and more accurate than using the mean vector.

Table 3. The GPO's translation errors in meter. (Path length is in parenthesis).

		Test1 (3.8832)	Test2 (2.2896)	Test3 (5.4559)	Average
GPO (Mean Vector)	min	0.0018	0.0115	0.0533	0.0222
	max	0.1694	0.1736	0.1194	0.1541
	avg	0.0435	0.0876	0.0886	0.0732
GPO (Median Vector)	min	0.0019	0.0040	0.0182	0.0080
	max	0.0787	0.0329	0.0731	0.0615
	avg	0.0337	0.0165	0.0485	0.0329

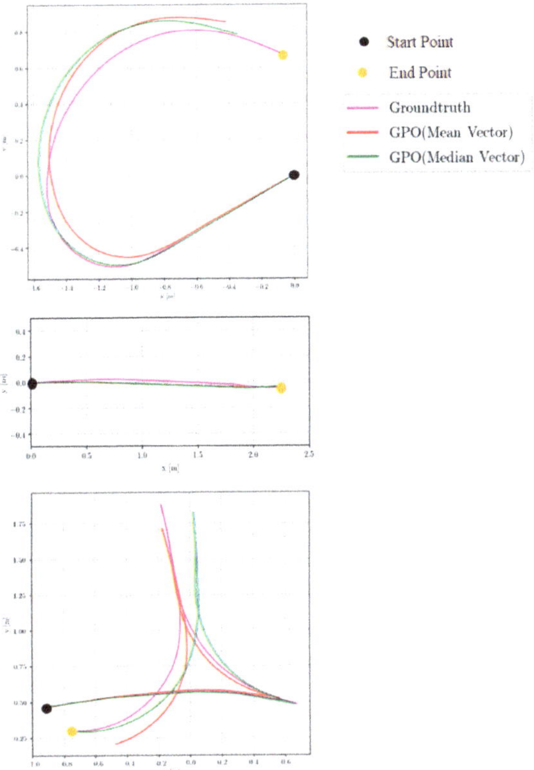

Figure 18. Comparison of median vector and mean vector method in GPO. From top, trajectories of Test1, Test2, and Test3.

4.3. Odometry Trajectory Evaluation

Figure 19 shows the comparison of three different odometry methods, T265 odometry, proposed GPO odometry, and EKF of T265 and GPO. For the evaluation of the EKF method, the T265 odometry and the GPO results are loosely coupled. To do this, the odometry outputs of the first two methods are filtered using the EKF algorithm in ROS. As shown in the figure, the EKF results best follow the ground truth trajectories. Table 4 is the translation error of three methods compared with the ground truth. In the table, GPO is better than the T265 odometry, and the EKF performs best. In addition, we also compare the GPO performance using two different ground plane images. In this table, using an ortho-rectified GPI is better than using the original GPI. This implies that ortho-rectified GPI can ideally generate the virtual top-view image from the fisheye camera.

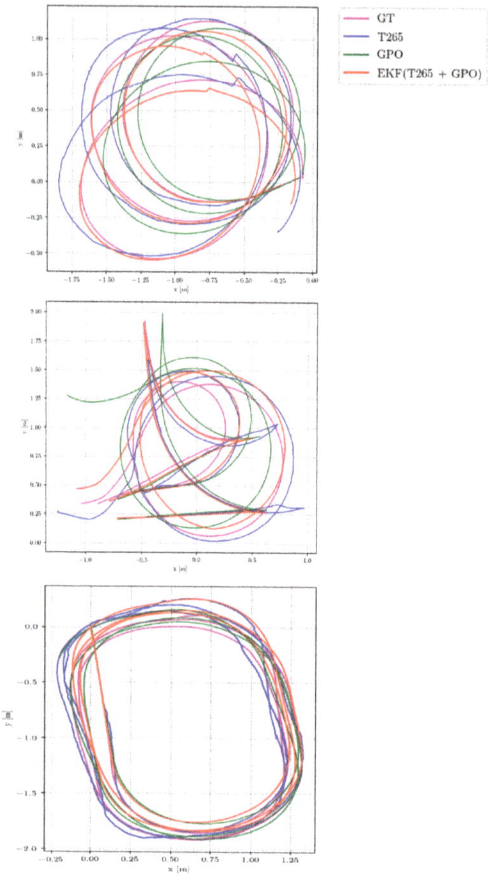

Figure 19. Comparison of three odometry methods. From top, trajectories of Test4, Test5, and Test6.

Table 4. Translation error comparison of three odometry methods. (Path length is in parenthesis and unit is in meters).

Image Type	Odometry	Test4 (12.7826)	Test5 (14.6013)	Test6 (17.0111)	Average
GPI	T265	0.118618	0.226798	0.151477	0.1656
	GPO	0.077336	0.183521	0.097809	0.1195
	EKF (GPO+T265)	0.047038	0.103502	0.068983	0.0731
Ortho-rectified GPI	T265	0.118618	0.226798	0.151477	0.1656
	GPO	0.049830	0.217686	0.085550	0.1177
	EKF (GPO+T265)	**0.038309**	**0.067267**	**0.068169**	**0.0579**

5. Conclusions

In this study, we propose a visual odometry method for a low-profile pallet robot using the ground plane images obtained from a fisheye camera. To overcome the inherent radial distortion problem of the fisheye image, we introduce a Ground Plane Image (GPI), which is generated from two faces of the cubemap. Initially, we generate an original GPI and transform the image again to an ortho-rectified GPI to obtain a perfect top-view image from the mobile robot. Consequently, the median of the motion vectors in the ortho-rectified GPI can represent the real and accurate motion at the front head of the robot. The bicycle motion model is employed to estimate the odometry of the robot. The performance of the proposed Ground Plane Odometry (GPO) is compared with a stereo-based VIO method of a T265 camera. The proposed GPO shows better performance than T265 odometry, and the EKF of GPO and T265 achieves the best performance. In addition, it is shown that using the ortho-rectified GPI is better than using the original GPI. The proposed GPO can be used in any low-profile robot equipped with fisheye cameras. In addition, GPO can be used by combining with other robot sensors for better performance. In future studies, GPO will be employed to multi-view fisheye cameras in the pallet robot for more accurate visual odometry performance.

Author Contributions: Conceptualization, S.-Y.P. and S.-H.B.; methodology, software, validation, formal analysis, U.-G.L. and S.-H.B.; sensor calibration U.-G.L.; investigation, S.-Y.P. and S.-H.B.; writing—original draft preparation, U.-G.L. and S.-Y.P.; writing—review and editing, S.-Y.P.; supervision, S.-Y.P.; project administration, S.-Y.P.; funding acquisition, S.-Y.P. All authors have read and agreed to the published version of the manuscript.

Funding: This work was supported by Basic Science Research Program through the National Research Foundation of Korea (NRF) funded by the Ministry of Education (No. 2021R1A6A1A03043144).

Institutional Review Board Statement: Not applicable.

Informed Consent Statement: Not applicable.

Data Availability Statement: The data presented in this study are available upon request from the corresponding author.

Conflicts of Interest: The authors declare no conflict of interest.

References

1. Wang, Y.; Cai, S.; Li, S.J.; Liu, Y.; Guo, Y.; Li, T.; Cheng, M.M. CubemapSLAM: A Piecewise-Pinhole Monocular Fisheye SLAM System. In *Proceedings of the Lecture Notes in Computer Science (Including Subseries Lecture Notes in Artificial Intelligence and Lecture Notes in Bioinformatics)*; Springer International Publishing: Cham, Switzerland, 2019; Volume 11366.
2. Barron, J.L.; Fleet, D.J.; Beauchemin, S.S. Performance of Optical Flow Techniques. *Int. J. Comput. Vis.* **1994**, *12*, 43–77. [CrossRef]
3. Polack, P.; Altche, F.; DAndrea-Novel, B.; de La Fortelle, A. The Kinematic Bicycle Model: A Consistent Model for Planning Feasible Trajectories for Autonomous Vehicles? In Proceedings of the IEEE Intelligent Vehicles Symposium, Los Angeles, CA, USA, 11–14 June 2017.
4. Julier, S.J.; Uhlmann, J.K. Unscented Filtering and Nonlinear Estimation. *Proc. IEEE* **2004**, *92*, 401–422. [CrossRef]
5. Intel T265 Tracking Camera. Available online: https://www.intelrealsense.com/tracking-camera-t265/ (accessed on 4 August 2023).
6. Flores, M.; Valiente, D.; Cebollada, S.; Reinoso, Ó.; Payá, L. Evaluating the Influence of Feature Matching on the Performance of Visual Localization with Fisheye Images. In Proceedings of the International Conference on Informatics in Control, Automation and Robotics, Paris, France, 6–8 July 2021.
7. Rublee, E.; Rabaud, V.; Konolige, K.; Bradski, G. ORB: An Efficient Alternative to SIFT or SURF. In Proceedings of the IEEE International Conference on Computer Vision, Barcelona, Spain, 6–13 November 2011.
8. Bay, H.; Ess, A.; Tuytelaars, T.; Van Gool, L. Speeded-Up Robust Features (SURF). *Comput. Vis. Image Underst.* **2008**, *110*, 346–359. [CrossRef]
9. Rosten, E.; Drummond, T. Machine Learning for High-Speed Corner Detection. In Proceedings of the Lecture Notes in Computer Science (Including Subseries Lecture Notes in Artificial Intelligence and Lecture Notes in Bioinformatics), Graz, Austria, 7–13 May 2006; Volume 3951.
10. Alcantarilla, P.F.; Bartoli, A.; Davison, A.J. KAZE Features. In *Proceedings of the European Conference on Computer Vision*; Springer: Berlin/Heidelberg, Germany, 2012.
11. Zhang, Z.; Rebecq, H.; Forster, C.; Scaramuzza, D. Benefit of Large Field-of-View Cameras for Visual Odometry. In Proceedings of the IEEE International Conference on Robotics and Automation, Stockholm, Sweden, 16–21 May 2016; Volume 2016.
12. Mur-Artal, R.; Montiel, J.M.M.; Tardos, J.D. ORB-SLAM: A Versatile and Accurate Monocular SLAM System. *IEEE Trans. Robot.* **2015**, *31*, 1147–1163. [CrossRef]
13. Cyganek, B.; Siebert, J.P. *An Introduction to 3D Computer Vision Techniques and Algorithms*; Wiley: Hoboken, NJ, USA, 2009.
14. Swank, A.J. *Localization Using Visual Odometry and a Single Downward-Pointing Camera*; NASA: Washington, DC, USA, 2012.
15. Zhang, J.; Singh, S.; Kantor, G. Robust Monocular Visual Odometry for a Ground Vehicle in Undulating Terrain. In *Proceedings of the Springer Tracts in Advanced Robotics*; Springer: Berlin/Heidelberg, Germany, 2014; Volume 92.

16. Jordan, J.; Zell, A. Kinematic Model Based Visual Odometry for Differential Drive Vehicles. In Proceedings of the 2017 European Conference on Mobile Robots, ECMR, Paris, France, 6–8 September 2017.
17. Zhou, D.; Dai, Y.; Li, H. Ground-Plane-Based Absolute Scale Estimation for Monocular Visual Odometry. *IEEE Trans. Intell. Transp. Syst.* **2020**, *21*, 791–802. [CrossRef]
18. Jordan, J.; Zell, A. Ground Plane Based Visual Odometry for RGBD-Cameras Using Orthogonal Projection. *IFAC-PapersOnLine* **2016**, *49*, 108–113. [CrossRef]
19. Malis, E. Improving Vision-Based Control Using Efficient Second-Order Minimization Techniques. In Proceedings of the IEEE International Conference on Robotics and Automation, New Orleans, LA, USA, 26 April–1 May 2004; Volume 2004.
20. Ouyang, M.; Cao, Z.; Guan, P.; Li, Z.; Zhou, C.; Yu, J. Visual-Gyroscope-Wheel Odometry with Ground Plane Constraint for Indoor Robots in Dynamic Environment. *IEEE Sens. Lett.* **2021**, *5*, 6000504. [CrossRef]
21. Chen, X.; Vempati, A.S.; Beardsley, P. StreetMap—Mapping and Localization on Ground Planes Using a Downward Facing Camera. In Proceedings of the IEEE International Conference on Intelligent Robots and Systems, Madrid, Spain, 1–5 October 2018.
22. Lin, H.Y.; Chung, Y.C.; Wang, M.L. Self-Localization of Mobile Robots Using a Single Catadioptric Camera with Line Feature Extraction. *Sensors* **2021**, *21*, 4719. [CrossRef] [PubMed]
23. Lee, U.-G.; Park, S.-Y. Visual odometry of a mobile palette robot using ground plane image from a fisheye camera. *Int. Arch. Photogramm. Remote Sens. Spat. Inf. Sci.* **2022**, *XLIII-B1-2*, 431–436. [CrossRef]
24. Lucas, B.D.; Kanade, T. *Iterative Image Registration Technique with an Application to Stereo Vision*; Carnegie Mellon University: Pittsburgh, PA, USA, 1981; Volume 2.
25. Khomutenko, B.; Garcia, G.; Martinet, P. An Enhanced Unified Camera Model. *IEEE Robot. Autom. Lett.* **2016**, *1*, 137–144. [CrossRef]
26. Rehder, J.; Nikolic, J.; Schneider, T.; Hinzmann, T.; Siegwart, R. Extending Kalibr: Calibrating the Extrinsics of Multiple IMUs and of Individual Axes. In Proceedings of the IEEE International Conference on Robotics and Automation, Stockholm, Sweden, 16–21 May 2016; Volume 2016.
27. Furgale, P.; Rehder, J.; Siegwart, R. Unified Temporal and Spatial Calibration for Multi-Sensor Systems. In Proceedings of the IEEE International Conference on Intelligent Robots and Systems, Tokyo, Japan, 3–7 November 2013.
28. Furgale, P.; Tong, C.H.; Barfoot, T.D.; Sibley, G. Continuous-Time Batch Trajectory Estimation Using Temporal Basis Functions. *Int. J. Robot. Res.* **2015**, *34*, 1688–1710. [CrossRef]
29. Maye, J.; Furgale, P.; Siegwart, R. Self-Supervised Calibration for Robotic Systems. In Proceedings of the IEEE Intelligent Vehicles Symposium, Gold Coast, QLD, Australia, 23–26 June 2013.
30. Oth, L.; Furgale, P.; Kneip, L.; Siegwart, R. Rolling Shutter Camera Calibration. In Proceedings of the Proceedings of the IEEE Computer Society Conference on Computer Vision and Pattern Recognition, Portland, OR, USA, 23–28 June 2013.
31. Lee, U.G.; Choi, K.J.; Park, S.Y. The Design and Implementation of Autonomous Driving Pallet Robot System Using ROS. In Proceedings of the International Conference on Ubiquitous and Future Networks, ICUFN, Jeju Island, Republic of Korea, 17–20 August 2021.
32. OptiTrack V120. Available online: https://optitrack.com/cameras/v120-trio/ (accessed on 4 August 2023).
33. Zhang, Z.; Scaramuzza, D. A Tutorial on Quantitative Trajectory Evaluation for Visual(-Inertial) Odometry. In Proceedings of the IEEE International Conference on Intelligent Robots and Systems, Madrid, Spain, 1–5 October 2018.

applied sciences

MDPI

Article

Advancing Image Object Detection: Enhanced Feature Pyramid Network and Gradient Density Loss for Improved Performance

Ying Wang [1], Qinghui Wang [1], Ruirui Zou [1], Falin Wen [1], Fenglin Liu [1], Yihang Zhang [2], Shaoyi Du [3] and Wei Zeng [1,*]

[1] School of Physics and Mechanical and Electrical Engineering, Longyan University, Longyan 364012, China; lyxyy_w@126.com (Y.W.); wqh0597@126.com (Q.W.); zray_zou@126.com (R.Z.); wenfalin2008@163.com (F.W.); liufenglin45@126.com (F.L.)
[2] School of Software Engineering, Xi'an Jiaotong University, Xi'an 710049, China; 18729358895@139.com
[3] Institute of Artificial Intelligence and Robotics, Xi'an Jiaotong University, Xi'an 710049, China; dushaoyi@xjtu.edu.cn
* Correspondence: zw0597@126.com

Abstract: In the era of artificial intelligence, the significance of images and videos as intuitive conveyors of information cannot be overstated. Computer vision techniques rooted in deep learning have revolutionized our ability to autonomously and accurately identify objects within visual media, making them a focal point of contemporary research. This study addresses the pivotal role of image object detection, particularly in the contexts of autonomous driving and security surveillance, by presenting an in-depth exploration of this field with a focus on enhancing the feature pyramid network. One of the key challenges in existing object detection methodologies lies in mitigating information loss caused by multi-scale feature fusion. To tackle this issue, we propose the enhanced feature pyramid, which adeptly amalgamates features extracted across different scales. This strategic enhancement effectively curbs information attrition across various layers, thereby strengthening the feature extraction capabilities of the foundational network. Furthermore, we confront the issue of excessive classification loss in image object detection tasks by introducing the gradient density loss function, designed to mitigate classification discrepancies. Empirical results unequivocally demonstrate the efficacy of our approach in enhancing the detection of multi-scale objects within images. When evaluated across benchmark datasets, including MS COCO 2017, MS COCO 2014, Pascal VOC 2007, and Pascal VOC 2012, our method achieves impressive average precision scores of 39.4%, 42.0%, 51.5%, and 49.9%, respectively. This performance clearly outperforms alternative state-of-the-art methods in the field. This research not only contributes to the evolving landscape of computer vision and object detection but also has practical implications for a wide range of applications, aligning with the transformative trends in the automotive industry and security technologies.

Keywords: deep learning; object detection; feature pyramid network; multi-scale features; gradient density loss

1. Introduction

Object detection holds a central position in the realm of computer vision, forming the bedrock for tasks such as image segmentation and object tracking. With the relentless progression of computer hardware capabilities, object detection has pervaded numerous domains, encompassing security, healthcare, and autonomous driving. It has notably catalyzed urban development by curtailing expenses and elevating societal efficacy. In everyday life, applications like facial recognition-based transactions, access control, and text recognition have seamlessly integrated into our routines. In the medical arena, object detection technology lends critical support to physicians, aiding in diagnoses and accurately pinpointing potential anomaly regions. Within industrial production, object detection methods discern product defects, facilitating automated oversight and serving as a pivotal component in assembly lines. In the military domain, these techniques offer reconnaissance potential by surveilling adversary targets through remote sensing imagery. In agriculture, object detection assumes the role of pest infestation monitoring and constitutes a cornerstone in automated harvest equipment. The ongoing evolution of object detection methodologies has instigated transformative revolutions across multifarious sectors, yielding enhancements in efficiency, precision, and automation.

The objective of object detection encompasses the identification of object categories within input images and the precise delineation of their positions and boundaries using rectangular bounding boxes, coupled with associated confidence scores. While the human eye may find recognizing object categories and positions intuitive, for computers, an image translates to a set of data points. Deriving object category insights and accurately localizing objects from these data points constitutes a formidable computational task. Variables like object dimensions, orientations, illumination, and background disturbances all contribute to the intricacies of object detection performance. Hence, object detection confronts a multitude of challenges, demanding precise object localization and classification. Attaining these objectives with a harmonious interplay of accuracy and speed underscores the core of successful object detection. Furthermore, objects manifest in diverse scales, driving the pursuit of equilibrium in performance across these scales as a pressing research endeavor. The incorporation of adept attention mechanisms further augments object detection efficiency, requiring the construction of streamlined attention mechanisms that adeptly balance efficacy and computational efficiency.

Conventional approaches to image object detection employ sliding window methodologies to systematically navigate through entire images. These methods are complemented by manually engineered features that possess the ability to remain invariant to factors like lighting conditions, rotation, translation, and scaling. These intricately designed features are tailored to accommodate common fluctuations in lighting and shapes. The subsequent phase involves training classifiers on these features. Viola and Jones [1] pioneered the Viola–Jones (VJ) detector, which extracts Haar features from images to automate facial detection through a sliding window mechanism. Dalal and Triggs [2] introduced the concept of histogram of oriented gradient (HOG) features, effectively balancing between the invariance to translation and scaling. Felzenszwalb et al. [3] innovated the deformable part model (DPM), which dissects object detection into various constituent detection tasks. Prominent classifiers employed include support vector machines (SVMs) and Bayesian classifiers.

Conventional methodologies for image object detection are fraught with several challenges. These include the substantial computational load of sliding window mechanisms and the intricate development of handcrafted features. Deep-learning-driven object detection algorithms present remarkable advantages over their traditional counterparts, excelling in terms of precision and efficiency, thereby positioning themselves as the leading approach in the field of object detection. The integration of deep learning into object detection has led to effective resolutions for many of these challenges. In 2012, Hinton et al. [4] introduced the groundbreaking AlexNet model, achieving a substantial breakthrough in the ImageNet image recognition competition. Following this milestone, convolutional neural networks (CNNs) have gradually gained prominence in the realm of object detection, forming the cornerstone of computer vision models. With each passing year, numerous researchers have contributed to the development of diverse network models that exhibit enhanced performance in both accuracy and speed for image recognition tasks. Deep-learning-based object detection algorithms predominantly fall into two distinct categories: one-stage algorithms and two-stage algorithms. Among these, the two-stage algorithms initially extract features from the input image, subsequently conducting object classification and bounding box regression on selected candidate regions.

The inception of the two-stage algorithm era was marked by the introduction of the R-CNN (regions with CNN feature), pioneered by Girshick et al. [5], which marked a pioneering breakthrough in the realm of object detection. The R-CNN approach ushered in the integration of convolutional neural networks (CNNs) into object detection, reshaping the landscape of this domain. The R-CNN method incorporated the Selective Search algorithm to generate a set of candidate regions. Subsequently, these regions were resized to a standardized dimension and fed into a neural network for feature extraction. This network was fine-tuned based on a pretrained ImageNet model tailored to the specific object detection task. Despite achieving commendable outcomes when compared to traditional methodologies of that time, the R-CNN approach faced certain limitations. Notably, the presence of fully connected layers hindered its accuracy when confronted with deformations, and its adaptability to varying image sizes remained constrained.

The challenges were addressed by He et al. [6] through the introduction of the SPPNet (Spatial Pyramid Pooling Network), a pioneering approach that brought about the concept of spatial pyramid pooling. This innovation facilitated the model's ability to seamlessly adapt to images of varying sizes. Subsequently, Girshick [7] presented the Fast R-CNN framework, leveraging the VGG network as the foundational architecture to enhance both the computational speed and precision of object detection. Further building upon the Fast R-CNN foundation, Faster R-CNN, as introduced by Ren et al. [8], incorporated the Region Proposal Network (RPN) for the generation of candidate bounding boxes. This innovation significantly improved both the accuracy and speed of the approach by replacing the resource-intensive selective search step with a more efficient candidate box generation mechanism. The Faster R-CNN methodology employed a shared feature map to simultaneously handle the tasks of classification and regression, integrating it seamlessly into the RPN for the generation of candidate bounding boxes.

Cai and Vasconcelos [9] introduced the Cascade R-CNN, a novel approach that involved training with varying thresholds to enhance the accuracy of object detection. It is important to note that while two-stage algorithms typically offer superior accuracy by initially predicting potential object locations based on input features and subsequently conducting object classification and bounding box regression for more precise classification and positioning, they often exhibit slower processing speeds. Recognizing this challenge, one-stage algorithms such as YOLO (You Only Look Once), proposed by Redmon et al. [10], execute both classification and regression tasks in a single step, thus boosting detection speed, albeit at the cost of a slight reduction in accuracy compared to Faster R-CNN.

Liu et al. [11] introduced the single-shot multibox detector (SSD), which leverages features from various layers to enhance feature extraction and achieve greater accuracy than YOLO. YOLOv3, presented by Redmon and Farhadi [12], incorporated the feature pyramid network (FPN) [13] with multi-scale features to enhance accuracy in detecting small objects. Bochkovskiy et al. [14] proposed YOLOv4, which employed the CSPDarkNet as the backbone, boasting increased parameters and utilizing the spatial pooling layer (SPP) to expand the receptive field of the backbone network, ultimately leading to superior performance. Dai et al. [15] adopted the Transformer architecture for object detection, eliminating the requirement for anchor boxes and introducing a direct set-based prediction approach. This work demonstrated the potential of Transformers in computer vision tasks beyond their traditional application in natural language processing.

This study addresses the imperative need for enhancing the performance of current object detection methodologies and introduces improvements in the accuracy of image object detection tasks through the utilization of computer vision techniques such as feature pyramids. The proposed approach introduces an object detection methodology founded upon an augmented feature pyramid network. This approach specifically targets two principal limitations encountered in prevailing image object detection techniques. Firstly, it addresses the challenge of inadequate object detection across various scales by incorporating an augmented feature pyramid network to extract multi-scale features. This ameliorates the issue of significant information loss in top-level features during the fusion of multi-scale features in conventional methods, resulting in superior-quality multi-scale information. Secondly, the method tackles the limitation of commonly employed focal classification loss functions in object detection models, which tend to overly emphasize the classification features of intricate samples, leading to delayed model convergence. This is mitigated by refining the classification loss function to align more effectively with the attributes of the classification task, rendering it more proficient in accurate classification based on real-world conditions. The final experimental section encompasses diverse assessments, including ablation experiments conducted on the COCO2017 dataset and performance comparisons with 14 existing object detection algorithms on four widely recognized publicly accessible object detection datasets. The consistently observed outcomes of these experiments underscore the effectiveness of the proposed approach, showcasing higher detection accuracy in comparison to established baseline image object detection methods across these extensively utilized object detection datasets.

In today's data-driven world, the role of images and videos as intuitive vehicles for transmitting information has taken on unprecedented significance. With the dawn of the AI era, computer vision techniques, particularly those grounded in deep learning, have enabled the autonomous and precise identification of objects within visual media. This technological advancement has propelled image and video analysis to the forefront of contemporary research, ushering in transformative possibilities across various domains. Our empirical findings not only substantiate the effectiveness of our approach but also establish it as a promising solution for discerning multi-scale objects within images and videos. Importantly, our research aligns with the quest for safer and more reliable autonomous driving systems, as it contributes to the evolving landscape of computer vision and object detection, directly impacting the future of mobility and security.

2. Problem Description

Object detection constitutes a ubiquitous domain within the realm of computer vision. Throughout its evolutionary trajectory, a central hurdle has involved crafting more potent feature extraction modules rooted in deep learning. In the nascent phases of object detection networks, comparatively straightforward convolutional neural networks were integrated as feature extraction modules within the bedrock network architecture. As a result, the foundational network architecture was constrained to extracting features solely at a singular scale. While these features showcased robust information representation capacities at a macroscopic level, the escalating requisites for heightened object detection performance accentuated the inadequacy of these features operating at a solitary scale. In the course of convolutional pooling, information pertaining to diminutive objects experienced a gradual erosion within the confines of the one-scale feature maps.

Subsequently, the significance of multi-scale feature extraction networks gained prominence, exemplified by models such as UNet [16] and FPN [17]. At present, the extensively employed FPN architecture has exhibited enhanced performance in object detection, as demonstrated in the case of Sparse R-CNN [18]. Through the fusion of multi-scale feature maps, FPN facilitates the reciprocal exploitation of both high-level and low-level features. High-level features augment the semantic representation of low-level features, while low-level features amplify intricate details grounded in high-level features. Nevertheless, considering FPN's multi-scale fusion principle, directly amalgamating features from distinct levels featuring pronounced semantic disparities may not yield optimal results. Moreover, during the utilization of FPN, direct fusion of features spanning the highest and lowest levels could potentially engender information loss.

The proposed enhancements in this study are predicated on the Sparse R-CNN object detection algorithm, which represents a distinctive approach. Sparse R-CNN, functioning as a two-stage object detection technique, deviates from the conventional paradigm of such algorithms. Notably, its Region Proposal Network (RPN) engenders a fixed ensemble of 100 candidate boxes. Subsequently, these designated boxes traverse a Dynamic Instance Interactive Head (DIIHead) module. For every individual candidate box, the RoIAlign algorithm orchestrates the extraction of region-specific features, subsequently facilitating the execution of tasks pertaining to target classification and positional regression. The processed boxes subsequently undergo a succession of six iterations through the DIIHead module, progressively refining the initial 100 candidate boxes and culminating in the derivation of the ultimate output.

Of notable significance, the Sparse R-CNN method circumvents the necessity for a post-processing step such as nonmaximum suppression (NMS). This sets it apart from conventional two-stage object detection techniques like Faster R-CNN, which generate a voluminous array of candidate boxes exceeding hundreds of thousands during the Region Proposal Network (RPN) phase. In stark contrast, Sparse R-CNN judiciously employs a sparse collection of 100 candidate boxes within the output feature map, each imbued with learnable attributes. This limited set of 100 candidates places a higher reliance on the extraction of high-level features. Notably, the feature pyramid network (FPN) encounters a diminution of high-level features as a consequence of fusion. Consequently, the direct adoption of FPN as the multi-scale feature extraction module for Sparse R-CNN remains poised for enhancements.

Within Sparse R-CNN, the ultimate goal is to derive conclusive outcomes from the sparse set of 100 learnable candidate boxes. To ensure a closer alignment between the ascertained classes and the actual target classes, an improved classification loss function becomes imperative. Such a function would yield a more accurate assessment of the proximity between identified and genuine target classes. In its classification loss, Sparse R-CNN employs focal loss. The introduction of focal loss was primarily aimed at mitigating the challenge of imbalanced hard and easy samples. Given the abundance of candidate boxes generated in detection, coupled with the paucity of positive samples, a class imbalance dilemma emerges. Nonetheless, incorporating focal loss in Sparse R-CNN gives rise to certain issues. Notably, focal loss tends to excessively concentrate on intricately classifiable samples. This predicament becomes pronounced when the model attains advanced training stages, resulting in a decline in performance.

To tackle the abovementioned challenges, this paper presents an image object detection methodology grounded in a feature pyramid framework. This approach rectifies the deficiencies in the multi-scale feature extraction network employed by Sparse R-CNN, consequently elevating the expressive capacity of multi-scale features within Sparse R-CNN. Furthermore, the classification loss incorporated in this methodology leverages a gradient density loss function. This function is adept at providing a more accurate assessment of classification loss concerning detection boxes.

In this work, we make the following key contributions and introduce novel aspects:

- Enhanced feature pyramid network: We propose an advanced feature pyramid network that significantly enhances multi-scale feature extraction and fusion capabilities. This innovation reduces information loss during feature fusion, leading to improved feature representations.
- Gradient density loss function: Our introduction of the gradient density loss function, as a replacement for the traditional focal loss, offers a more precise assessment of object detection classification losses, particularly for challenging samples.
- Comprehensive evaluation: We rigorously evaluate our method on four well-established public object detection datasets, including MS COCO 2017, MS COCO 2014, Pascal VOC 2007, and Pascal VOC 2012. Substantial improvements in average precision are achieved compared to existing methods.
- Enhanced small-scale target detection: The application of the enhanced feature pyramid network results in superior detection capabilities for small-scale objects, while maintaining high accuracy for larger and medium-sized targets.
- Visual comparisons: We present visual comparisons to demonstrate the practical effectiveness of our enhancements.

3. Method

Multi-scale feature extraction networks play a crucial role in object detection tasks. In widely employed datasets, small objects account for a smaller fraction of the entire image, while other objects necessitate the preservation of high-level large-scale features. Furthermore, the utilized loss function for assessing classification performance might not comprehensively cater to real-world scenarios. With these considerations in mind, this paper introduces an object detection approach founded on a feature pyramid network. The overarching structure of this methodology encompasses two primary constituents: an enhanced feature pyramid network module and a loss function tailored for object detection classification.

3.1. Enhanced Feature Pyramid Network

In preparation for feature fusion within the FPN, a 1×1 convolutional layer is employed. This step serves to transform the channel dimensions, facilitating the subsequent aggregation of feature maps from different scales. The outcome is a fused feature map endowed with enriched information. However, a commonly employed fusion strategy involves channel dimension reduction in higher-level features via convolutional layers, enabling their summation with lower-level feature maps. This process, unfortunately, introduces information loss within feature maps originating from varying scales. Additionally, the semantic gap between feature maps generated at different scales is substantial. If the model neglects these notable semantic disparities and directly applies 1×1 convolutions followed by addition for multi-scale information fusion, there remains room for improvement within this fusion strategy. It is essential to acknowledge that the fundamental design principle of the FPN involves fusing high-level features with their low-level counterparts, thereby empowering high-level features to enhance low-level feature representation by leveraging semantic insights from larger scales. Nevertheless, the feature map located at the highest level lacks a corresponding higher-level, larger-scale feature map for fusion. Consequently, the original network architecture generates a feature map at the highest level, corresponding to the largest scale, devoid of input from features at other scales. Instead, this map is directly subjected to channel dimension reduction before being merged with feature maps from lower scales. This approach seems inequitable to the highest-scale feature map, especially considering that lower-scale feature maps have already undergone fusion with features from more elevated scales.

The approach proposed in this paper effectively tackles the limitations of the FPN utilized in the Sparse R-CNN network to extract multi-scale features. It incorporates mechanisms for enhancing both top-level and bottom-level features, thereby facilitating the extraction of multi-scale image features that encompass both fine-grained and coarse-grained details. This approach better satisfies the demands of object detection. Given an input image, the process of multi-scale feature extraction benefits from the top-level and progressive feature enhancement mechanisms, resulting in the extraction of more finely fused and higher-quality multi-scale feature maps with minimized information loss. Subsequently, the detection network processes these multi-scale features to generate object classification scores and bounding box coordinates, which collectively constitute the foundation for the final object detection outcomes.

The architecture of the feature extraction network is depicted in Figure 1. It encompasses two principal stages: top-level feature enhancement and progressive feature fusion across multiple scales. During the first stage, top-level feature enhancement leverages the multi-scale features obtained from the backbone network. These features are integrated into the ultimate output network features. By employing a reduction in channel dimensionality, the extent of information loss in top-level features is ameliorated. The subsequent stage involves progressive feature fusion across multiple scales during the multi-scale feature fusion phase. This approach effectively diminishes the semantic gap existing between feature maps of varying resolutions within the feature pyramid's multi-scale feature fusion process. Consequently, this strategy contributes to more refined multi-scale feature fusion. The combination of these two stages culminates in the creation of well-integrated multi-scale feature maps.

During the stage of multi-scale feature extraction, the initial input image undergoes processing through the backbone network, which is constructed based on the ResNet-50 architecture. Convolutional operations are executed across four residual convolutional modules (referred to as layer_1 to layer_4), generating feature maps at corresponding scales: $256 \times H \times W$, $512 \times H/2 \times W/2$, $1024 \times H/4 \times W/4$, and $2048 \times H \times W$. This design allows the feature extraction network to progressively gather insights from both diminutive-scale and substantial-scale features spanning diverse scales. Notably, subsequent to each residual convolutional module, the feature map is proportionally downsampled by factors of 0.5, 0.25, and 0.125, respectively. Within the FPN network of Sparse R-CNN, subsequent to the extraction of feature maps at the four distinct scales, a process of reversed feature fusion is initiated. To facilitate this fusion, a 1×1 convolutional operation is applied to the feature map of each layer. This operation ensures uniformity in output channel dimensions across layers, thus simplifying the ensuing feature fusion process. Typically, fusion is executed through element-wise addition.

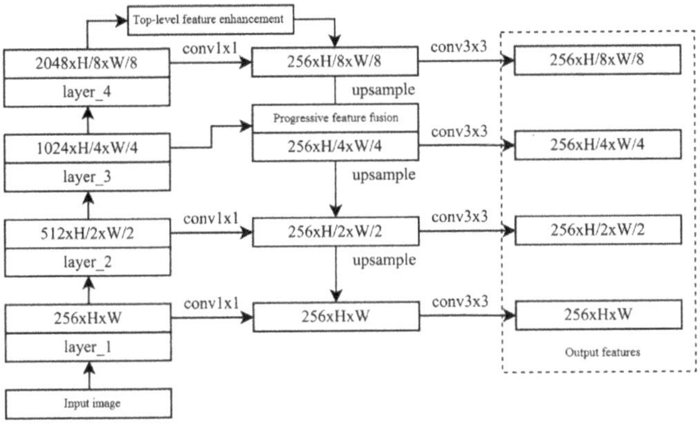

Figure 1. Feature extraction network.

In the ultimate top-level feature map at the 0.125× scale (depicted in Figure 2), a module for top-level feature enhancement is incorporated. The module's output undergoes fusion with the original top-level features, thereby generating enhanced top-level features. This augmentation effectively counteracts the loss of top-level feature data that can arise during the ensuing process of reversed multi-scale feature fusion. Consequently, this addresses the constraint attributed to the absence of higher-level features available for fusion with the top-level counterparts.

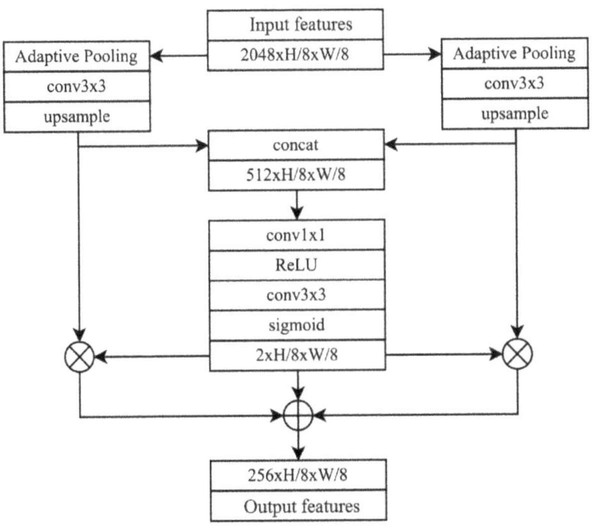

Figure 2. Top-level feature fusion module.

To be more precise, the top-level feature enhancement module incorporates an adaptive pooling mechanism. In this process, the aspect ratio of the feature map remains constant, while the pooling ratio can be altered. This permits the execution of adaptive pooling using multiple pooling ratios. Consequently, the model operates at diverse scales, generating target features of varying scales based on the original top-level input features. This approach enables progressive feature fusion across distinct channels, and the resultant feature scale ς after extraction can be expressed as shown in Equation (1).

$$\varsigma = (a_1 \times S, a_2 \times S, \ldots, a_n \times S) \tag{1}$$

where S represents the initial feature scale, corresponding to the scale of the top-level feature. The symbols a_i, where $i = 1, 2, \ldots, n$, signify the designated pooling ratios. Following this, a 3×3 convolutional kernel is applied to condense the multi-channel features to 256. Subsequently, an upsampling procedure is engaged to restore the dimensions to the original scale S in preparation for subsequent fusion. The technique of bilinear interpolation is harnessed in the upsampling phase to compute new pixel values. For given coordinates of four points, $(x_1, y_1), (x_2, y_2), (x_3, y_3)$, and (x_4, y_4), the interpolation formula employed to ascertain the value at a novel coordinate (x, y) is as follows:

$$f(x, y_1) \approx \frac{x_2 - x}{x_2 - x_1} f(Q_{11}) + \frac{x - x_1}{x_2 - x_1} f(Q_{21}) \tag{2}$$

$$f(x, y_2) \approx \frac{x_2 - x}{x_2 - x_1} f(Q_{12}) + \frac{x - x_1}{x_2 - x_1} f(Q_{22}) \tag{3}$$

$$f(x, y) \approx \frac{y_2 - y}{y_2 - y_1} f(x, y_1) + \frac{y - y_1}{y_2 - y_1} f(x, y_2) \tag{4}$$

where $Q_{11} = (x_1, y_1), Q_{12} = (x_1, y_2), Q_{21} = (x_2, y_1), Q_{22} = (x_2, y_2)$.

In this process, following the upsampling step, the feature maps that underwent pooling with varying ratios in the preceding stage have their scales equalized. These feature maps are subsequently concatenated. Recognizing the semantic disparity among these feature maps with distinct original ratios, each of them undergoes a sequence consisting of a 1×1 convolutional layer, an ReLU layer, and a 3×3 convolutional layer. This series of operations generates weight maps for each multi-scale feature map after the concatenation. Finally, these weight maps are element-wise multiplied with their corresponding feature maps from before the concatenation on each channel, followed by summation. This sequence constitutes the output of the top-level feature enhancement module.

The rationale behind the capability of the top-level enhancement module to improve the representation of the highest-level features lies in its utilization of adaptive pooling at the highest scale. This process effectively extracts features one level higher on the top-level feature map, which contains richer semantic content. Subsequently, this enhanced feature map is integrated with the original top-level feature map. Moreover, the paper introduces two adaptive pooling branches with distinct scaling ratios to combine high-level semantic features with even higher-level, diverse semantic information. These adaptive pooling branches can be extended to encompass multiple ratios, further enhancing the feature representation potential of the highest-level features.

Following the operations of the two adaptive pooling branches, bilinear interpolation is employed to upsample the feature maps. These upsampled maps are then concatenated, forming a cascaded representation of multi-scale enhanced features based on the top-level features. Subsequent to a 1×1 convolutional layer that reduces the number of channels, the result is multiplied with the original feature map that has not been concatenated. This procedure yields the final enhanced top-level features.

In the reverse progressive feature fusion process of the pyramid network, the prevalent approach involves upsampling each layer's feature map, resulting in uniform dimensions and channel numbers for the feature maps at every level. This facilitates direct merging during the fusion with lower-level features. While nearest neighbor interpolation is frequently used for upsampling in FPN implementation, this method lacks consideration for semantic differences among features of different scales. Hence, the use of bilinear interpolation during upsampling generates smoother multi-scale feature maps, thereby minimizing the semantic gap between distinct feature maps. Moreover, when enhancing top-level features and conducting multi-scale feature fusion, direct application of nearest neighbor interpolation during upsampling might lead to information loss during feature fusion at the top level. The enhanced progressive feature fusion module effectively mitigates the decline in quality of multi-scale features caused by significant semantic gaps across different levels.

3.2. Gradient Density Loss Function

In conventional classification tasks, the cross-entropy loss function is widely utilized. The equation for the binary cross-entropy loss function is expressed as follows:

$$L_{\text{cross-entropy}} = \begin{cases} -log(p), & \text{if } T = 1 \\ -log(1-p), & \text{if } T = 0 \end{cases} \tag{5}$$

where $T \in \{0,1\}$ signifies the true class label of the sample, and the corresponding probability is denoted as $p \in [0,1]$.

Next, let us denote *"out"* as the output of the model. In this scenario, we have $p = \text{sigmoid}(out)$. Now, by differentiating the aforementioned cross-entropy loss function with respect to the model's output, we arrive at

$$\frac{\partial L_{\text{cross-entropy}}}{\partial \text{out}} = \begin{cases} p-1, & \text{if } T = 1 \\ p, & \text{if } T = 0 \end{cases} \tag{6}$$

At this juncture, the magnitude of the gradient is defined as *"gradnorm"*, which can be represented as

$$gradnorm = |p - T| = \begin{cases} 1-p, & \text{if } T = 1 \\ p, & \text{if } T = 0 \end{cases} \tag{7}$$

where the value of *"gradnorm"* indicates the level of difficulty posed by a sample and its impact on the overall gradient. To address the challenge stemming from uneven gradient distribution due to varying sample quantities, the concept of the "gradient density loss function" is introduced. This strategy involves partitioning the gradient into distinct ranges, counting the number of samples within each range, and evaluating the distribution within these ranges. Subsequently, using the sample count in each range and the length of the range, the *"gradientdensity"* (GD) is defined. This metric reflects the number of samples per unit magnitude.

$$GD(gradnorm) = \frac{1}{l_\varepsilon(gradnorm)} \sum_{k=1}^{N} \delta_\varepsilon(gradnorm_k, gradnorm) \tag{8}$$

$$l_\varepsilon(gradnorm) = min(gradnorm + \frac{\varepsilon}{2}, 1) - max(gradnorm - \frac{\varepsilon}{2}, 0) \tag{9}$$

$$\delta_\varepsilon(x,y) = \begin{cases} 1, & y - \frac{\varepsilon}{2} \le x < y + \frac{\varepsilon}{2} \\ 0, & otherwise \end{cases} \tag{10}$$

where ε represents the magnitude within the range. Therefore, the gradient density $GD(gradnorm)$ can be expressed as the number of samples within a certain interval $[gradnorm - \frac{\varepsilon}{2}, gradnorm + \frac{\varepsilon}{2}]$. Subsequently, the gradient density parameter is defined as follows:

$$\beta_i = \frac{N}{GD(gradnorm_i)} \tag{11}$$

In the equation, N represents the total number of samples. This parameter is used to weigh the cross-entropy loss classification function, resulting in a new gradient density loss function. This function is designed to reduce the weight of larger gradient density intervals, thereby diminishing the impact of challenging samples on the model. This adjustment helps mitigate the influence of difficult outliers on the final accuracy. For better comprehension, let us express the above equation in an alternative form:

$$\beta_i = \frac{1}{GD(gradnorm_i)/N} \tag{12}$$

As can be discerned, $GD(gradnorm_i)/N$ signifies the proportion of sample gradients within the gradient partitioning interval concerning the total sample count. If the sample count within the gradient interval is uniformly distributed, then $\beta_i = 1$ for every interval, implying that the gradient density parameter in that interval would hold no sway over the result. Conversely, for demanding samples characterized by higher gradient densities and larger counts, this value would diminish. This characteristic serves to counterbalance the impact of a multitude of challenging samples on the model's precision.

Hence, the substitution of the classification loss function focal loss [19] with the gradient density loss function in Sparse R-CNN can be mathematically expressed as follows:

$$L_{gcls} = \frac{1}{N}\sum_{i=1}^{N} \beta_i L_{cross\ entropy}(p_i, p_i^{gt}) = \sum_{i=1}^{N} \frac{L_{cross\ entropy}(p_i, p_i^{gt})}{GD(gradnorm_i)} \tag{13}$$

In this equation, p_i^{gt} represents the actual classification label, p_i represents the predicted label, and $GD(gradnorm_i)$ represents the count of samples falling within the ith interval after dividing the gradient magnitudes into N intervals using the parameter N. The term $L_{cross\ entropy}(p_i, p_i^{gt})$ denotes the cross-entropy classification loss function.

Figure 3 displays the gradient magnitude distribution of a converged model. Along the horizontal axis, the gradient magnitude is depicted, with the right side representing more challenging samples and the left side representing simpler ones. The vertical axis, on a logarithmic scale, illustrates the corresponding number of samples at each gradient magnitude. The logarithmic scale is chosen due to the substantial variance in quantities between simple and moderately difficult samples in practical scenarios. The graph highlights that the largest number of samples falls into the simple category, while moderately difficult samples are fewer in number. Remarkably, the count of extremely challenging samples increases notably, surpassing even the quantity of moderately difficult ones by a substantial margin. As a consequence, these highly challenging samples can potentially lead to a reduction in model accuracy. This is due to the fact that these challenging samples, acting as outliers, often possess gradient directions significantly distinct from those of other simple and moderately difficult samples. Forcing a nearly converged model to learn from these exceedingly challenging exceptional samples can result in inaccurate classification for a considerable number of other samples.

After applying the gradient density loss function to weight the cross-entropy loss function, the relationship between the gradient magnitudes of samples and their corresponding quantities is illustrated in Figure 4.

In Figure 4, the horizontal axis represents the initial gradient magnitudes, while the vertical axis illustrates the gradient magnitudes after employing distinct loss functions. The observation reveals that within the region associated with simple and moderately difficult samples, situated on the left side of Figure 3, the tendencies of the gradient density loss and focal loss align. This correspondence implies that these two loss functions demonstrate coherent behavior in discriminating between simple and moderately difficult samples. Conversely, for challenging samples situated on the right side of the graph, the gradient density loss effectively diminishes the gradient magnitudes. This adjustment serves to stabilize the model when confronted with difficult and exceptional samples, thereby amplifying training efficiency.

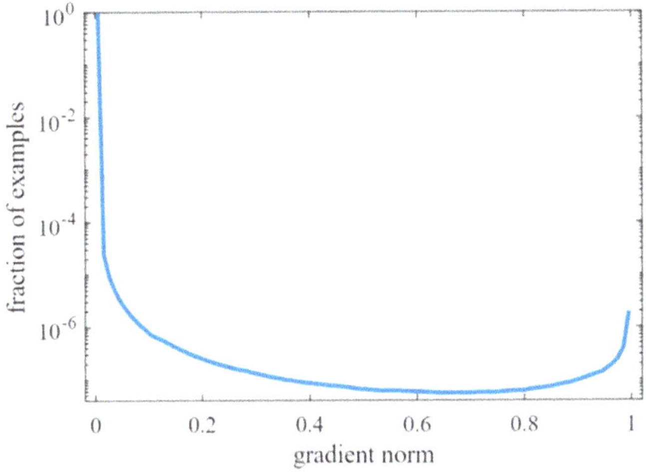

Figure 3. Distribution of sample gradient magnitudes and corresponding quantities [20].

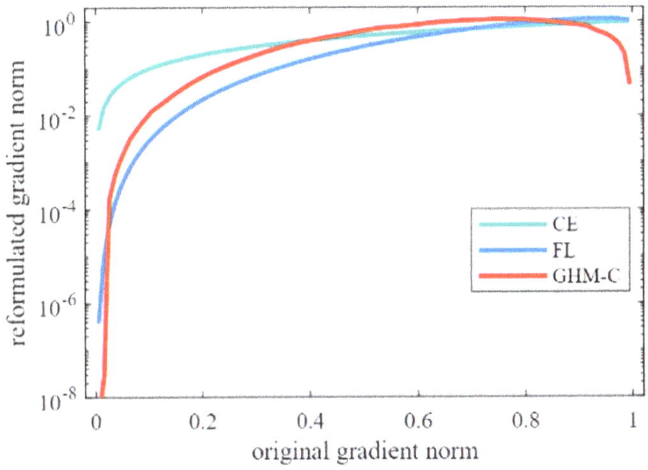

Figure 4. Gradient magnitudes with different weighted loss functions [20], where gradient density loss is represented by the red line and focal loss is represented by the blue line. Here, FL represents the focal loss, CE represents the cross-entropy, and GHM-C represents the gradient harmonizing mechanism classification.

After integrating the enhanced loss function, the comprehensive loss function for the image object detection method is formulated as

$$Loss = \lambda_1 Loss_{cls} + \lambda_2 Loss_{L1} + \lambda_3 Loss_{giou} \tag{14}$$

where λ_1, λ_2, and λ_3 denote the coefficients allocated to the classification loss, bounding box regression loss, and intersection over union (IoU) loss, correspondingly.

The FL is a well-known loss function introduced to address the class imbalance problem in object detection. It assigns higher weights to hard-to-classify examples, reducing the contribution of easily classifiable examples to the loss. While FL effectively handles class imbalance, it does not explicitly consider the density of object instances within an image, potentially leading to suboptimal performance in scenarios where objects are densely packed.

Class-agnostic focal loss (CF) extends FL by making it class-agnostic, thus focusing on the overall detection quality without distinguishing between different object classes. This modification enhances the detection of objects irrespective of their classes, which is particularly useful when dealing with diverse or unknown objects. However, CF may not effectively address the issue of object density when objects of varying classes are densely distributed.

GHM-C introduces gradient harmonization to address the imbalance between easy and hard examples during training. It effectively balances the learning process between different examples based on the gradient distribution, enhancing the network's ability to focus on challenging examples. However, similar to FL, GHM-C does not explicitly consider object density.

In contrast, our proposed gradient density loss is specifically designed to address the challenge of object detection in scenarios with varying object densities. It combines the benefits of FL and GHM-C by incorporating density-based weighting into the loss calculation. Gradient density loss assigns higher weights to challenging examples, both in terms of classification difficulty and object density. By integrating these aspects, it aims to enhance the detection of objects in densely populated scenes, contributing to improved overall detection performance.

In summary, the proposed gradient density loss introduces an innovative approach that combines the strengths of FL, CF, and GHM-C while explicitly considering object density. This approach aims to offer a more comprehensive solution for object detection, particularly in challenging scenarios. Our comparative analysis shows that gradient density loss provides a valuable addition to the field of object detection, especially in cases where object density varies significantly across images.

3.3. The Overall Framework

The model proposed in this paper follows the overall architecture of Sparse R-CNN while introducing improvements on top of it, which is depicted in Figure 5.

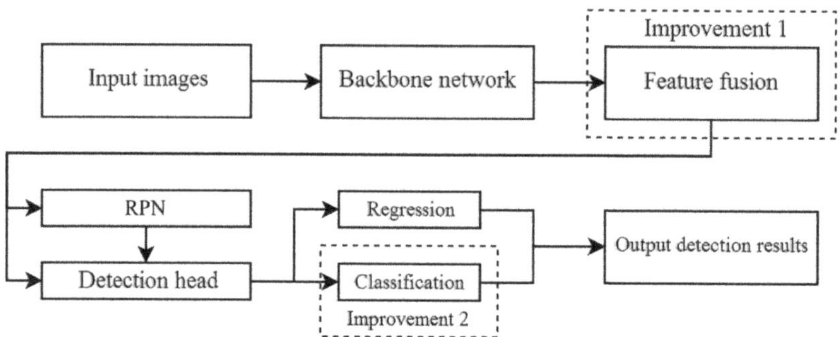

Figure 5. The overall framework.

In the section pertaining to the backbone network, the model employs a ResNet-50 network that has been pretrained on the ImageNet dataset. Subsequent enhancements are applied to the feature pyramid network (FPN) module of the Sparse R-CNN network using the proposed enhanced feature pyramid network (FPN) for multi-layer feature fusion. Given that the input and output scales, as well as the channel numbers, of the enhanced FPN remain consistent with those of the Sparse R-CNN network, which utilizes feature vectors from the four stages of the ResNet-50 backbone network (Stage 1–4), these improvements do not affect other components of the network.

Within the enhanced FPN's module for multi-scale fusion, the top-level features receive initial enhancement. A dedicated top-level feature enhancement module is constructed based on the output of the Stage 4 residual layer of ResNet-50. The convolutional layers within the top-level feature enhancement module utilize Xavier initialization. After producing higher-level feature vectors through this module, subsequent stages involve the utilization of bilinear interpolation for both upsampling and fusion. This stands in contrast to the nearest-neighbor interpolation employed in the fusion stage of the original Sparse R-CNN algorithm. This refined approach mitigates top-level feature degradation, a concern in the original Sparse R-CNN algorithm's multi-scale features. Moreover, due to the smoother feature fusion between different scales, the semantic gap between these features is reduced, leading to superior quality features and ultimately augmenting the algorithm's accuracy.

Post feature extraction, the process proceeds by employing the Region Proposal Network (RPN) to generate an embedding matrix for each of the fixed 100 detection boxes. These matrices encapsulate the coordinates and classification features of the detection boxes and possess dimensions of 100×4 and 100×256, respectively. The initial coordinates of the detection boxes are assigned using a random distribution. Subsequently, RoIAlign is utilized to extract features from the corresponding positions of the input feature maps for each detection box. Employing the multi-scale feature maps from the four layers, each layer maps the embedding matrices of the detection boxes to the extracted multi-scale feature maps. Consequently, feature vectors linked to each detection box are retrieved.

In relation to the label assigner within the detection head, a threshold of 0.5 is defined as the criterion for categorizing positive and negative samples. Classification and regression tasks are accomplished via fully connected layers. The regression loss employs the L1 loss from the original Sparse R-CNN. The GIoU loss is employed for IoU loss. Additionally, an enhancement is introduced via the proposed gradient density classification loss to modify the original Sparse R-CNN's focal loss. The classification loss weight is set to 2.0, the regression loss weight to 2.0, and the IoU loss weight to 5.0. Upon completion of the Sparse R-CNN detection process, the customary nonmaximum suppression (NMS) post-processing step is eschewed. Instead, the model directly outputs the 100 detection boxes. During visualization, confidence thresholding can be applied to display pertinent detection boxes.

4. Experiments and Analysis

4.1. Datasets and Evaluation Metrics

This section outlines the datasets employed in the experiments and the evaluation criteria utilized.

MS COCO 2017: The Microsoft Common Objects in Context (MS COCO) dataset [21] stands as a cornerstone in computer vision tasks, including object detection and semantic segmentation. It gained prominence following the discontinuation of the ImageNet competition in 2017. COCO has attracted contributions from global tech giants like Google, Microsoft, Facebook, and esteemed research institutions. It boasts a comprehensive object detection dataset drawn from intricate everyday scenes, featuring over 200,000 images and 80 categories. The training set of COCO 2017 comprises 118,287 images, and the validation set contains 5000 images.

MS COCO 2014: Introduced in 2014 [21], the COCO dataset of 2014 diverges in content from its 2017 counterpart. It incorporates a training set of 82,783 images, a validation set of 40,504 images, and a test set encompassing 40,775 images.

Pascal VOC 2007 [22]: Pascal VOC (Pattern Analysis, Statistical Modeling, and Computational Learning Visual Object Classes) serves as a benchmark dataset for object detection, encompassing image classification, object detection, and semantic segmentation tasks. The 2007 edition comprises 20 categories and a total of 9963 annotated images. The dataset is partitioned into a training set of 5011 images and a test set of 4952 images.

Pascal VOC 2012: Representing the concluding year of this esteemed computer vision challenge, Pascal VOC 2012 features image classification, object detection, object segmentation, and action recognition as its principal tasks. It encompasses 11,530 images with 27,450 object annotations and 6929 segmentation annotations.

Evaluation Criteria: For object detection evaluation, the COCO dataset's evaluation metrics are employed due to the conversion of Pascal VOC datasets into COCO format for experimentation. The primary evaluation metric for object detection in the COCO dataset is the mean average precision (mAP), a measure of average accuracy. Given that object detection involves generating detection boxes, calculating the average precision entails a series of steps. The average AP values are determined for detecting diverse classes of objects prior to assessing classification accuracy.

Throughout COCO evaluation, multiple intersection over union (IoU) thresholds are set, often spanning from 0.5 to 0.95 with increments of 0.05. For each IoU threshold, the average precision is computed for each object class. Ultimately, the obtained AP values across various IoU thresholds are averaged to yield the comprehensive mAP score.

When computing the mean mAP, a crucial metric at play is intersection over union (IoU), employed to gauge the intersection extent between a detected bounding box and its corresponding ground truth bounding box. While assessing the COCO dataset, IoU thresholds, frequently set at levels like 0.5 and 0.75, come into play. If the intersection over union value between the network's output detection box and the actual ground truth bounding box exceeds the threshold, COCO deems the detection as accurate. A comprehensive illustration of the IoU calculation process is depicted in Figure 6.

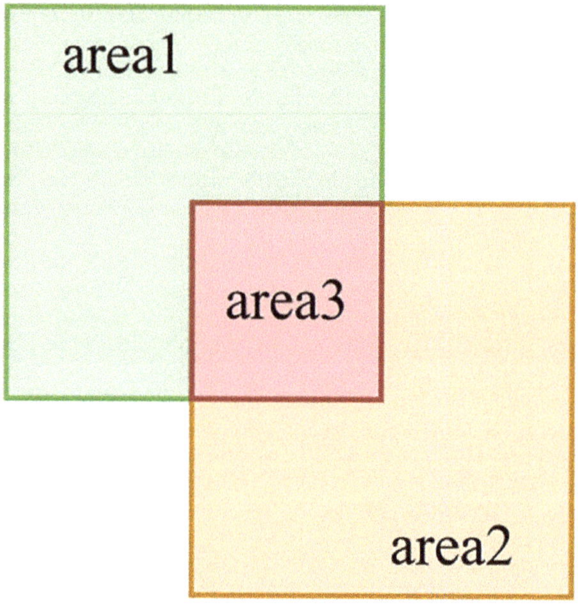

Figure 6. Calculation of IoU for bounding boxes.

IoU stands as a pivotal metric in evaluating the efficacy of object detection, as it gauges the alignment between the predicted detection box and the true ground truth box, thereby indicating the accuracy of the detection. This assessment is based on the comparison of the predicted detection box, denoted as $area_1$, and the actual annotated boundary box for the target, referred to as $area_2$, with their intersection represented as $area_3$. Mathematically, IoU is computed as follows:

$$IoU = \frac{S_{area3}}{S_{area1} + S_{area2} - S_{area3}} \tag{15}$$

In tandem with IoU calculation, the evaluation process also incorporates classification attributes, encompassing the counts of true positives (TPs), false positives (FPs), false negatives (FNs), and true negatives (TNs). These metrics capture the four potential scenarios that arise between the classification outcome and the actual ground truth object: (1) TPs: Detection boxes with an IoU greater than the threshold accurately represent detected objects. (2) FPs: Detection boxes lacking corresponding ground truth annotations are deemed false positives. (3) FNs: Ground truth annotations without corresponding detection boxes are classified as false negatives. (4) TNs: Instances where neither ground truth annotations nor detection boxes are present.

Drawing from these scenarios, the evaluation of precision and recall becomes feasible. Precision, also referred to as the positive predictive value, assesses the proportion of correctly identified positive instances out of all instances predicted as positive. On the other hand, recall, also known as sensitivity, measures the proportion of correctly identified positive instances out of all actual positive instances. The formulas for precision and recall computation are as follows:

$$Precision = \frac{TP}{TP + FP} \tag{16}$$

$$Recall = \frac{TP}{TP + FN} \tag{17}$$

Accuracy signifies the proportion of correct outcomes among all the targets predicted by the network, whereas recall indicates the proportion of correctly predicted detection boxes output by the network in comparison to all the true target annotations. In the COCO dataset, each annotated detection box is accompanied by a class confidence score. By varying the confidence score thresholds, different detection results can be generated, yielding varying accuracy and recall values, which are graphically represented in the precision–recall (P–R) curve. The preeminent metric, mean average precision (mAP), is calculated through a series of IoU thresholds spanning from 0.5 to 0.95, with increments of 0.05. The determination of mAP entails computing the precision of detection at each IoU threshold, culminating in the final mAP value.

Within the ambit of COCO's evaluation metrics, there exists an element that computes precision based on target size. The COCO dataset classifies targets into three size categories: large, medium, and small, contingent on the area of their authentic annotation boxes. Targets featuring an annotation pixel area exceeding 96 × 96 are designated as large, while those below 32 × 32 are denoted as small, and those encompassed between 32 × 32 and 96 × 96 are categorized as medium. In the ultimate assessment, the COCO dataset computes precision separately for these three target sizes, denoted as AP_{small}, AP_{medium}, and AP_{large}.

4.2. Experimental Environment

The method presented in this paper is implemented using the PyTorch deep learning framework, and all experiments are carried out on an NVIDIA RTX 1080Ti GPU with 11 GB of memory. The initial learning rate for training is configured at 0.000025, and the training process employs the SGD optimizer with a momentum value of 0.9. The learning rate is subjected to an equal interval reduction strategy. The model undergoes training for a cumulative 150 epochs across the complete dataset, employing a batch size of 2 for training images.

4.3. Ablation Experiments

In contrast to existing methodologies, this paper introduces two novel modules: the enhanced feature pyramid network and the gradient density object detection classification loss function. To meticulously assess the actual efficacy of these modules in enhancing the performance of the baseline method in real-world scenarios, this study undertakes ablation experiments. The comparative reference point for these evaluations is the baseline method derived from the official implementation of Sparse R-CNN.

The results in Table 1 highlight the enhancements achieved by incorporating the enhanced feature pyramid network and gradient density object detection classification loss function modules into the initial baseline method, Sparse R-CNN. The integration of the enhanced feature pyramid network module contributes to a 0.4% increase in average precision compared to the original Sparse R-CNN baseline. This improvement is consistent across various object scales, encompassing large, medium, and small objects. By replacing focal loss with the gradient density loss function in the baseline approach, no substantial changes are observed in the overall metrics, yet the average precision experiences a slight elevation of 0.1%. Notably, when both the enhanced feature pyramid network module and gradient density loss function are simultaneously applied to the baseline, the average precision advances by 1.1%. This translates to a 2.4% boost in precision for small-scale objects, and 0.9% and 0.3% enhancements for medium and large objects, respectively.

The adoption of the enhanced feature pyramid network module brings about a marked improvement in feature quality, positively impacting detection performance across various object scales. Furthermore, the incorporation of the gradient density loss function during training effectively reduces the model's susceptibility to challenging small objects, resulting in enhanced detection accuracy for such objects, without compromising the performance for medium and large objects. In terms of computational efficiency, both Improvement 1 and Improvement 2 had a minor effect on the training and inference speeds of the model. Specifically, the model's inference rate slightly decreased by 1.1 images per second, and the training time per iteration saw a marginal increase of 0.0159 s.

Table 1. The results of the ablation experiments on the COCO 2017 dataset.

Model	$AP_{0.5:0.95}$	$AP_{0.5}$	$AP_{0.75}$	AP^s	AP^m	AP^1	AP^{100}	AR^s	AR^m	AR^1	Inference (img/s)	Training (s/iter)
Baseline	0.384	0.560	0.414	0.206	0.409	0.540	0.578	0.340	**0.621**	0.772	**27.5**	**0.2982**
Baseline + Improvement 1	0.388	0.566	0.418	0.209	0.413	**0.547**	0.579	0.341	**0.621**	0.782	26.6	0.3095
Baseline + Improvement 2	0.385	0.564	0.413	0.208	0.407	0.540	0.579	0.342	**0.621**	0.781	26.7	0.3100
Baseline + Improvement 1 + Improvement 2	**0.395**	**0.578**	**0.423**	**0.232**	**0.418**	0.543	**0.581**	**0.367**	**0.621**	**0.788**	26.4	0.3141

4.4. Comparative Experiments with Existing Methods

To validate the practical effectiveness of the proposed enhancements, a series of comparative experiments were conducted by incorporating the proposed modules into the baseline method. Subsequently, these enhanced methods were systematically compared against several state-of-the-art object detection approaches, all evaluated on the four widely used benchmark datasets. The obtained validation results unequivocally demonstrate that the proposed method surpassed the performance of these existing advanced methods.

Table 2 presents the performance comparison on the COCO 2017 dataset, where the enhancements proposed in this study were rigorously assessed against prominent 2021 methods, including YOLOX, YOLOF, and Deformable DETR. The observed improvements in average precision were substantial, with increments of 0.3%, 0.1%, and 0.4% for the aforementioned methods, respectively. Furthermore, Table 3 reveals the outcome of the enhancements on the COCO 2014 dataset. The proposed improvements exhibited remarkable average precision boosts of 0.7% and 0.6% when juxtaposed with the YOLOX and YOLOF methods proposed in 2021. Additionally, a 0.1% enhancement was achieved when compared to Deformable DETR. These comparisons illustrate the potency of the introduced enhancements in enhancing object detection accuracy.

The superiority of the proposed method is evident from the results displayed in Table 4 on the Pascal VOC 2007 dataset. Against Deformable DETR, YOLOF, and YOLOX, the method introduced in this paper exhibited remarkable average precision enhancements of 0.4%, 0.6%, and 0.7%, respectively. Notably, in terms of accuracy for detecting large objects, the proposed method outperformed YOLOX by a substantial margin of 1.9%. Moreover, when benchmarked against Dynamic R-CNN, DETR, and CentriPetalNet, the enhancements showcased their prowess by delivering substantial average precision improvements of 2.4%, 2.3%, and an impressive 4.0%, respectively. This compelling comparison demonstrates the considerable advancements introduced by the proposed method across multiple evaluation metrics.

Table 2. Comparative experiments were conducted on the COCO 2017 dataset.

Model	$AP_{0.5:0.95}$	$AP_{0.5}$	$AP_{0.75}$	AP^{s}	AP^{m}	AP^{1}	AP^{100}	AR^{s}	AR^{m}	AR^{1}
Faster RCNN [8]	0.306	0.511	0.324	0.150	0.336	0.389	0.482	0.179	0.530	0.618
Focal loss [19]	0.307	0.502	0.330	0.142	0.335	0.390	0.478	0.175	0.528	0.623
CornerNet [23]	0.302	0.444	0.316	0.137	0.320	0.389	0.480	0.170	0.522	0.619
ATSS [24]	0.315	0.479	0.340	0.176	0.348	0.406	0.489	0.202	0.532	0.628
carafe [25]	0.308	0.208	0.331	0.184	0.342	0.377	0.487	0.201	0.524	0.611
Cascade RCNN [9]	0.342	0.518	0.369	0.188	0.375	0.447	0.509	0.312	0.585	0.706
CenterNet [26]	0.295	0.461	0.314	0.102	0.329	0.467	0.485	0.228	0.517	0.728
AutoAssign [27]	0.348	0.528	0.377	0.184	0.385	0.449	0.506	0.309	0.599	0.709
CentriPetal Net [28]	0.358	0.522	0.378	0.188	0.382	0.470	0.517	0.320	0.608	0.720
DETR [29]	0.391	0.577	0.409	0.174	0.424	0.587	0.571	0.350	0.617	**0.790**
Dynamic R-CNN [30]	0.389	0.576	0.427	0.221	0.419	0.517	0.570	0.354	0.621	0.779
Deformable DETR [31]	0.391	0.581	0.424	**0.239**	0.423	0.541	0.585	0.369	0.631	**0.790**
YOLOF [32]	0.393	0.583	0.427	0.238	0.421	0.542	0.582	0.368	0.620	0.788
YOLOX [33]	0.390	0.580	**0.430**	0.237	**0.440**	0.506	0.583	0.369	0.638	0.772
Ours	**0.394**	**0.588**	0.423	0.232	0.418	**0.543**	**0.586**	**0.370**	**0.639**	0.788

Table 3. Comparative experiments were conducted on the COCO 2014 dataset.

Model	$AP_{0.5:0.95}$	$AP_{0.5}$	$AP_{0.75}$	AP^{s}	AP^{m}	AP^{1}	AP^{100}	AR^{s}	AR^{m}	AR^{1}
Faster RCNN [8]	0.324	0.528	0.345	0.178	0.359	0.403	0.501	0.195	0.552	0.638
Focal loss [19]	0.327	0.527	0.347	0.182	0.352	0.404	0.506	0.198	0.554	0.642
CornerNet [23]	0.324	0.468	0.335	0.157	0.348	0.408	0.504	0.195	0.541	0.632
ATSS [24]	0.336	0.498	0.361	0.195	0.368	0.428	0.528	0.224	0.554	0.648
carafe [25]	0.324	0.228	0.354	0.201	0.368	0.394	0.521	0.224	0.548	0.637
Cascade RCNN [9]	0.362	0.538	0.375	0.204	0.392	0.467	0.527	0.334	0.607	0.724
CenterNet [26]	0.314	0.485	0.338	0.125	0.348	0.483	0.502	0.248	0.538	0.743
AutoAssign [27]	0.362	0.548	0.395	0.204	0.405	0.468	0.542	0.328	0.617	0.725
CentriPetal Net [28]	0.380	0.546	0.392	0.207	0.415	0.498	0.547	0.337	0.608	0.747
DETR [29]	0.410	0.604	0.425	0.195	0.442	0.601	0.595	0.372	0.638	0.809
Dynamic R-CNN [30]	0.402	0.594	0.448	0.243	0.440	0.534	0.591	0.372	0.641	0.799
Deformable DETR [31]	0.424	0.594	0.447	0.259	**0.442**	0.561	0.607	**0.389**	0.651	0.814
YOLOF [32]	0.418	0.608	0.448	**0.268**	0.448	0.567	0.604	0.382	0.620	0.801
YOLOX [33]	0.419	0.601	0.439	0.258	0.431	0.529	0.602	0.378	0.655	0.795
Ours	**0.425**	**0.609**	**0.449**	**0.268**	**0.442**	**0.568**	**0.609**	**0.389**	**0.658**	**0.819**

Table 4. Comparative experiments were conducted on the Pascal VOC 2007 dataset.

Model	$AP_{0.5:0.95}$	$AP_{0.5}$	$AP_{0.75}$	AP^{s}	AP^{m}	AP^{1}	AP^{100}	AR^{s}	AR^{m}	AR^{1}
Faster RCNN [8]	0.418	0.619	0.418	0.275	0.429	0.508	0.598	0.284	0.624	0.729
Focal loss [19]	0.416	0.614	0.420	0.267	0.425	0.511	0.596	0.280	0.618	0.722
CornerNet [23]	0.428	0.548	0.429	0.284	0.427	0.518	0.618	0.297	0.633	0.728
ATSS [24]	0.440	0.599	0.481	0.299	0.459	0.518	0.628	0.314	0.642	0.740
carafe [25]	0.428	0.358	0.438	0.315	0.485	0.499	0.618	0.328	0.649	0.735
Cascade RCNN [9]	0.471	0.642	0.474	0.308	0.402	0.575	0.631	0.425	0.709	0.815
CenterNet [26]	0.423	0.574	0.427	0.234	0.468	0.591	0.603	0.359	0.637	0.832
AutoAssign [27]	0.452	0.658	0.499	0.305	0.496	0.557	0.651	0.417	0.729	0.834
CentriPetal Net [28]	0.485	0.644	0.483	0.308	0.524	0.599	0.658	0.428	0.708	0.851
DETR [29]	0.502	0.700	0.528	0.286	0.561	0.623	0.684	0.483	0.728	0.895
Dynamic R-CNN [30]	0.501	0.692	0.538	0.344	0.548	0.625	0.689	0.483	0.752	0.899
Deformable DETR [31]	0.521	0.703	**0.548**	0.354	0.554	0.672	0.699	0.481	**0.755**	**0.910**
YOLOF [32]	0.519	0.700	**0.548**	0.353	0.542	0.661	0.702	0.482	0.731	0.900
YOLOX [33]	0.518	0.701	0.547	0.351	0.561	0.655	0.700	0.472	0.751	0.895
Ours	**0.525**	**0.708**	**0.548**	**0.358**	**0.568**	**0.674**	**0.704**	**0.485**	**0.755**	0.899

Table 5 reveals the compelling performance of the proposed method on the Pascal VOC 2012 dataset. Compared to YOLOX, YOLOF, and Deformable DETR, the method introduced in this paper excelled, with average precision improvements of 0.2%, 0.7%, and 0.1%, respectively. When placed against Dynamic R-CNN, DETR, and CentriPetalNet, the proposed method boasted impressive enhancements of 1.9%, 2.4%, and a notable 3.7% in terms of average precision. The results illustrate that the proposed method not only exhibited slight yet consistent improvements in average detection precision and detection recall metrics for various object sizes when compared to advanced methods like Deformable DETR and YOLOX, but it also outperformed Dynamic R-CNN, DETR, and CentriPetalNet, reaffirming its superiority across diverse evaluation metrics.

Table 5. Comparative experiments were conducted on the Pascal VOC 2012 dataset.

Model	$AP_{0.5:0.95}$	$AP_{0.5}$	$AP_{0.75}$	AP^{s}	AP^{m}	AP^{1}	AP^{100}	AR^{s}	AR^{m}	AR^{1}
Faster RCNN [8]	0.395	0.592	0.392	0.254	0.402	0.486	0.532	0.263	0.608	0.701
Focal loss [19]	0.394	0.589	0.389	0.251	0.397	0.485	0.530	0.251	0.610	0.698
CornerNet [23]	0.402	0.521	0.405	0.261	0.404	0.496	0.598	0.275	0.612	0.704
ATSS [24]	0.421	0.572	0.462	0.271	0.432	0.492	0.604	0.294	0.624	0.723
carafe [25]	0.401	0.331	0.418	0.295	0.465	0.472	0.592	0.301	0.624	0.712
Cascade RCNN [9]	0.458	0.621	0.450	0.281	0.381	0.551	0.612	0.402	0.698	0.792
CenterNet [26]	0.401	0.557	0.449	0.215	0.448	0.571	0.605	0.334	0.624	0.819
AutoAssign [27]	0.431	0.635	0.471	0.285	0.474	0.539	0.621	0.394	0.700	0.804
CentriPetal Net [28]	0.462	0.621	0.468	0.281	0.504	0.574	0.638	0.402	0.684	0.831
DETR [29]	0.475	0.675	0.501	0.263	0.541	0.682	0.659	0.462	0.705	0.876
Dynamic R-CNN [30]	0.480	0.672	0.512	0.322	0.524	0.602	0.664	0.462	0.730	0.870
Deformable DETR [31]	0.498	0.682	0.521	0.335	0.532	0.651	0.681	0.462	0.732	0.871
YOLOF [32]	0.492	0.688	0.521	**0.342**	0.528	0.642	0.681	0.468	0.710	0.870
YOLOX [33]	0.497	0.682	0.521	0.335	0.538	0.601	0.681	0.474	0.731	0.872
Ours	**0.499**	**0.689**	**0.525**	0.304	**0.547**	**0.652**	**0.682**	**0.478**	**0.741**	**0.874**

In summary, our characterization of "superior performance" is based on improvements over the baseline model's results within the specific domain of image object detection.

(1) Benchmark and context: When referring to our results as "superior performance", we take into account the benchmark and context in which our evaluation is conducted. Our research aims to enhance image object detection methods, and our proposed techniques achieve significant improvements over existing approaches. While the absolute value of 51% AP may not seem extraordinarily high in a broader context, it represents a noteworthy advancement within the specific domain of image object

detection, which typically involves numerous challenges, including diverse object sizes, scales, and varying degrees of complexity.

(2) Superiority over baseline: We compare our approach with a baseline model, which is often a standard practice in the field of computer vision. By demonstrating an improvement from the baseline model's performance, we highlight the effectiveness of our proposed methodologies. Our experiments are carried out in the same training/testing data setting, ensuring a fair evaluation.

(3) Impact of data variability: Data variability can indeed affect model performance. To address this concern, our experimentation is designed to include a diverse range of object classes, sizes, and complexities, which often reflect real-world scenarios. We understand that by changing the dataset, the performance might vary, and it could be an interesting direction for future research.

4.5. Visualization Results

This subsection presents the visualization results obtained from the COCO 2017, COCO 2014, Pascal VOC 2007, and Pascal VOC 2012 datasets. Among them, Figure 7 provides a visual representation of the outcomes achieved by this work on the COCO 2017 dataset. In the first row of visual comparison results, the baseline method succeeds in detecting human targets and three bird targets, albeit missing the handbag. However, with the improved model, the detection of the handbag is accomplished without any adverse impact on the detection of other objects. Moving to the second row of visual comparison results, the baseline method can spot train cars and traffic lights, yet it overlooks the small distant bird target. Interestingly, the improved model successfully captures this overlooked target. In the third row of visual comparison results, the baseline method incorrectly identifies the camera in the bottom left corner as a water bottle and also mistakes the charger for a cellphone. However, upon evaluation with the improved model, these misclassifications are effectively rectified.

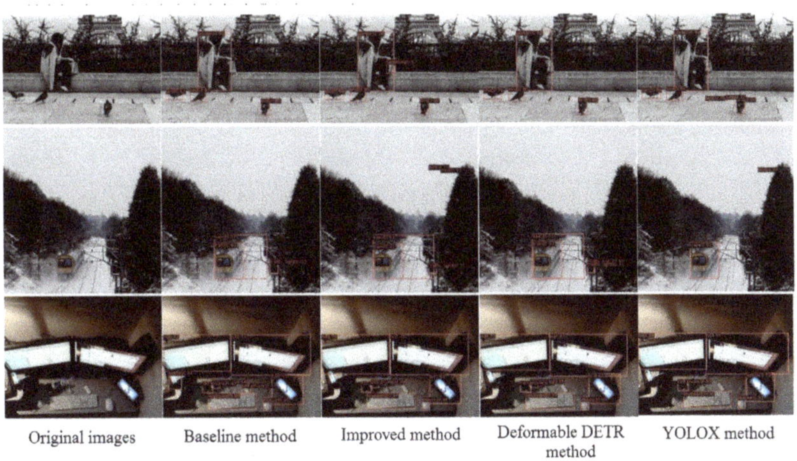

Figure 7. Visualization results on MS COCO 2017 dataset.

Figure 8 illustrates the visualization results of this work on the MS COCO 2014 dataset.

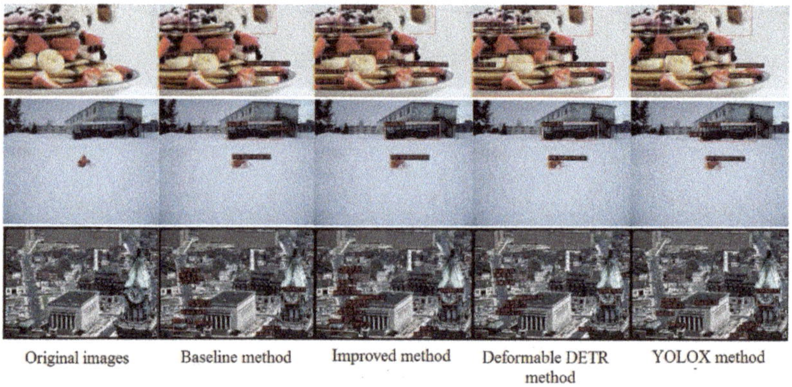

Figure 8. Visualization results on MS COCO 2014 dataset.

In the first row of results, the baseline method successfully detects the occluded teddy bear target and the sliced banana target, yet it falls short in detecting another occluded banana target. The improved method's detection results, on the other hand, manage to successfully identify the occluded banana target. However, there are a few instances of misclassifications, such as pies being incorrectly labeled as a bowl and banana targets. Moving to the second row of results, the baseline method effectively detects the bus target and the fire hydrant target, but struggles with misclassifying several pillars behind the bus as humans. The improved method's results rectify this misclassification. In the third row, a scenario involving extremely small targets is presented, showcasing multiple car targets. Here, the baseline method misses a substantial number of car targets, while the improved method detects more car targets when compared to the baseline. Moreover, the improved method also manages to detect the clock target that the baseline's results display, albeit both methods exhibiting some misclassifications of bus targets. This clearly demonstrates that the incorporation of the enhanced feature pyramid module enhances the detection capability for small-scale targets while maintaining accuracy for larger and medium-sized targets.

Figure 9 provides a visual comparison of results on the VOC 2007 dataset. In the first row of results, the baseline method successfully detects the car target and the person driving the car. However, it misclassifies the billboard located next to the race track as a train target. Upon implementing the improved model, this misclassification is rectified. Shifting to the second row of results, the baseline method produces inaccurate detections, which are corrected when the improved model is utilized. Nevertheless, some misclassifications still persist, such as the incorrect labeling of a gift-wrapped target as a handbag target. Finally, in the third row of results, the baseline method encounters a notable misclassification scenario where a substantial number of fruit and biscuit targets are erroneously labeled as cake targets.

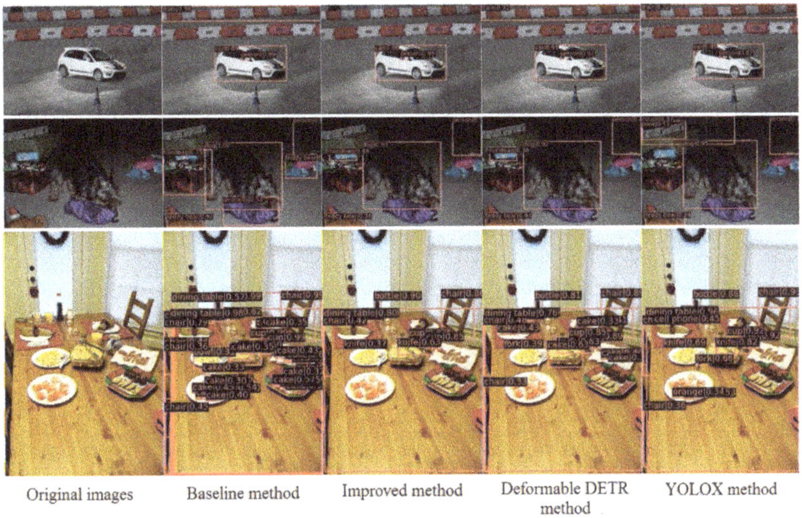

Figure 9. Visualization results on VOC 2007 dataset.

Figure 10 provides a visual comparison of results on the VOC 2012 dataset. In the first row of examples, the baseline method successfully detects the dining table, person, and chair targets. However, the car detection result in the top right corner is incorrect. The improved model corrects this misclassification and identifies additional targets, such as mobile phones, blankets, bowls, and more chairs, showcasing superior performance compared to the baseline method. Transitioning to the second row of results, the baseline method misclassifies a distant car as a bus and erroneously labels a prominent piece of clothing as a person. These misclassifications are rectified in the output of the improved model. Moving on to the third row of results, in scenarios featuring smaller distant person targets, the improved model demonstrates improved detection performance compared to the baseline method, successfully capturing more of the smaller targets.

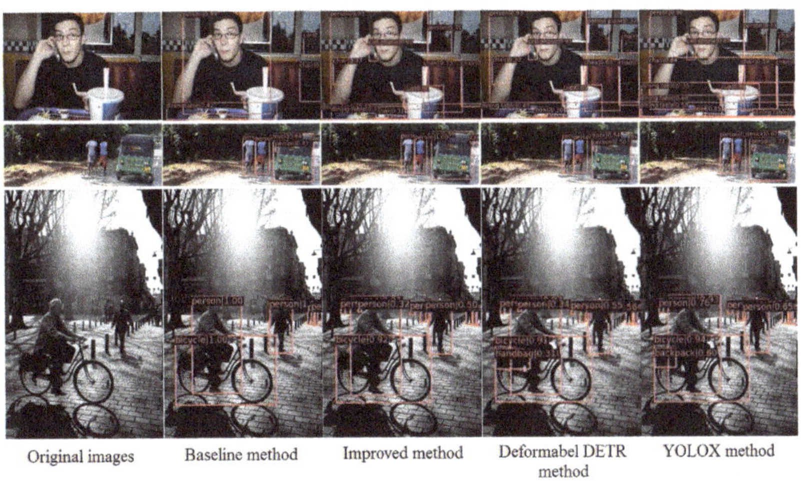

Figure 10. Visualization results on Pascal VOC 2012 dataset.

Here, we present a comprehensive analysis of the performance variations observed when applying our proposed method to different datasets. These variations are crucial for understanding the adaptability and limitations of our approach and shedding light on the intricate interplay of dataset characteristics, specific challenges, data distribution, model adaptability, and generalization capabilities.

(1) Dataset characteristics: The COCO 2017 dataset and the COCO 2014 dataset differ in terms of the number of images, object categories, and scene complexity. COCO 2017 is more extensive, while COCO 2014 is a slightly smaller dataset. These differences in dataset characteristics can influence the performance due to variations in object diversity, object sizes, and scene complexities. The Pascal VOC datasets (2007 and 2012) are relatively smaller and contain fewer object categories. The smaller dataset size can make it more challenging for models to generalize, potentially affecting performance.
(2) Data distribution: The distribution of object sizes and object densities within images can vary significantly between datasets. COCO datasets are known for their diverse object sizes and complex scenes, while Pascal VOC datasets may have different size and density characteristics. Our method, which takes into account object density using the gradient density loss, may show variations in performance based on these distributions. Class imbalance, where some classes have more instances than others, can also impact performance. Some datasets may exhibit more pronounced class imbalances than others, leading to variations in detection performance across object categories.
(3) Specific challenges: Each dataset may come with its unique challenges. For example, COCO datasets may involve instances of small objects or objects with heavy occlusions, while Pascal VOC datasets may have specific challenges related to class distribution and object appearance. These dataset-specific challenges can affect detection performance.
(4) Adaptability of model: Our proposed approach, including the enhanced feature pyramid network and gradient density loss, is designed to improve object detection under varying conditions. However, its adaptability to different datasets depends on the specific characteristics of those datasets. The adaptability of our approach may also be influenced by the selection of hyperparameters. Fine-tuning hyperparameters for specific datasets can further enhance performance.
(5) Model generalization: The ability of our model to generalize across datasets is a crucial factor. Generalization depends on the diversity of the training data, the robustness of the network architecture, and the effectiveness of the proposed loss function.

The proposed image object detection approach brings several novel elements to the field:

(1) Enhanced feature pyramid network: The introduction of an enhanced feature pyramid network stands as a key innovation. This enhancement improves the baseline's multi-scale feature extraction and fusion capabilities, which is crucial for accurate object detection. The reduction in information loss during feature fusion enhances the network's ability to represent features effectively.
(2) Gradient density loss function: The replacement of the conventional focal loss function with the gradient density loss function is a novel contribution. This novel loss function provides a more accurate assessment of object detection classification loss, mitigating the influence of challenging samples and leading to improved detection performance.

The contributions of this work can be summarized as follows:

(1) Enhanced detection accuracy: The integration of the enhanced feature pyramid network and the gradient density loss function leads to a significant improvement in detection accuracy across a range of object scales and complexities. This contributes to the advancement of object detection methodologies.
(2) Improved multi-scale features: The proposed network architecture enhances the extraction and fusion of multi-scale features, which results in higher-quality features

and overall improved detection performance. This contribution addresses a crucial aspect of object detection tasks.

(3) Novel loss function: The introduction of the gradient density loss function provides a novel approach to handling classification loss in object detection. By accounting for the influence of challenging samples, this loss function enhances the model's capability to accurately classify objects.

While the proposed approach shows promising results, there are certain limitations to be considered:

(1) Computational overhead: Although the introduced enhancements have a minor impact on training and inference speeds, there might still be computational overhead associated with the proposed method. Further optimization might be necessary for real-time applications.
(2) Generalization to diverse datasets: While the proposed approach performs well on the tested benchmark datasets, its generalization to more diverse and challenging datasets remains to be validated. Different datasets may present unique challenges that could affect the method's performance.

In the pursuit of further advancement, several avenues for future research open up:

(1) Domain adaptation: Exploring the adaptability of the proposed approach to domain-specific challenges, such as specific object categories or environmental conditions, could enhance its practical applicability.
(2) Efficiency optimization: Continued efforts to optimize the computational efficiency of the method will be valuable for real-time applications and large-scale deployments.
(3) Robustness testing: Evaluating the proposed approach's robustness against occlusions, lighting variations, and other real-world challenges could provide insights into its practical viability.
(4) Exploration of loss functions: Further investigation into alternative loss functions or modifications to the existing ones could contribute to even more accurate and stable training processes.
(5) Interpretability: Exploring methods to enhance the interpretability of the model's decisions could improve the transparency and trustworthiness of the proposed approach.

5. Conclusions

This paper presents an innovative image object detection approach based on the pyramid network, aimed at enhancing the detection performance of current object detection methods. The key contributions and innovations of this research can be summarized in three aspects. Firstly, an enhanced feature pyramid network is introduced to enhance the baseline method. This enhancement significantly strengthens the capabilities of multi-scale feature extraction and fusion. By minimizing information loss during the fusion of multi-scale features, the enhanced network effectively enhances the feature representation potential. Secondly, the object detection classification loss function of the baseline method is enhanced by introducing the gradient density loss function, replacing the focal loss function used in the baseline. This incorporation of the gradient density loss function results in a more precise evaluation of object detection classification loss. It enriches the landscape of classification loss evaluation, a critical component not only for our method but also for the evolving paradigms in autonomous driving and security surveillance. The pursuit of higher classification accuracy aligns seamlessly with the demands of real-world systems where safety and reliability are paramount. Lastly, comprehensive experiments validate the effectiveness of the proposed method across four widely used public object detection datasets: MS COCO 2017, MS COCO 2014, Pascal VOC 2007, and Pascal VOC 2012. The achieved average precision on these datasets reaches 39.4%, 42.0%, 51.5%, and 49.9% respectively. These outcomes surpass the performance of existing methods in terms of precision metrics. In summary, this approach introduces a novel strategy for image object detection, combining an enhanced feature pyramid network with a gradient density object detection classification loss function. The experimental results underscore its superior performance compared to existing methods on diverse benchmark datasets.

Author Contributions: Conceptualization, W.Z.; methodology, Y.W.; software, Q.W., F.W. and F.L.; validation, R.Z. and Y.Z.; writing, Y.W. and Y.Z.; writing—review and editing, F.W.; project administration, S.D.; funding acquisition, S.D. and W.Z. All authors have read and agreed to the published version of the manuscript.

Funding: This work was supported by the Natural Science Foundation of Fujian Province (Grant Nos. 2021J011086, 2023J01964, 2023J01965, 2023J01966), by the Fujian Province Chinese Academy of Sciences STS Program Supporting Project (Grant no. 2023T3084) and by the Qimai Science and Technology Innovation Project of Wuping Country (Grant no. WPQM001).

Institutional Review Board Statement: Not applicable.

Informed Consent Statement: Not applicable.

Data Availability Statement: All the datasets used in this manuscript are publicly available datasets, already in the public domain.

Conflicts of Interest: The authors declare no conflict of interest.

References

1. Viola, P.; Jones, M. Rapid object detection using a boosted cascade of simple features. In Proceedings of the 2001 IEEE Computer Society Conference on Computer Vision and Pattern Recognition (CVPR 2001), Kauai, HI, USA, 8–14 December 2001.
2. Dalal, N.; Triggs, B. Histograms of oriented gradients for human detection. In Proceedings of the 2005 IEEE Computer Society Conference on Computer Vision and Pattern Recognition (CVPR'05), San Diego, CA, USA, 20–25 June 2005; pp. 886–893.
3. Felzenszwalb, P.F.; Girshick, R.B.; McAllester, D.; Ramanan, D. Object detection with discriminatively trained part-based models. *IEEE Trans. Pattern Anal. Mach. Intell.* **2009**, *32*, 1627–1645. [CrossRef] [PubMed]
4. Krizhevsky, A.; Sutskever, I.; Hinton, G.E. Imagenet classification with deep convolutional neural networks. *Adv. Neural Inf. Process. Syst.* **2012**, *25*, 1097–1105 . [CrossRef]
5. Girshick, R.; Donahue, J.; Darrell, T.; Malik, J. Region-based convolutional networks for accurate object detection and segmentation. *IEEE Trans. Pattern Anal. Mach. Intell.* **2015**, *38*, 142–158. [CrossRef] [PubMed]
6. He, K.; Zhang, X.; Ren, S.; Sun, J. Spatial pyramid pooling in deep convolutional networks for visual recognition. *IEEE Trans. Pattern Anal. Mach. Intell.* **2015**, *37*, 1904–1916. [CrossRef] [PubMed]
7. Girshick, R. Fast r-cnn. In Proceedings of the IEEE International Conference on Computer Vision, Santiago, Chile, 7–13 December 2015; pp. 1440–1448.
8. Ren, S.; He, K.; Girshick, R.; Sun, J. Faster r-cnn: Towards real-time object detection with region proposal networks. *Adv. Neural Inf. Process. Syst.* **2015**, *28*, 1137–1149 . [CrossRef] [PubMed]
9. Cai, Z.; Vasconcelos, N. Cascade r-cnn: Delving into high quality object detection. In Proceedings of the IEEE Conference on Computer Vision and Pattern Recognition, Salt Lake City, UT, USA, 18–23 June 2018; pp. 6154–6162.
10. Redmon, J.; Divvala, S.; Girshick, R.; Farhadi, A. You only look once: Unified, real-time object detection. In Proceedings of the IEEE Conference on Computer Vision and Pattern Recognition, Las Vegas, NV, USA, 27–30 June 2016; pp. 779–788.
11. Liu, W.; Anguelov, D.; Erhan, D.; Szegedy, C.; Reed, S.; Fu, C.Y.; Berg, A.C. Ssd: Single shot multibox detector. In Proceedings of the Computer Vision-ECCV 2016: 14th European Conference, Amsterdam, The Netherlands, 11–14 October 2016; pp. 21–37.
12. Redmon, J.; Farhadi, A. Yolov3: An incremental improvement. *arXiv* **2018**, arXiv:1804.02767.
13. Lin, T.Y.; Dollár, P.; Girshick, R.; He, K.; Hariharan, B.; Belongie, S. Feature pyramid networks for object detection. In Proceedings of the IEEE Conference on Computer Vision and Pattern Recognition, Honolulu, HI, USA, 21–26 July 2017; pp. 2117–2125.
14. Bochkovskiy, A.; Wang, C.Y.; Liao, H.Y.M. Yolov4: Optimal speed and accuracy of object detection. *arXiv* **2020**, arXiv:2004.10934.
15. Dai, X.; Chen, Y.; Yang, J.; Zhang, P.; Yuan, L.; Zhang, L. Dynamic detr: End-to-end object detection with dynamic attention. In Proceedings of the IEEE/CVF International Conference on Computer Vision, Montreal, BC, Canada, 11–17 October 2021; pp. 2988–2997.
16. Guan, S.; Khan, A.A.; Sikdar, S.; Chitnis, P.V. Fully dense UNet for 2-D sparse photoacoustic tomography artifact removal. *IEEE J. Biomed. Health Inform.* **2019**, *24*, 568–576. [CrossRef] [PubMed]
17. Ghiasi, G.; Lin, T.Y.; Le Q.V. Nas-fpn: Learning scalable feature pyramid architecture for object detection. In Proceedings of the IEEE/CVF Conference on Computer Vision and Pattern Recognition, Long Beach, CA, USA, 15–20 June 2019; pp. 7036–7045.
18. Lopez-Marcano, S.; LJinks, E.; Buelow, C.A.; Brown, C.J.; Wang, D.; Kusy, B.; Connolly, R.M. Automatic detection of fish and tracking of movement for ecology. *Ecol. Evol.* **2021**, *11*, 8254–8263. [CrossRef] [PubMed]
19. Lin , T.Y.; Goyal, P.; Girshick, R.; He, K.; Dollár, P. Focal loss for dense object detection. In Proceedings of the IEEE International Conference on Computer Vision, Venice, Italy, 22–29 October 2017; pp. 2980–2988.

20. Li, B.; Liu, Y.; Wang, X. Gradient harmonized single-stage detector. In Proceedings of the AAAI Conference on Artificial Intelligence, Honolulu, HI, USA, 27 January–1 February 2019; pp. 8577–8584.
21. Lin, T.Y.; Maire, M.; Belongie, S.; Hays, J.; Perona, P.; Ramanan, D.; Zitnick, C.L. Microsoft coco: Common objects in context. In Proceedings of the Computer Vision-ECCV 2014: 13th European Conference, Zurich, Switzerland, 6–12 September 2014; pp. 740–755.
22. Shetty, S. Application of convolutional neural network for image classification on Pascal VOC challenge 2012 dataset. *arXiv* **2016**, arXiv:1607.03785.
23. Law, H.; Deng, J. Cornernet: Detecting objects as paired keypoints. In Proceedings of the European Conference on Computer Vision, Munich, Germany, 8–14 September 2018; pp. 734–750.
24. Zhang, S.; Chi, C.; Yao, Y.; Lei, Z.; Li, S.Z. Bridging the gap between anchor-based and anchor-free detection via adaptive training sample selection. In Proceedings of the IEEE/CVF Conference on Computer Vision and Pattern Recognition, Seattle, WA, USA, 13–19 June 2020; pp. 9759–9768.
25. Wang, J.; Chen, K.; Xu, R.; Liu, Z.; Loy, C.C.; Lin, D. Carafe: Content-aware reassembly of features. In Proceedings of the IEEE/CVF International Conference on Computer Vision, Seoul, Korea, 27 October–2 November 2019; pp. 3007–3016.
26. Duan, K.; Bai, S.; Xie, L.; Qi, H.; Huang, Q.; Tian, Q. Centernet: Keypoint triplets for object detection. In Proceedings of the IEEE/CVF International Conference on Computer Vision, Seoul, Korea, 27 October–2 November 2019; pp. 6569–6578.
27. Zhu, B.; Wang, J.; Jiang, Z.; Zong, F.; Liu, S.; Li, Z.; Sun, J. Autoassign: Differentiable label assignment for dense object detection. *arXiv* **2020**, arXiv:2007.03496.
28. Dong, Z.; Li, G.; Liao, Y.; Wang, F.; Ren, P.; Qian, C. Centripetalnet: Pursuing high-quality keypoint pairs for object detection. In Proceedings of the IEEE/CVF Conference on Computer Vision and Pattern Recognition, Seattle, WA, USA, 13–19 June 2020; pp. 10519–10528.
29. Carion, N.; Massa, F.; Synnaeve, G.; Usunier, N.; Kirillov, A.; Zagoruyko, S. End-to-end object detection with transformers. In Proceedings of the 16th European Conference on Computer Vision, Glasgow, UK, 23–28 August 2020; pp. 213–229.
30. Zhang, H.; Chang, H.; Ma, B.; Wang, N.; Chen, X. Dynamic R-CNN: Towards high quality object detection via dynamic training. In Proceedings of the 16th European Conference on Computer Vision, Glasgow, UK, 23–28 August 2020; pp. 260–275.
31. Zhu, X.; Su, W.; Lu, L.; Li, B.; Wang, X.; Dai, J. Deformable detr: Deformable transformers for end-to-end object detection. *arXiv* **2020**, arXiv:2010.04159.
32. Chen, Q.; Wang, Y.; Yang, T.; Zhang, X.; Cheng, J.; Sun, J. You only look one-level feature. In Proceedings of the IEEE/CVF Conference on Computer Vision and Pattern Recognition, Nashville, TN, USA, 20–25 June 2021; pp. 13039–13048.
33. Ge, Z.; Liu, S.; Wang, F.; Li, Z.; Sun, J. Yolox: Exceeding yolo series in 2021. *arXiv* **2021**, arXiv:2107.08430.

applied sciences

MDPI

Article

Self-Learning Robot Autonomous Navigation with Deep Reinforcement Learning Techniques

Borja Pintos Gómez de las Heras *, Rafael Martínez-Tomás and José Manuel Cuadra Troncoso

Department of Artificial Intelligence, National Distance Education University, Juan del Rosal 16, 28040 Madrid, Spain; rmtomas@dia.uned.es (R.M.-T.); jmcuadra@dia.uned.es (J.M.C.T.)
* Correspondence: bpintosgomezheras@gmail.com

Abstract: Complex and high-computational-cost algorithms are usually the state-of-the-art solution for autonomous driving cases in which non-holonomic robots must be controlled in scenarios with spatial restrictions and interaction with dynamic obstacles while fulfilling at all times safety, comfort, and legal requirements. These highly complex software solutions must cover the high variability of use cases that might appear in traffic conditions, especially when involving scenarios with dynamic obstacles. Reinforcement learning algorithms are seen as a powerful tool in autonomous driving scenarios since the complexity of the algorithm is automatically learned by trial and error with the help of simple reward functions. This paper proposes a methodology to properly define simple reward functions and come up automatically with a complex and successful autonomous driving policy. The proposed methodology has no motion planning module so that the computational power can be limited like in the reactive robotic paradigm. Reactions are learned based on the maximization of the cumulative reward obtained during the learning process. Since the motion is based on the cumulative reward, the proposed algorithm is not bound to any embedded model of the robot and is not being affected by uncertainties of these models or estimators, making it possible to generate trajectories with the consideration of non-holonomic constrains. This paper explains the proposed methodology and discusses the setup of experiments and the results for the validation of the methodology in scenarios with dynamic obstacles. A comparison between the reinforcement learning algorithm and state-of-the-art approaches is also carried out to highlight how the methodology proposed outperforms state-of-the-art algorithms.

Keywords: autonomous robots; deep reinforcement learning; dynamic environment; comfort driving; self-learning

1. Introduction

Autonomous driving is a very wide field of research that gathers many different methodologies. These methodologies are usually targeted to a very specific problem within the autonomous driving domain. The methodology presented in this work proposes a solution for a robot driving in a scenario with spatial restrictions (limited drivable space), interacting with static and dynamic obstacles, and fulfilling at all times safety, legal, and comfort requirements. The safety requirements ensure that the robot moves within the drivable space, avoiding collision with any other obstacle and always fulfilling the non-holonomic constrains that the robot might have. The legal requirements define the maximum speed limit and the comfort requirements establish limitations on the robot's acceleration and jerk to maximize comfort. The state-of-the-art solutions that attempt to solve this autonomous driving problem are characterized by overly complicated algorithms that demand a lot of computational resources. This article uses deep reinforcement learning techniques to overcome these issues. Specifically, the proposed methodology contains the following contributions:

(1) The methodology relies on the definition of simple, multiple, and intuitive reward functions. The reward functions are closely related to the feeling of pain or pleasure that we might receive after performing a series of actions. In contrast to a cost function, the reward function does not evaluate how good or bad it is to take a specific action in a given state.
(2) The perception, motion planning, and motion control software modules are not embedded in a complex deep neural network architecture. Only the motion planning software module is implemented using a deep neural network model, making training easier.
(3) A continuous multi-action space is defined to solve complex dynamic scenarios.
(4) The comfort requirements are easily included through the definition of simple reward functions. Since all reward functions are normalized between -1 and 0, a weighted sum of the reward functions is performed to prioritize safety requirements over legal or comfort requirements.

The rest of the paper is structured as follows. The Materials and Methods section reviews the relevant literature and gives a detailed explanation of the proposed methodology. The design of experiments for validation of the methodology and the simulation results are also discussed in the Discussion section and compared with state-of-the-art algorithms. Finally, the main conclusions and advantages of the proposed methodology are listed in the Conclusions section.

2. Materials and Methods

To review the related work, we went through all robotic paradigms, always keeping in mind the autonomous driving problem that we were attempting to solve. The definition of this autonomous driving problem has basically 4 challenging points: first, the accomplishment of the safety and legal requirements under dynamic scenarios, i.e., avoiding dynamic obstacles and driving the robot within the drivable space, controlling its maximum linear speed. Some methods in the literature struggle to accomplish these requirements under dynamic scenarios. Second, the fulfillment of comfort requirements. Some robotic paradigms are less suitable than others to fulfill comfort requirements. Third, the generation of feasible trajectories, i.e., the generation of trajectories that are capable of being realized by the motion control. This point represents a headache for many state-of-the-art algorithms, specifically when facing driving situations close to the limit (grip limit). Finally, the proposed algorithm must have a reasonable computational load. State-of-the-art algorithms are well known for demanding very high computational resources, making them hard to run on microcontrollers.

A. Reactive Paradigm

Methods based on the reactive paradigm perform well with holonomic robots in spatially restricted scenarios with static obstacles while fulfilling safety and legal requirements. Common methods are the potential field method [1], velocity obstacle method [2], dynamic window approach [3], or partial center of area method [4]. However, non-holonomic robots might be difficult to integrate with the reactive paradigm since they need to plan their movement ahead in order to avoid collision with obstacles. Moreover, the control based on reactions to certain sensorial inputs (behaviors) without any local planning can lead to aggressive and long trajectories [5], making it difficult to integrate comfort requirements. On the other hand, the reactive paradigm demands low computational resources, which enables the usage of non-expensive microcontrollers.

B. Deliberative Paradigm

The robotic paradigm with a local planification stage is the deliberative paradigm. The state-of-the-art solutions that attempt to solve the previously described autonomous driving problem are mainly grouped in this robotic paradigm. The methods within the deliberative paradigm can be divided into three different subgroups: graph-based methods, sampling-based methods, and model predictive control methods.

Within the graph-based methods group, the Dijkstra algorithm [6] performs very well at avoiding obstacles and finding the shortest path. However, it is difficult to use with large grids because it is computationally very expensive. The A* algorithm [7] followed the Dijkstra algorithm to reduce the computational load by including heuristic functions to find the solution faster. However, the Dijkstra and A* algorithms have also an additional and very important disadvantage: they do not consider the robot dynamics in the motion planning, making the method invalid for non-holonomic robots. Kinematic or dynamic embedded models can be considered to make the method valid for non-holonomic robots. A good example is the hybrid A* algorithm [8]. However, these methods rely on embedded models that are affected by uncertainties, producing trajectories that might be unfeasible to control, specifically if the robot faces driving situations close to the limit.

With the aim of reducing the computational cost, sampling-based methods were created. Sampling-based methods can work on a continuous space where only specific discrete waypoints are sampled. Common methods are rapidly exploring random trees (RRTs) [9] or the probabilistic roadmap method (PRM) [10]. Once the sampling process is completed, a path considering non-holonomic constrains can be generated: [11] used maximum and minimum limitations of a path's curvature and the momentum of the vehicle and [12] used an embedded dynamic model of the robot. Eventually, the final trajectory was selected based on the score given by a cost function. Since non-holonomic constrains were considered, the cost function only needed to evaluate aspects such as robot collisions, comfort, or total travel time. However, [11] did not consider all non-holonomic constrains and [12] relied on embedded models that could differ from reality. Driving situations close to the limit are also challenging, even when using complex embedded robot dynamic models. Feasible trajectories can be easily generated under scenarios of low velocity, such as parking [13] or scenarios with moderate accelerations [14]. But, scenarios close to the limit are difficult to consider in this approach. Some other approaches based on pure mathematical functions can be applied to connect the sampled waypoints with a smooth path profile, such as Bezier curves [15], splines curves [16], or the lattice planner with polynomial curves for the path's curvature [17]. These approaches can have the advantage of not embedding a dynamic model of the robot into the logic, which can simplify the implementation. However, since some of the trajectories generated might not fulfill the non-holonomic constrains, the cost function must consider these constrains to reject non-feasible trajectories.

The predictive control model attempts to generate feasible trajectories considering all non-holonomic constrains, without having to completely discretize (graph-based) or partially sample (sampling-based) the continuous space. An embedded dynamic model of the robot is included in the logic. A cost function defines the goals of the autonomous driving problem and an optimizer iteratively solves this problem, minimizing the cost function and generating the corresponding control actions over a future horizon. The predictive control model also addresses the problem of planning feasible trajectories under driving situations close to the limit. In [18], the parameters of the embedded model were updated over the horizon in situations close to the limit based on a predictive friction estimate to obtain even more reliable and feasible trajectories. However, there was still a reliance on models or estimators that could have differences with respect to reality and the computational consumption was still very high due to the usage of an iterative optimizer.

In general, the main problems of the deliberative paradigm are the challenging design of the cost function, the high computational cost, and the uncertainties of embedded dynamic models, no matter whether the algorithm falls under a graph-based, sampling-based, or predictive control model. In [8], the challenges of designing an effective cost function for the hybrid A* algorithm were addressed. In [19], a comparison of motion planning algorithms was performed in terms of computational cost, highlighting the high computational cost of graph-based and sampling-based methods.

C. Sense–Think–Act Paradigm

The problem of the uncertainties of embedded models, which eventually compromises the feasibility of the generated trajectory, is addressed by the robotic paradigm sense–think–act. In this paradigm, a model, usually a deep neural network, is trained with real trajectories (supervised learning) to create a path planner algorithm based on artificial intelligence [20]. Thus, it avoids the usage of embedded dynamic models. However, the main drawbacks of this paradigm are the need for big databases and the labeling of these databases. The labeling of databases is not only a very time-consuming task, but it can also introduce bias to the results, specifically if the labeling task is done manually.

D. Sense–Explore–Act–Learn Paradigm

The methods based on reinforcement learning techniques attempt to bring a solution to the previously mentioned drawbacks. The drawbacks and the solution proposed by the reinforcement learning approach are summarized in Table 1.

Table 1. Strengths of reinforcement learning-based algorithms for autonomous driving navigation.

State-of-the-Art Drawbacks	Reinforcement Learning Strengths
Cost function design	Cost function is automatically learned
Performance under dynamic environments	Agent learns from dynamic environments
Uncertainties of embedded dynamic models	Agent free from embedded dynamic models
High computational cost	Low computational cost avoiding iterative motion planning modules
Integration of comfort requirements	Easy integration with simple reward functions
Databases must be available and labeled	Data are generated on-the-go and automatically labeled based on a reward value

Due to the advantages summarized in the previous table, reinforcement learning techniques have been the subject of extensive study within the autonomous driving domain. In [21], Q-learning was used to find the shortest path from the starting pose to the destination pose. Reinforcement learning techniques with continuous actions and states were introduced in [22], with very promising results. A deep deterministic policy gradient (DDPG) was introduced in [23] to learn from a real scenario how to drive and keep the vehicle inside the road boundaries by providing just a single camera image of the road as the input. However, this work only implemented partial safety requirements (driving within the drivable space without obstacle avoidance) and did not consider any comfort or legal requirements. Additionally, the perception, motion planning, and motion control modules were embedded in the deep neural networks of the deep reinforcement learning algorithm. In [24], the autonomous driving algorithm relied not only on input images of the cameras as in [23] but also on post-processed data as velocities or relative distances. In [25], comfort requirements were included for lane change maneuvers as a simple reward function. In contrast to [23], it separated the perception module from the reinforcement learning task, and the agent consumed already post-processed signals from the perception module, such as robot velocities or relative distances, instead of images. The algorithm presented in this work keeps the perception, motion planning, and motion control modules separated from each other. Images are not directly provided to the agent. Instead, all the relevant post-processed information from the perception module is provided to the agent so that it can learn the best possible trajectory based only on variables that have an important impact on motion planning. Regarding the definition of reward functions, [26] included positively and negatively weighted rewards. The methodology proposed only included negatively weighted rewards to exclude policies that end up in endless loops of collecting permanently positive rewards instead of reaching the destination. In terms of requirements, ref. [26] included vehicle speed (legal requirements), collisions (safety requirements), and travel distance towards the destination (task-oriented requirements), but comfort requirements were not included. In [27], comfort, legal, and task-oriented requirements (lane keeping) were included, but safety requirements were not, which are needed for, for example, overtaking maneuvers in dynamic scenarios. In [28], all requirements were included. However, the comfort requirements were not expressed in terms of acceleration and jerk but in terms of limiting the steering wheel angle (thus only penalizing lateral acceleration but excluding lateral jerk and longitudinal acceleration and jerk). In [28], a mix of sparse and continuous reward functions was used. The methodology proposed only used continuous reward functions to better guide the reinforcement learning algorithm to find the most valuable rewards.

The methodology proposed is based on the deep deterministic policy gradient (DDPG) algorithm. The DDPG method, like the Q-learning method, is a model-free algorithm. That is, it does not use any embedded model of the robot inside the agent logic. Instead, it learns the optimal policy directly from the environment. The DDPG method is intended for environments with continuous states and actions. This is possible by approximating the actor model (which computes the action a_t) with deep neural network μ of weights θ as a function of state s at time t (1).

$$a_t = \mu(s_t|\theta) = \mu_\theta(s_t) \tag{1}$$

The DDPG algorithm is an actor–critic method and, therefore, the critic function is also approximated with another deep neural network Q of weight ϕ as a function of state s and the action a at time t (2).

$$Q_\phi(s_t, a_t) = Q_\phi(s_t, \mu_\theta(s_t)) \tag{2}$$

Actor–critic methods combine the benefits of policy-based and value-based methods. They benefit from the good convergence properties of policy-based methods and they also benefit from the sample efficiency of value-based methods, which tend to find an optimal solution faster.

The deterministic policy gradient theorem (3) updates the weight θ of the actor model μ_θ in the direction of the maximum value of the critic model Q_ϕ:

$$\nabla_\theta J(\mu_\theta) = \mathbb{E}_{s\sim\mathcal{D}}\left[\nabla_a Q_\phi(s,a)\big|_{a=\mu_\theta(s)} \nabla_\theta \mu_\theta(s)\right] \tag{3}$$

On the other hand, the Bellman equation is used to iteratively update the weight ϕ of the critic network Q_ϕ (4) and eventually learn the optimal critic and actor networks.

$$Q_\phi(s_t, a_t) = \mathbb{E}_{s\sim\mathcal{D}}\left[r(s_t, a_t) + \gamma Q_\phi(s_{t+1}, \mu_\theta(s_{t+1}))\right] \tag{4}$$

Additionally, this actor–critic method uses the replay buffer and target networks to bring more stability to the learning process. Random observations are selected from the replay buffer to avoid training the neural networks with consecutive and correlated observations. The weights ϕ' and θ' of the target networks $\mu'_{\theta'}(s)$ and $Q'_{\phi'}(s,a)$ are updated after the weights of the actor and critic networks with a soft factor τ (5). This lag brings stability to the learning.

$$\begin{aligned} \theta' &= \tau\theta + (1-\tau)\theta' \\ \phi' &= \tau\phi + (1-\tau)\phi' \end{aligned} \tag{5}$$

The neural network architecture of actor and critic models considers several hidden layers (Figure 1). The variables in brackets denote the number of neurons for that layer.

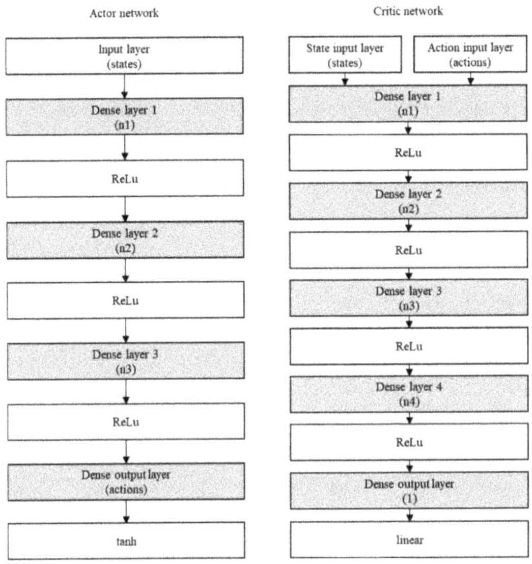

Figure 1. Actor and critic deep neural network architecture.

Finally, the exploration is essential to find the optimal policy. For that purpose, a white noise signal of zero mean value is added to the actor–network output (6). The amplitude of the exploration is controlled with the variance σ of the normal distribution $\mathcal{N}$.

$$a_t = \mu_\theta(s_t) + \mathcal{N}(0,\sigma) \tag{6}$$

As the learning process progresses and the actor and critic networks learn the optimal values, less exploration is needed. Therefore, a decay of the exploration was programmed. This decay also helped to stabilize the learning process. The DDPG pseudocode is shown in Algorithm 1:

Algorithm 1. Deep Deterministic Policy Gradient

Init $\mu_\theta(s)$ and ${\mu'}_{\theta'}(s)$ with weights θ and $\theta' = \theta$
Init $Q_\phi(s,a)$ and ${Q'}_{\phi'}(s,a)$ with weights ϕ and $\phi' = \phi$
Init discount factor γ and soft update τ
for episode $\in \{1 \ldots episode_{max}\}$ do
    $s_t = s_0$
    while True
        $a_t = \mu_\theta(s_t) + \boldsymbol{\mathcal{N}}(0,\sigma)$
        Execute a_t and observe next state s_{t+1}, reward r_t,
            and termination flag T_t
        Record transition (s_t, a_t, r_t, s_{t+1}) in the buffer
        Sample a batch of transitions $\boldsymbol{\mathcal{D}}$ from the buffer
        Compute $y_i = r_i + \gamma {Q'}_{\phi'}(s_{i+1}, \mu'_{\theta'}(s_{i+1}))$
        Update ϕ by minimizing the loss function L:
            $L = \frac{1}{N} \sum_{i \in \mathcal{D}} (y_i - Q_\phi(s_i, \mu_\theta(s_i)))^2$
        Update θ:
            $\nabla_\theta J(\mu_\theta) = \mathbb{E}_{s\sim\mathcal{D}}\left[\nabla_a Q_\phi(s,a)\big|_{a=\mu_\theta(s)} \nabla_\theta \mu_\theta(s) \right]$
        Update ϕ' and θ':
            $\theta' = \tau\theta + (1-\tau)\theta'$
            $\phi' = \tau\phi + (1-\tau)\phi'$
        if T_t == True
            break

Before reaching the maximum number of learning episodes, the learning process can be terminated if some specific conditions are met. When reaching these conditions, it is considered that the policy is mature enough to fulfill the requirements. There are mainly 2 conditions: first, the number of total episodes must be greater than the minimum number. With this condition, we ensured that the exploration rate decayed until almost 0 to avoid very noisy and unstable policies. Secondly, no termination due to non-compliance of the safety, legal, or comfort requirements must be detected for a minimum number of episodes to ensure proper learning. In order words, the termination of the learning episode must be only related to task-oriented requirements, such as if the robot reaches the destination pose. If the episode gets terminated because the robot crashes into an obstacle, the counter to consider that the policy is mature enough gets reset to 0.

The agent, which is based on the DDPG algorithm, interacts with the environment. Important concepts of the environment are the state variables, reward functions, and motion control.

A. State Variables

The state variables S are essential during the reinforcement learning process (7). One important property is that the state variables must be capable of generalization so that the policy obtained after the learning process can solve unseen scenarios. Therefore, absolute variables like the absolute distance traveled by the robot are avoided. A set of relative distances $(d_1, \ldots, d_n)$ with respect to the robot's position, the relative orientation to the destination (θ_{dest}), and dynamic variables such as robot linear and angular velocities (v_{rob}, w_{rob}) and accelerations $(\dot{v}_{rob}, \dot{w}_{rob})$ are selected. The relative distances come from a lidar sensor, which covers 180 degrees of the robot's front view. To smoothly reduce the linear velocity before reaching the destination, the relative distance between the robot and the destination (d_{dest}) is also included.

$$S = \{d_1, \ldots, d_n, v_{rob}, w_{rob}, \dot{v}_{rob}, \dot{w}_{rob}, \theta_{dest}, d_{dest}\} \tag{7}$$

Figure 2 shows an image of the previously mentioned state variables:

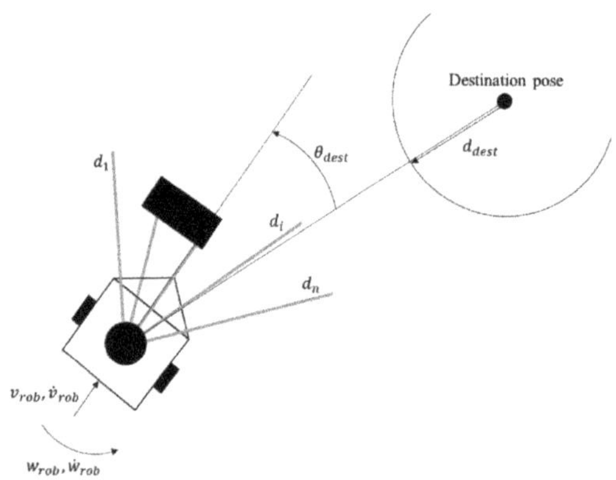

Figure 2. State variables.

For the learning process to be stable, the state variables must fulfill the Markov property. The Markov property is formulated in (8):

$$p(s_{t+1}|s_t) = p(s_{t+1}|s_1, s_2, \ldots, s_t) \tag{8}$$

According to (8), the future state s_{t+1} only depends on the current state s_t. This means that all the information from the history $(s_1, s_2, \ldots, s_{t-1})$ is irrelevant for computing the future state s_{t+1}. To make it possible for the state variables to fulfill the Markov property, they must include not only relative distances but also dynamic variables like the linear and angular velocities and accelerations. Therefore, they are included in the list of state variables in (7).

Finally, it is necessary to make sure that all states variables can be measured with sensors. The relative distances come from a lidar sensor mounted on the top of the robot, covering 180 degrees of the front view. The linear and angular velocities and accelerations of the robot can be measured with encoders mounted on the driven wheels and IMU sensors. The distance and orientation to the destination pose can be calculated based on camera images. The destination pose can be recognized by using some special color that the camera recognizes. Based on that recognition and after proper calibration of the camera parameters, the distance to the destination pose and the relative orientation of the robot to the destination pose can be calculated.

B. Reward Functions

The reward functions are the key to the reinforcement learning algorithm. The reward function does not define which action to take. Instead, the reward functions are a measurement of how good or bad it is to end up in a new state after an action is taken. This measurement of good or bad is closely related to the sensation of pain or pleasure. Based on the reward functions, the reinforcement learning algorithm automatically learns a complex critic function to navigate the robot autonomously. In contrast to the reward function, the critic function defines, for a given state, how good an action is, before the robot performs such an action. Comparing the critic function with the state-of-the-art algorithms like the lattice planner or the predictive control model, the critic function is equivalent to the cost function. In these state-of-the-art algorithms, the cost function is usually a single and very complex function, which needs to be manually designed to fulfill the requirements of the autonomous driving problem. On the other hand, the reinforcement learning algorithm automatically learns this cost function (critic function) based on simple, multiple, and very intuitive reward functions. Therefore, the critic function gathers all the experience seen during the learning process to build a complex function that fulfills all requirements. To learn this complexity, usually, deep neural networks are selected as policy and critic approximation functions.

There are some important rules to follow before the definition of the reward functions:

- In problems with continuous actions and states, or in case of too large discrete spaces, it is very convenient to shape the reward functions instead of using sparse rewards. The reason behind this is that the robot learns by means of exploration. If the robot starts randomly exploring, the probability of the robot seeing this sparse reward can be very low, leading to either an unsuccessful learning process because the robot does not find the reward or to an extremely long learning process. On the other hand, in case of shaped rewards (rewards with a smooth continuous gradient), the robot can adapt its actions in the direction of maximizing the cumulative reward right from the beginning.
- It is better to avoid purely negative rewards. If the reward functions are always negative, excluding the 0, the reinforcement learning might find a solution by terminating the learning episode as soon as possible rather than keeping on driving and therefore accumulating more negative rewards.
- Positive rewards are also tricky, and they must be treated carefully. If the robot finds a positive reward, the logic might end up in a loop solution.
- The learning scenarios must always be carefully designed to facilitate finding rewards from a beneficial starting situation.

Regarding the type of rewards, they are divided into safety, legal, comfort, and task-oriented requirements. The safety, legal, and comfort requirements are based on the pain that the robot might suffer if they are not fulfilled. On the other hand, task-oriented requirements are based on the sensation of pleasure that is obtained by fulfilling the proposed tasks. Therefore, depending on the robot's mission, task-oriented requirements can be built following different criteria.

(1) Safety requirements

Safety requirements implement a penalization or negative reward if the robot crashes into obstacles. Situations where the robot drives too close to an obstacle are also penalized even when no contact between the robot and the obstacle happens. For that reason, a safety margin around the obstacle footprint of distance d_{limit} is defined, as represented in Figure 3. A very simple reward function (9) is therefore defined: 0 if the robot drives outside the safety margin, and a linear interpolation between 0 and −1 if the robot drives inside the safety margin. The variable d is the relative distance from the robot to the obstacle.

$$r_{safety} = \begin{cases} 0 & , \ d > d_{limit} \\ \frac{d - d_{limit}}{d_{limit}} & , \ d \leq d_{limit} \end{cases} \quad (9)$$

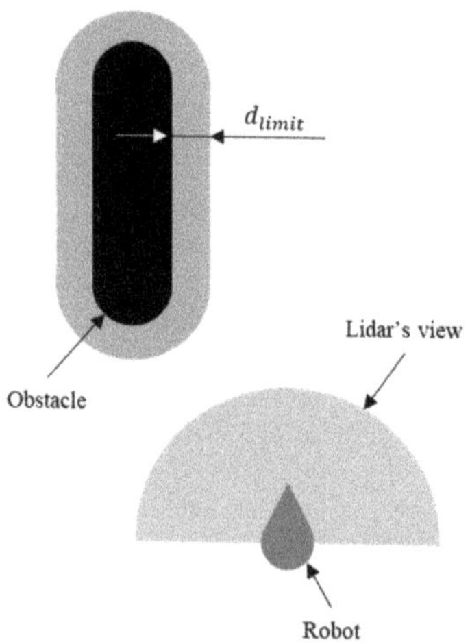

Figure 3. Safety margin around the obstacle footprint. If the robot enters the area delimited by the safety margin, a negative reward will be forwarded to the agent during the learning process.

(2) Legal requirements

Legal requirements limit the maximum speed that the robot can travel. The robot must control its linear speed v_{rob} so that it does not exceed this maximum value. The implementation of this reward function is very simple as well (10). There are two speed thresholds (v_1 and v_2). If the speed does not exceed the lower threshold v_1, the reward will be 0. The reward will be linearly interpolated between 0 and −1 between the first and second thresholds.

$$r_{legal} = \begin{cases} 0 & , \ v_{rob} < v_1 \\ -\frac{v_{rob} - v_1}{v_2 - v_1} & , \ v_1 \leq v_{rob} \leq v_2 \\ -1 & , \ v_{rob} > v_2 \end{cases} \quad (10)$$

If the legal maximum velocity is v_{legal}, v_1 can be equal to v_{legal} and v_2 can be set a little bit higher than v_{legal}.

(3) Comfort requirements

Comfort can be defined as the pleasant feeling of the people traveling inside the robot. Feelings of discomfort can come from different sources. The three main sources that can decrease the feeling of comfort are listed below:

- High acceleration values (longitudinal and lateral/angular)
- High jerk values (longitudinal and lateral/angular)
- Low-frequency acceleration over long periods of time (feeling of sickness)

In this work, the first source of discomfort is implemented. The reward function (11) defines a window given two acceleration values, $\dot{w}_1$ and $\dot{w}_2$. Accelerations below $\dot{w}_1$ are not penalized. Accelerations within $\dot{w}_1$ and $\dot{w}_2$ are penalized using linear interpolation. The maximum penalization value is clipped to -1 if the acceleration is higher than $\dot{w}_2$.

$$r_{comfort} = \begin{cases} 0 & , \dot{w}_{rob} < \dot{w}_1 \\ -\frac{\dot{w}_{rob} - \dot{w}_1}{\dot{w}_2 - \dot{w}_1} & , \dot{w}_1 \leq \dot{w}_{rob} \leq \dot{w}_2 \\ -1 & , \dot{w}_{rob} > \dot{w}_2 \end{cases} \tag{11}$$

Reward functions for longitudinal and angular accelerations are added together to build the final comfort function.

(4) Task-oriented requirements

It is possible to define multiple and specific tasks to be accomplished during autonomous navigation. In this work, the robot is intended to reach a specific destination pose. For that purpose, the robot is rewarded when facing the right direction towards the destination pose. The robot will be also rewarded when traveling at the maximum legal speed.

For the first task, if the direction to the destination pose θ_{target} (determined by visual sensors) is not the same as the robot's heading θ_{rob}, the robot is penalized according to (12).

$$r_{destination} = -\frac{|\theta_{target} - \theta_{rob}|}{\pi} \tag{12}$$

If the destination cannot be detected by visual sensors, the destination reward is set to 0 and the robot moves forward within the drivable space.

For the second task, the robot is penalized if the linear speed is not the maximum legal one (13). Since legal requirements are also implemented, if the robot exceeds the maximum legal speed, it is penalized according to (10).

$$r_{speed} = \begin{cases} -\frac{v_{target} - v_{rob}}{v_{target}} & , \quad v_{rob} < v_{target} \\ 0 & , \quad v_{rob} \geq v_{target} \end{cases} \tag{13}$$

In the previous reward function, the target linear velocity v_{target} is equal to v_{legal}.

Finally, the reward due to the task-oriented requirements is obtained by adding the single functions (14):

$$r_{task} = r_{destination} + r_{speed} \tag{14}$$

(5) Final reward function

The final reward function can be simply built by adding the single reward functions (15). Each reward function is normalized between -1 and 0. A weighted addition of the single reward functions is performed to prioritize the requirements:

$$\begin{aligned} r = K_{safety} * r_{safety} + K_{legal} * r_{legal} + \\ K_{comfort} * r_{comfort} + K_{task} * r_{task} \end{aligned} \tag{15}$$

In general, the order of importance within the requirements is as follows: safety, legal, comfort, and task-oriented (16):

$$K_{safety} \geq K_{legal} \geq K_{comfort} \geq K_{task} \tag{16}$$

C. Motion Control

The motion control strategy depends on the type of robot to be controlled. The commanded signals from the agent are the linear and angular velocities. The motion control takes care of the actuator control to fulfill the commanded velocities. In this work, a differential robot was used for the validation. Considering that a differential robot traveling at a constant linear speed must fulfill non-holonomic constrains as well, the differential robot is a simple but valid approach to validate the methodology.

The commanded linear and angular velocities of the robot were converted in wheel rotational speed by means of the kinematic equations of the differential robot (17) and (18).

$$w_{left} = \frac{2v_{rob} + w_{rob}D}{2r_{left}} \tag{17}$$

$$w_{right} = \frac{2v_{rob} - w_{rob}D}{2r_{right}} \tag{18}$$

The differential robot geometry and kinematic variables are shown in Figure 4:

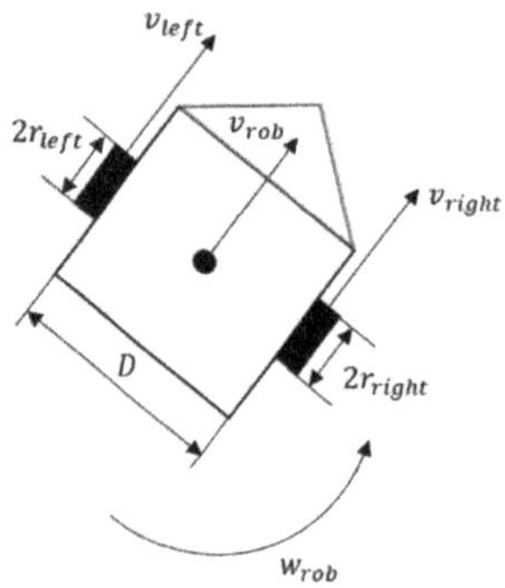

Figure 4. Kinematic variables for a differential robot.

Finally, two independent PI controllers control the commanded velocities by adapting the duty cycles of the electrical motors connected to the driven wheels. Figure 5 shows the final cascade control of the robot's motion. An external motion planning loop based on deep reinforcement learning techniques generates the requested linear and angular speeds of the robot, depending on the robot's state measured from the environment. These requested linear and angular speeds are forwarded to an internal control loop based on classical PID controllers. The internal loop measures the linear and angular actual speeds of the robot for closed-loop control.

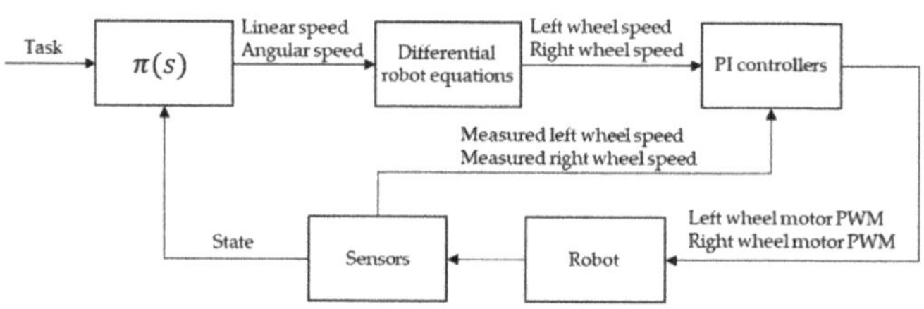

Figure 5. Motion control diagram. Cascade control for the position of the robot.

3. Results

All the experiments were carried out in the simulator CoppeliaSim (Figure 6). A Python script controlled the connection with the simulation by means of APIs. The main libraries used to build the reinforcement learning algorithm were TensorFlow and Gym. Tensorflow and Keras provide the necessary functions to build actor and critic neural network models. Gym provides the necessary tools to implement and compare reinforcement learning algorithms. The actor neural network (policy) produced linear and angular speed set points, which were converted into left and right wheel speeds of the differential robot. These wheel speeds were sent over to the simulator via APIs. The simulator implemented PID controllers to control in a closed loop these speeds.

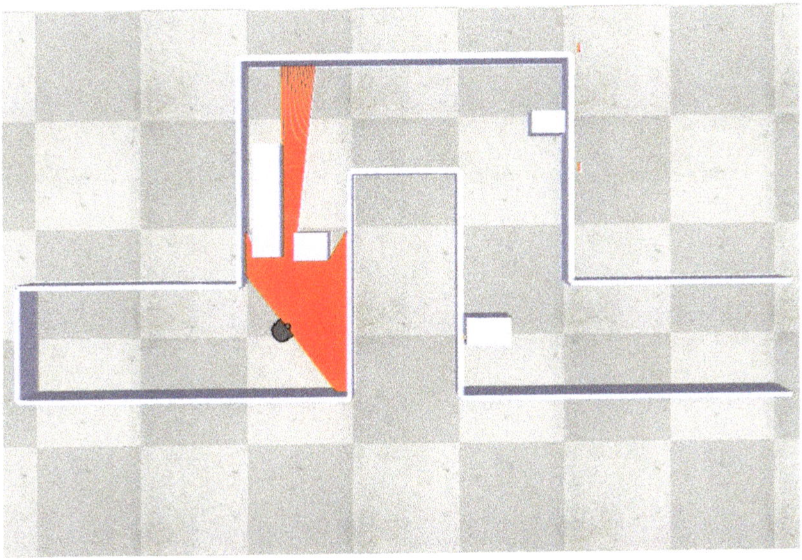

Figure 6. CoppeliaSim simulation platform where all experiments were carried out.

The experiments defined within this section can be divided into two main groups. First, an experiment or set of experiments used for the learning process. Second, an experiment or set of experiments used for the validation process. It was important to use different experiments for the validation process to prove that the methodology proposed can be generalized to new unseen scenarios after the learning process is completed.

For the validation purpose, an experiment including a complex scenario with dynamic and static obstacles was set up. The scenario (see Figure 7) was carefully chosen to validate the following maneuvers:

(1) Follow-up
(2) Overtaking
(3) Lane keeping
(4) Lane change

Before the robot could perform in this validation experiment, it needed to learn how to drive autonomously. For the learning purpose, less complex experiments were set up. In this case, it was important to carefully select the experiments and scenarios so that the robot could easily find rewards and learn a policy that would enable the robot to generalize and drive autonomously in other unseen and more complex scenarios. Therefore, the learning scenario (Figure 8) was specially adapted to see reward values early in the learning process and better guide the policy to learn the right strategy.

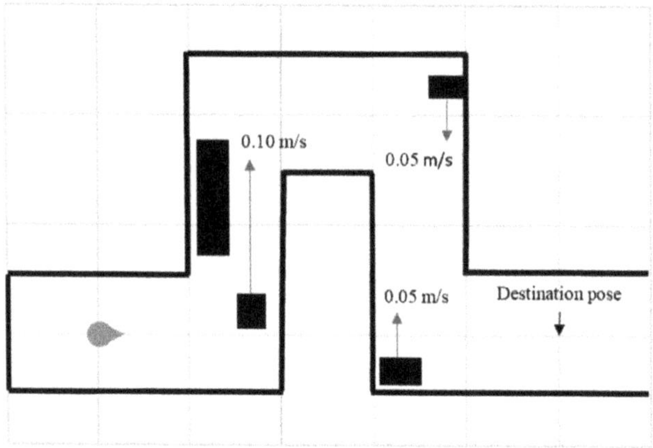

Figure 7. Validation scenario. First dynamic obstacle blocked the way, forcing the robot to follow up the obstacle. Second and third dynamic obstacles were driving at opposite speeds. The robot was forced to perform lane change maneuvers to overtake the obstacles.

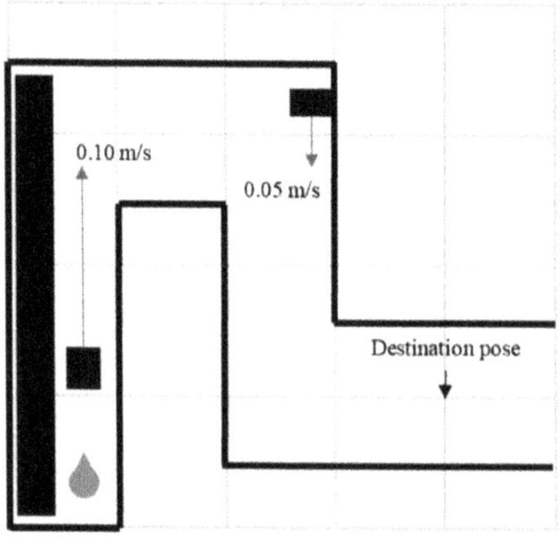

Figure 8. Learning scenario. Left static obstacle reduced the drivable space so that the robot received rewards right from the beginning of the learning process.

The static obstacle of the left hand reduced the drivable space of the robot. This allowed the robot to receive a negative reward right from the beginning if it got too close to the walls, which delimited the boundaries of the circuit. Additionally, the starting pose of the robot was not too far away from the first dynamic obstacle. This helped the robot to quickly see negative rewards while approaching the dynamic obstacle and adapt the policy and critic functions to follow up the obstacle instead of crashing into it. The resulting robot's trajectory after the learning process is shown in Figure 9. The robot's linear velocity was reduced to follow up the dynamic obstacle at the beginning of the circuit and the commanded angular velocity was always smooth to maximize comfort (Figure 10).

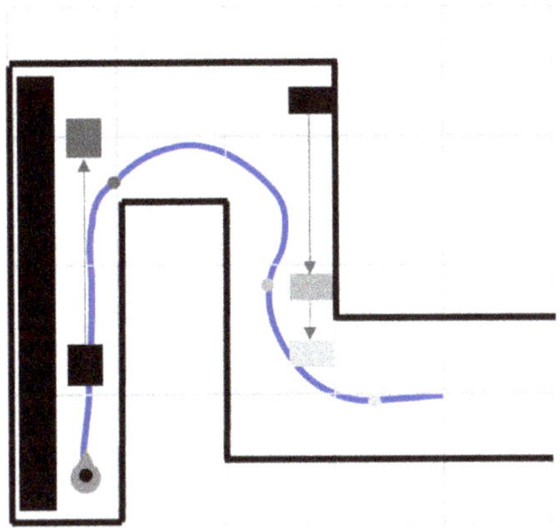

Figure 9. Traveled path by the robot in the learning scenario. Same colors indicate instant positions of the dynamic obstacles and the robot in the traveled path.

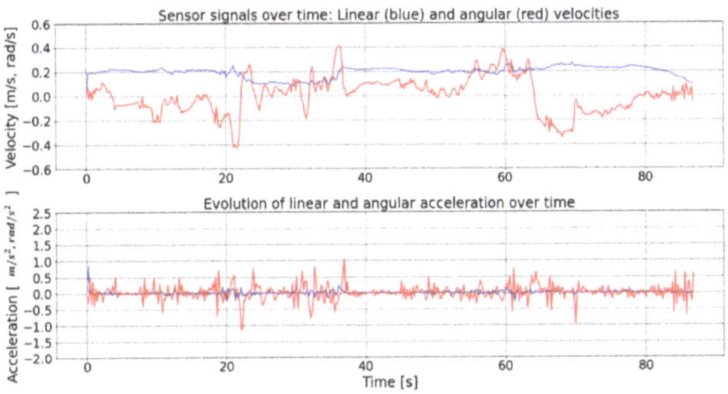

Figure 10. Top graph: linear (blue) and angular (red) velocities of the robot in the learning scenario. Bottom graph: linear (blue) and angular (red) accelerations. The maximum and minimum levels of the linear and angular accelerations were limited to maximize comfort.

Figure 11 shows the episodic reward as a function of the learning episodes. The episodic reward increased towards 0 after each learning episode. There was even room for improvement as the episodic reward did not seem to be completely flat after more than 200 learning episodes. It is interesting to note the recovery process in the episodic reward plot, that is, the episodic reward suddenly went down and then up again in the next episodes. This was due to two different factors: first, the exploration ratio sometimes led the robot to collide with the obstacles. In the beginning, when the policy was still unmature and the exploration ratio was high, the recovery peaks could be seen more frequently. Second, when updating the policy following the gradient descent approach, the step or learning rate could be too large, producing a policy that crashes the robot into the obstacles. For these two reasons, a decay for both exploration and learning rate was implemented. This is the reason why as the learning process went on, the recovery peaks were less frequent. The total time needed to automatically learn the policy was 6.57 h. The system used for the learning process was a desktop computer with an AMD Ryzen 5 2600 Six-Core Processor of 3.40 GHz and 8 Gb RAM installed (no GPU installation).

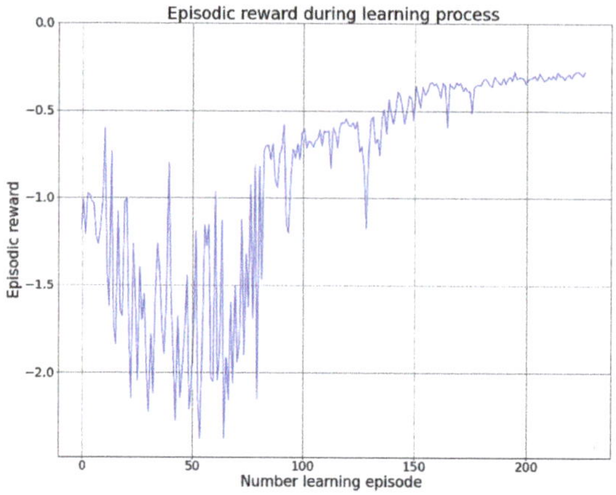

Figure 11. Episodic reward versus learning episodes of the learning scenario.

Finally, the traveled path of the robot in the validation scenario is shown in Figure 12. The first dynamic obstacle blocked the way, and the robot needed to adapt its linear speed to follow up the obstacle. Once the robot detected enough free space, it overtook the obstacle. The second and third obstacles traveled at opposite speeds. Lane change maneuvers were needed to overtake both obstacles. The linear and angular velocities and accelerations are shown in Figure 13. The robot safely avoided all static and dynamic obstacles of the circuit traveling at the maximum legal linear speed of 0.2 m/s (except in the follow-up sector). In addition, the longitudinal and angular accelerations did not exceed the maximum comfort levels defined for this scenario ($0.5\ \mathrm{m/s^2}$ and $1\,\frac{\mathrm{rad}}{\mathrm{s^2}}$, respectively). Finally, the robot reached the destination pose and ramped down the linear velocity to a standstill position. All of this was happening in a scenario never used during a learning process, concluding that the policy learned during the learning process can be generalized to new and unseen scenarios.

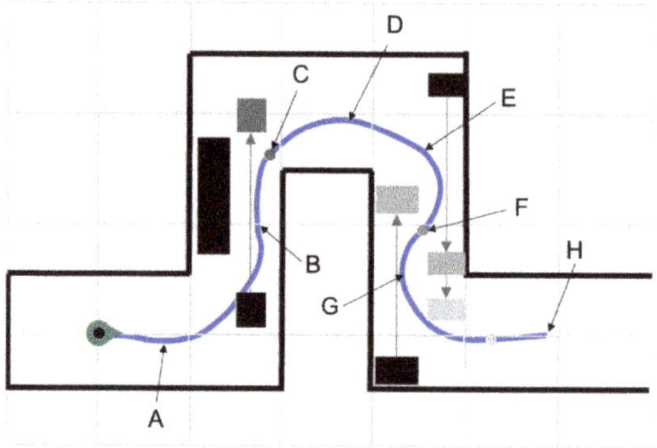

Figure 12. Traveled path by the robot in the validation scenario. Same colors indicate instant positions of the dynamic obstacles and the robot in the traveled path. A, D and G marks indicate a preparation of the robot before the curve to maximize comfort feeling. B mark is the follow up section. C, E and F marks show the moment when the robot found enough free space to overtake. Finally, mark H is the destination pose.

There are several key points marked with capital letters in Figures 12 and 13. These points are explained in detail in Table 2.

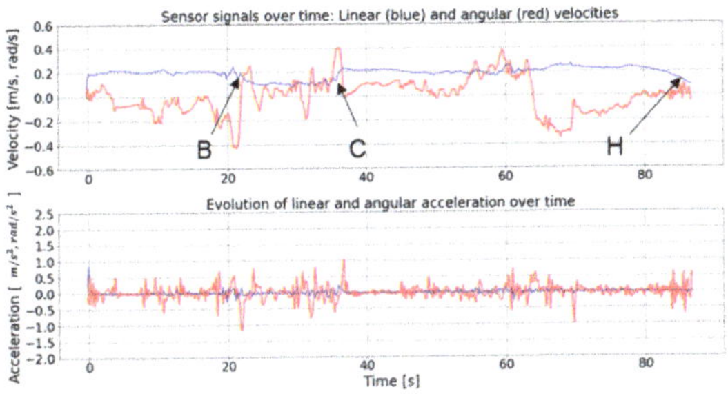

Figure 13. Top graph: linear (blue) and angular (red) velocities of the robot in the validation scenario. Bottom graph: linear (blue) and angular (red) accelerations. The maximum and minimum levels of the linear and angular accelerations were limited to maximize comfort. Between marks B and C is the follow up section, where the robot slowed down its linear speed. Mark H shows the stopping procedure.

Table 2. Explanation of the key points of the traveled path by the robot in the validation scenario.

Mark	Description
A	The robot drove outside of the track before entering the curve to reduce angular acceleration and maximize comfort.
B	The robot followed up the dynamic obstacle at a constant linear speed, keeping a fixed distance between the obstacle and the robot.
C	The robot detected enough free space to overtake the dynamic obstacle and increased its linear speed up to the legal one.
D	Similar to A, the robot drove outside of the track before entering the curve to reduce angular acceleration.
E	The robot detected free space and drove to the left (lane change) to overtake the second dynamic obstacle.
F	The robot detected free space and changed lane again to overtake the third dynamic obstacle.
G	Similar to A and D, the robot drove outside of the track before entering the curve to reduce angular acceleration.
H	The robot ramped down the linear velocity and stopped at the destination pose.

Another interesting autonomous driving use case is a robot exploring inside an open-world scenario. For this purpose, the experiment depicted in Figure 14 was used as a validation scenario to test the methodology in open-world scenarios.

Figure 14. Validation scenario for open-world use cases. The destination pose is marked with a red spot. The robot was intended to avoid all obstacles under safety, comfort, and legal conditions.

The scenario used for the learning process was a simplified open-world scenario to speed up the learning process. The results after the learning process was completed are shown in Figure 15 for different destination poses.

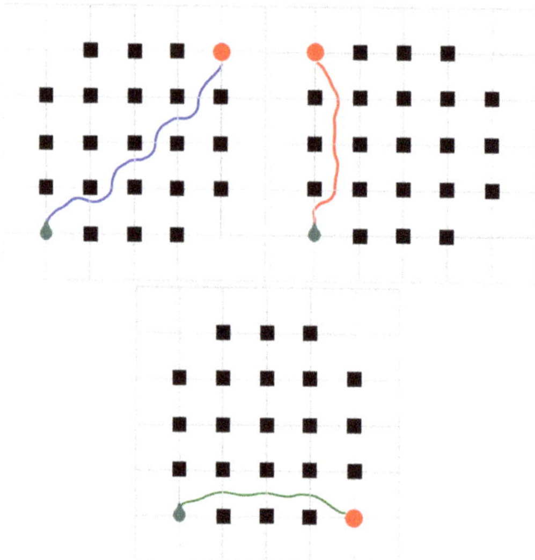

Figure 15. Traveled path (blue, red and green) by the robot in the validation scenario, considering different destination poses. The destination pose is marked with a red spot. The green spot shows the initial position of the robot.

In all three cases from Figure 15, the robot reached the destination pose under safety, legal, and comfort conditions. Nevertheless, the robot still entered at some points (beginning of blue path) the safety margin and got close to some obstacles. This was because the termination conditions of the learning process were met while the agent was still learning. Figure 16 shows precisely that the agent was still learning since the episodic reward was still not flat after more than 200 learning episodes (red line).

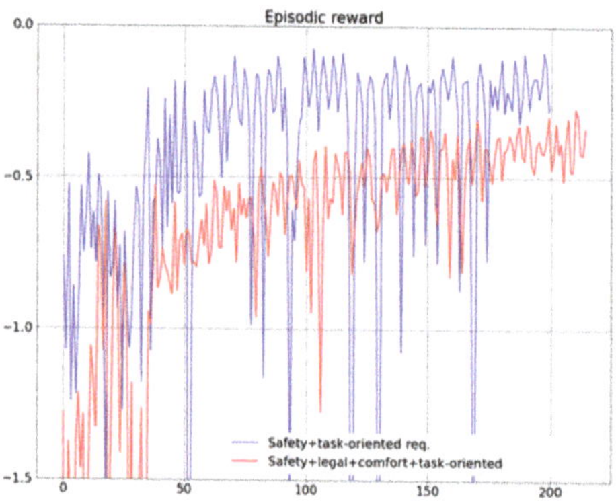

Figure 16. Episodic reward versus learning episodes of the learning scenario for the open-world scenario. The red line represents the case where all requirements (safety, legal, comfort, and task-oriented) were implemented. The agent was still learning after the termination condition was met since the slope of the curve was not flat. On the other hand, the blue line, representing the case where only safety and task-oriented requirements were implemented, was flat after the termination condition was met, ensuring that the agent had completed the learning process.

However, since the learning process took a long time, a relax termination condition needed to be defined to find a trade-off between the performance of the resulting policy and the total learning time. In this case, the total learning time went up to 6.1 h. In the case of using of a more powerful system with a server and GPUs, the termination condition could have been adapted to check that the episodic reward does not increase over the learning episodes and ensure in that way a complete learning process.

Finally, a comparison was performed between different strategies to create reward functions. First, a mix of positive (robot driving at desired speed) and negative (collision, comfort, legal, and direction towards the destination) reward functions was created, such as the ones used in many algorithms of the literature reviewed in the section Materials and Methods. Second, a variant with strictly negative reward functions, such as the one proposed in this paper, was chosen as a reference for comparison. The results are shown in Figure 17. The variant with positive and negative terms found a suboptimal solution where the robot keeps driving to collect more rewards.

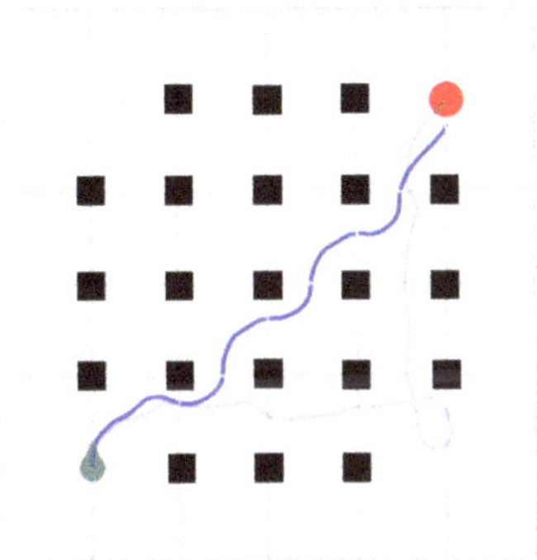

Figure 17. Comparison between different strategies when generating reward functions. Dark blue corresponds to strictly negative reward functions. Light blue corresponds to a mix of positive and negative reward functions.

These results show the effectiveness of the reinforcement learning techniques in dynamic scenarios, which are usually the most complex to solve, and open-world scenarios, which are common for robots whose task is to explore the surroundings. Maneuvers such as obstacle avoidance, overtaking, follow up, lane keeping, or lane change were successfully validated in the previous scenarios.

4. Discussion

The results of the previous section were compared with the dynamic window approach (DWA), which is a state-of-the-art algorithm that belongs to the reactive paradigm group. The DWA algorithm constructs a set of trajectories based on the velocity space. Only velocities that produce safe trajectories are used (i.e., velocities that allow the robot to stop without collision in case of facing a static obstacle). Then, the best trajectory is selected based on the score produced by a well-designed cost function. The results of the DWA algorithm in the dynamic environment are shown in Figure 18.

If the robot finds enough free space, the DWA algorithm succeeds in finding the path to the destination, as shown in mark C. The robot found here enough free space to get through the obstacles, despite being surrounded by dynamic obstacles. On the other hand, if the robot gets trapped between static and dynamic obstacles, the robot fails to find a feasible path and collides with obstacles. Mark A shows when the robot collided with the dynamic obstacle. As a consequence of that, the robot moved backwards to mark B. From mark B onwards, the robot moved forward again to reach the destination. The comparison shows how the reinforcement learning algorithm found a way to adapt the robot's linear velocity to follow up the dynamic obstacle until free space was detected again. This way, a collision was avoided and the total time to destination was reduced, outperforming the results obtained with the DWA algorithm.

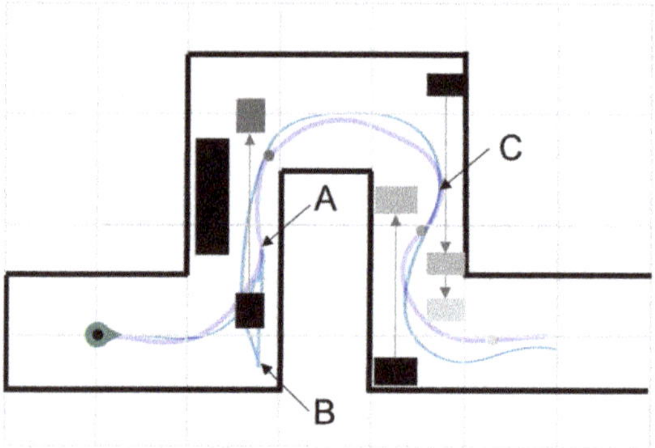

Figure 18. Traveled path by the robot for the reinforcement learning algorithm (light blue) and DWA algorithm (dark blue). Mark A shows a collision with the dynamic obstacle in the DWA algorithm, being forced to move the robot backwards to mark B. Mark C shows a successful overtake of the dynamic obstacles.

Scenarios with dynamic obstacles are very suitable for use cases involving passenger vehicles under real traffic conditions. Deep reinforcement learning models can learn the non-holonomic constrains of a front steering robot, even under conditions close to the grip limit, and solve complex driving scenarios, such as obstacle avoidance with dynamic obstacles, lane keeping, lane change, or overtaking maneuvers.

5. Conclusions

If we think of autonomous driving, many requirements need to be considered, making the development of software a challenging task. Not only do safety requirements need to be considered (lane keeping or avoiding static and dynamic obstacles) but comfort, legal (maximum speed limit), and task-oriented requirements are also essential to be fulfilled. The state-of-the-art solutions produce thousands of possible trajectories to finally select the best one based on the score of a well-designed cost function. However, there are many disadvantages with these state-of-the-art solutions. They are not only very computationally demanding but also very challenging to implement to fulfill all predefined requirements. The self-learning navigation algorithm introduced in this work comes up automatically with a policy that fulfills all predefined requirements at a reasonable computational cost. These requirements are defined by multiple, intuitive, and simple reward functions. Therefore, a highly mature autonomous driving logic or policy is obtained just by trial and error. This methodology has been proven by setting up some classical dynamic scenarios for the autonomous driving domain with maneuvers such as overtaking, following up, lane keeping, or lane changing. Within this section, Figures 12 and 13 show how the safety, legal, and comfort requirements were fulfilled along the entire trajectory driven by the robot. Additionally, the proposed methodology was compared with the dynamic window approach (DWA) algorithm (Figure 18), which is a classic state-of-the-art algorithm to drive robots autonomously while avoiding obstacles. The comparison shows how the reinforcement learning algorithm outperformed the DWA algorithm. Figure 15 also proves that the methodology works effectively in open-world scenarios, avoiding obstacles and driving up to the destination pose under safety, legal, and comfort conditions. The results also show that the learning process can be very time-consuming. However, it can be automated by simulation, loading multiple and complex dynamic driving scenarios. As the autonomous driving logic is self-learned, engineering time and costs can be saved through the software development process. Additionally, the reinforcement learning algorithm produces a policy considering the dynamic constrains of the non-holonomic robot. This is because the agent interacts and learns directly from the environment where the robot belongs, making it unnecessary to implement embedded models of robot dynamics like in the predictive control model approach.

Author Contributions: Conceptualization, B.P.G.d.l.H., R.M.-T. and J.M.C.T.; validation, B.P.G.d.l.H., R.M.-T. and J.M.C.T.; formal analysis, B.P.G.d.l.H., R.M.-T. and J.M.C.T.; writing—review and editing, B.P.G.d.l.H., R.M.-T. and J.M.C.T.; supervision, R.M.-T. and J.M.C.T. All authors have read and agreed to the published version of the manuscript.

Funding: This research received no external funding.

Institutional Review Board Statement: Not applicable.

Informed Consent Statement: Not applicable.

Data Availability Statement: The data that support the findings of this study are openly available in https://github.com/bpintos/smart-trajectory-planning (accessed on 20 December 2023).

Conflicts of Interest: The authors declare no conflicts of interest.

References

1. Warren, C.W. Multiple robot path coordination using artificial potential fields. In Proceedings of the IEEE International Conference on Robotics and Automation, Cincinnati, OH, USA, 13–18 May 1990.
2. Fiorini, P.; Shiller, Z. Motion Planning in Dynamic Environments Using Velocity Obstacles. *Int. J. Robot. Res.* **1998**, *17*, 760–772. [CrossRef]
3. Fox, D.; Burgard, W.; Thrun, S. The dynamic window approach to collision avoidance. *IEEE Robot. Autom. Mag.* **1997**, *4*, 23–33. [CrossRef]
4. Álvarez-Sánchez, J.; De la Paz López, F.; Troncoso, J.; Sierra, D. Reactive navigation in real environments using partial center of area method. *Robot. Auton. Syst.* **2010**, *58*, 1231–1237. [CrossRef]
5. Tobaruela, J.A.; Rodríguez, A.O. Rodríguez Reactive navigation in extremely dense and highly intricate environments. *PLoS ONE* **2017**, *12*, e0189008.
6. Dreyfus, S.E. An Appraisal of Some Shortest Path Algorithm. *Oper. Res.* **1969**, *17*, 395–412. [CrossRef]
7. Moses, B.; Jain, L.; Finn, R.; Drake, S. Multiple UAVs path planning algorithms: A comparative study. *Fuzzy Optim. Decis. Mak.* **2008**, *7*, 257–267.
8. Dolgov, D.; Thrun, S.; Montemerlo, M.; Diebel, J. Practical Search Techniques in Path Planning for Autonomous Driving. In *AAAI Workshop—Technical Report*. 2008; Available online: https://api.semanticscholar.org/CorpusID:16143233 (accessed on 29 December 2023).
9. LaValle, S.M.; Kuffner, J.J. Randomized Kinodynamic Planning. *Int. J. Robot. Res.* **1999**, *20*, 378–400. [CrossRef]
10. Kavraki, L.E.; Svestka, P.; Latombe, J.-C.; Overmars, M.H. Probabilistic roadmaps for path planning in high-dimensional configuration spaces. *IEEE Trans. Robot. Autom.* **1996**, *12*, 566–580. [CrossRef]
11. Bautista, D.G.; Pérez, J.; Milanes, V.; Nashashibi, F. A Review of Motion Planning Techniques for Automated Vehicles. *IEEE Trans. Intell. Transp. Syst.* **2015**, *17*, 1135–1145.
12. Pepy, R.; Lambert, A.; Mounier, H. Path Planning using a Dynamic Vehicle Model. In Proceedings of the 2006 2nd International Conference on Information & Communication Technologies, Damascus, Syria, 24–28 April 2006.
13. Paden, B.; Čáp, M.; Yong, S.Z.; Yershov, D.; Frazzoli, E. A Survey of Motion Planning and Control Techniques for Self-Driving Urban Vehicles. *IEEE Trans. Intell. Veh.* **2016**, *1*, 33–55. [CrossRef]
14. Pereira, G.C. Lateral Model Predictive Control for Autonomous Heavy-Duty Vehicles: Sensor, Actuator, and Reference Uncertainties. Ph.D. Thesis, KTH Royal Institute of Technology, Stockholm, Sweden, 2020.
15. González, D.S.; Pérez, J.; Lattarulo, R.; Montero, V.M.; Nashashibi, F. Continuous curvature planning with obstacle avoidance capabilities in urban scenarios. In Proceedings of the 17th International IEEE Conference on Intelligent Transportation Systems (ITSC), Qingdao, China, 8–11 October 2014; pp. 1430–1435.
16. Farouki, R.T.; Sakkalis, T. Pythagoreanhodograph space curves. *Adv. Comput. Math.* **1994**, *2*, 41–66. [CrossRef]
17. Xu, W.; Wei, J.; Dolan, J.; Zhao, H.; Zha, H. A Real-Time Motion Planner with Trajectory Optimization for Autonomous Vehicles. In Proceedings of the 2012 IEEE International Conference on Robotics and Automation, Saint Paul, MN, USA, 14–18 May 2012.
18. Svensson, L.; Bujarbaruah, M.; Karsolia, A.; Berger, C.; Törngren, M. Traction Adaptive Motion Planning at the Limits of Handling. *IEEE Trans. Control Syst. Technol.* **2020**, *30*, 1888–1904. [CrossRef]
19. Zhou, C.; Huang, B.; Fränti, P. A review of motion planning algorithms for intelligent robots. *J. Intell. Manuf.* **2022**, *33*, 387–424. [CrossRef]
20. Nair, R.S.; Supriya, P. Robotic Path Planning Using Recurrent Neural Networks. In Proceedings of the 2020 11th International Conference on Computing, Communication and Networking Technologies (ICCCNT), Kharagpur, India, 1–3 July 2020.

21. Hu, Y.; Yang, L.; Lou, Y. Path Planning with Q-Learning. *J. Phys. Conf. Ser.* **2021**, *1948*, 012038. [CrossRef]
22. Lillicrap, T.; Hunt, J.; Pritzel, A.; Heess, N.; Erez, T.; Tassa, Y.; Silver, D.; Wierstra, D. Continuous control with deep reinforcement learning. *arXiv* **2015**, arXiv:1509.02971.
23. Kendall, A.; Hawke, J.; Janz, D.; Mazur, P.; Reda, D.; Allen, J.-M.; Lam, V.-D.; Bewley, A.; Shah, A. Learning to Drive in a Day. In Proceedings of the 2019 International Conference on Robotics and Automation (ICRA), Montreal, QC, Canada, 20–24 May 2019.
24. Vitelli, M.; Nayebi, A. CARMA: A Deep Reinforcement Learning Approach to Autonomous Driving. 2016. Available online: https://anayebi.github.io/files/projects/CS_239_Final_Project_Writeup.pdf (accessed on 29 December 2023).
25. Wang, P.; Chan, C.-Y.; de La Fortelle, A. A Reinforcement Learning Based Approach for Automated Lane Change Maneuvers. In Proceedings of the 2018 IEEE Intelligent Vehicles Symposium (IV), Changshu, China, 26–30 June 2018.
26. Dosovitskiy, A.; Ros, G.; Codevilla, F.; Lopez, A.; Koltun, V. CARLA: An open urban driving simulator. In Proceedings of the 1st Annual Conference on Robot Learning, Mountain View, CA, USA, 13–15 November 2017; pp. 1–16.
27. Kardell, S.; Kuosku, M. Autonomous Vehicle Control via Deep Reinforcement Learning. Master's Thesis, Chalmers University of Technology, Gothenburg, Sweden, 2017.
28. Chen, J.; Yuan, B.; Tomizuka, M. Model-free deep reinforcement learning for urban autonomous driving. In Proceedings of the 2019 IEEE Intelligent Transportation Systems Conference (ITSC), Auckland, New Zealand, 27–30 October 2019; pp. 2765–2771.

applied sciences

Article

Adaptive Kalman Filter for Real-Time Visual Object Tracking Based on Autocovariance Least Square Estimation

Jiahong Li [1,2], Xinkai Xu [1,2], Zhuoying Jiang [2] and Beiyan Jiang [1,2,3,*]

1 Beijing Key Laboratory of Information Service Engineering, Beijing Union University, Beijing 100101, China; jqrjiahong@buu.edu.cn (J.L.); ldtxinkai@buu.edu.cn (X.X.)
2 College of Robotics, Beijing Union University, Beijing 100027, China; jiangzhuoying043@hotmail.com
3 State Key Lab of Automotive Safety and Energy, School of Vehicle and Mobility, Tsinghua University, Beijing 100084, China
* Correspondence: jqrbeiyan@buu.edu.cn

Abstract: Real-time visual object tracking (VOT) may suffer from performance degradation and even divergence owing to inaccurate noise statistics typically engendered by non-stationary video sequences or alterations in the tracked object. This paper presents a novel adaptive Kalman filter (AKF) algorithm, termed AKF-ALS, based on the autocovariance least square estimation (ALS) methodology to improve the accuracy and robustness of VOT. The AKF-ALS algorithm involves object detection via an adaptive thresholding-based background subtraction technique and object tracking through real-time state estimation via the Kalman filter (KF) and noise covariance estimation using the ALS method. The proposed algorithm offers a robust and efficient solution to adapting the system model mismatches or invalid offline calibration, significantly improving the state estimation accuracy in VOT. The computation complexity of the AKF-ALS algorithm is derived and a numerical analysis is conducted to show its real-time efficiency. Experimental validations on tracking the centroid of a moving ball subjected to projectile motion, free-fall bouncing motion, and back-and-forth linear motion, reveal that the AKF-ALS algorithm outperforms a standard KF with fixed noise statistics.

Keywords: visual object tracking; Kalman filter; autocovariance least-squares estimation; background subtraction; adaptive thresholding

1. Introduction

Visual object tracking (VOT) is a fundamental task in computer vision and has extensive applications including autonomous vehicles [1], video surveillance [2], robot vision [3], and human-computer interaction [4]. Specifically, in autonomous vehicle systems and robotics, robust and efficient VOT algorithms that identify and track nearby vehicles and pedestrians are essential for real-time navigation, obstacle avoidance, and environment perception, ensuring safe and efficient operation in dynamic scenarios. The essence of VOT is to accurately track the states (e.g., the centroid, the bounding box, or the contour) of an object of interest through successive frames within a video stream. This involves two key steps: object detection which identifies the presence and location of the object, and object tracking which associates the object and generates the trajectory of the object throughout the frame sequence.

The challenges faced in VOT are accentuated in the context of autonomous systems. These systems must perform reliably amidst non-stationary conditions caused by variable lighting, weather conditions, diverse terrains, and unstructured environments. Furthermore, in robotics, VOT is tested against robot motion, varying perspectives, and interactions with objects that can introduce additional complexities. The object may change poses, deform, become occluded, or disappear temporarily from the field of view [5], resulting in high model and measurement noise in the object tracking step. Regarding the uncertainty (e.g., sudden or drastic changes in object appearance) as the noise, the approach to the two above challenges depends heavily on the real-time accurate estimation of the noise statistics, especially the accurate noise covariance estimation. Therefore, the real-time VOT under uncertain noise covariance problem is addressed in this paper, which can account for the varying model and measurement noise covariance's influence on object detection and tracking.

1.1. Object Detection

The background subtraction (BS) technique is a widely used object detection technique for extracting moving foreground objects from video against a static background. In an autonomous vehicle or robot's vision systems, the choice of BS technique can significantly influence performance due to several factors, such as illumination variations, occlusion, shape deformation, scale variation, and background clutter. According to the model and computational complexity, the BS techniques can be classified into three broad categories: simple BS, statistical BS, and neural networks (NN) based BS [6–9].

- The simple BS techniques including the frame differencing [10], average filtering [11,12], median filtering [13,14], and histogram over time [15], model the background simply, often by just using the previous frames or an average/median of the recent frames. These are fast and easy to implement but lack adaptability to changes in the background.
- The statistical BS techniques including the mixture of Gaussians (MoG) [16,17], kernel density estimation [18–20], support vector models [21,22], and principal component analysis [23,24], build a statistical model of the background and classify pixels based on the model. They model the background more robustly using the history of pixels but are prone to the setting parameters and models.
- NN-based BS techniques including the radial basis function NN [25], self-organizing NN [26,27], the convolutional NN [28,29], and the generative adversarial networks (GAN) [30,31], learn the specialized NN architectures that can adapt to changes in the background model to detect foreground over time. These can learn complex representations of background appearance and maintain robust models of multi-modal backgrounds, but they require large training datasets and have expensive computation loads to train and run, which may not be available in a real-time environment.

Dynamic backgrounds with movement from objects can be erroneously marked as foreground if the model cannot adapt swiftly enough, leading to noisy segmentation and false detections. Therefore, the simple frame difference-based BS technique with the adaptive threshold (AT) method is appealing in the real-time VOT application due to its simplicity, efficiency, and robustness [32]. In this method, the threshold is adjusted based on the pixel's intensity variations over some time, e.g., the threshold should be smaller for regions with low contrast.

1.2. Object Tracking

Recent developments of VOT have been witnessed to solve the VOT problem from the representation (e.g., correlation filter) based tracker [33,34], discriminate and generative tracker [35], to deep Siamese networks based tracker [36,37]. Conventional representation-based VOT methods rely on target appearance models based on handcrafted visual features and are prone to inaccuracies from background interference and occlusion. The discriminative and deep Siamese networks-based VOT methods are more robust but may fail to update adequately to sudden variations of the object in real-time applications due to high computation complexity.

To achieve robust and real-time VOT, various recursive Bayesian filtering methods like the mean-shift (MS) tracker [38,39], the Kalman filter (KF) tracker [40–42], and particle filter (PF) tracker [43–45] are commonly developed. Among the above, the KF tracker has become one of the most preferred tracking filters because of the high computational efficiency and robustness to missed detections or occlusions, especially for linear and Gaussian systems.

The KF-based VOT method uses the background subtraction method for foreground object detection through background modeling and foreground segmentation and updates the object's state recursively by model prediction and measurement correlation.

For the linear time-invariant (LTI) system model with the LTI measurement system, KF is proven to be the optimal linear minimum variance estimator [46]. However, in practice, the model mismatches and inaccurate noise statistics result in sub-optimal filtering that outputs unbiased and even divergent estimates [47]. The adaptive KF is one effective approach and has been applied in many applications [48–50]. In the KF-based object tracking task, the prior knowledge of the process and measurement noise covariance matrices is hard to determine due to occlusions and changes in object appearance. Therefore, designing adaptive KF focusing on noise covariance estimation is of vital importance.

Existing noise covariance estimation methods typically utilize historical open-loop data, which can be categorized as Bayesian [51], maximum likelihood estimation (MLE) [52,53], covariance matching (CM) [54,55], and subspace identification (SI) [56], correlation [57–62]. Each of these methods has its strengths and weaknesses. For instance, while the Bayesian method provides optimal posterior estimation using Bayes' rule to update the prior knowledge of the noise covariance, it suffers from computational complexity and sensitivity to hyper-parameter choices [51]. Similarly, while the MLE method maximizes the likelihood of innovations without prior knowledge, it is computationally intensive and sensitive to model misspecification [53]. The CM method, although robust and computationally efficient, is sensitive to outliers and unsuitable for online scenarios [63]. The correlation method, and especially the auto-covariance least-squares (ALS) method has attracted quite considerable attention as it provides an unbiased estimate based on the autocovariance of the innovations with acceptable computational complexity and no specific noise model requirements [60].

Therefore, the autocovariance least square estimation (ALS) method is integrated into the KF-based VOT method, and the AKF-ALS algorithm is proposed to address the requirements of simplicity and robustness in real-time visual object tracking under uncertain noise covariance. Our contributions are threefold:

1. The proposed AKF-ALS algorithm provides a robust and efficient solution for noise covariance estimation in the visual object tracking problem.
2. A novel adaptive thresholding method based on the estimated process noise covariance that can predict sudden variations without heavy computations is proposed to improve the robustness of the BS method.
3. The experiments on tracking the centroid of a moving ball subjected to projectile motion, free-fall bouncing motion, and back-and-forth linear motion are conducted to show the efficiency and superiority of the proposed AKF-ALS algorithm.

The rest of the paper is organized as follows. Section 2 formulates the VOT problem and introduces the KF and the ALS method. In Section 3, the AKF-ALS algorithm is developed to deal with the visual object tracking task with uncertain noise covariance. It utilizes the ALS method to achieve real-time noise covariance tuning and proposes a novel adaptive thresholding method to improve the robustness of the BS method. The computation complexity of the AKF-ALS algorithm is derived and a numerical analysis is conducted to show its efficiency. In Section 4, the proposed AKF-ALS algorithm is validated to outperform the Kalman filter given constant noise covariance through the experiments on tracking the centroid of a moving ball subjected to projectile motion, free-fall bouncing motion, and back-and-forth linear motion. Section 5 provides concluding remarks and a discussion of future work.

2. Problem Formulation and Preliminaries

The visual object tracking problem consists of two main stages: background subtraction for object detection and Kalman filter for object tracking.

2.1. Background Subtraction

Background subtraction involves distinguishing the foreground (moving objects) from the background in video frames. We can formulate this problem mathematically as follows. The input video sequence is represented as a series of frames: $F_1, F_2, \ldots, F_t, \ldots, F_N$, where N is the total number of frames and t is the current time instance. Each frame F_t is a 2D array of pixels denoted as $F_t(i,j)$, where i and j represent the spatial coordinates of the pixel in the frame. Denote the input image at time t as $I(x,y,t)$ which is composed of a static background $B(x,y)$ and a moving foreground $F(x,y,t)$, such that:

$$I(x,y,t) = B(x,y) + F(x,y,t) \tag{1}$$

The goal of background subtraction is to create a binary mask $M(x,y)$ for each frame $F_t(i,j)$, where pixels belonging to moving objects are labeled as 1 (foreground), and the rest are labeled as 0 (background).

$$M(x,y) = \begin{cases} 1, & \text{if } |I(x,y,t_d) - I(x,y,t_f)| > T_t \\ 0, & \text{otherwise} \end{cases} \tag{2}$$

where T_t is the user-defined threshold, t_d and t_f represent the time instances of the dark and bright frames, respectively.

2.2. Kalman Filter

Kalman filter is a recursive estimator that provides the optimal solution under the assumptions of a linear system and Gaussian noise. The Kalman filter operates in a two-step process recursively: prediction and correlation. In the prediction step, the filter predicts the state and error covariance estimates to obtain the prior estimates for the next time step.

$$\begin{aligned} \hat{x}_{k+1} &= F\hat{x}_k + Bu_k \\ P_{k+1} &= FP_kF^T + Q \end{aligned} \tag{3}$$

where $\hat{x}_k$ is the state estimate at time k, P_k is the state covariance matrix, and Q is the process noise covariance.

In the correction step, the filter incorporates a new measurement into the prior estimate to obtain an improved posterior estimate.

$$\begin{aligned} K_{k+1} &= P_{k+1}H^T(HP_{k+1}H^T + R)^{-1} \\ \hat{x}_{k+1} &= \hat{x}_{k+1} + K_{k+1}(z_{k+1} - H\hat{x}_{k+1}) \\ P_{k+1} &= (I - K_{k+1}H)P_{k+1} \end{aligned} \tag{4}$$

where K_{k+1} is the Kalman gain, R is the measurement noise covariance.

2.3. Auto-Covariance Least-Squares (ALS) Method

The auto-covariance least-squares (ALS) method is introduced to estimate the unknown noise covariance in a linear state-space model in model-based estimation and control, which is crucial for improving the performance of a visual object tracking task.

The ALS method is based on the innovation sequence, denoted as e_k, which is defined as the difference between the observed measurement and the estimated measurement based on the predicted state in the prediction step:

$$e_k = z_k - H\hat{x}_{k|k-1}, \forall k. \tag{5}$$

Denote the residual, which is the difference between the actual state and the predicted state as $\varepsilon_k = x_k - \hat{x}_{k|k-1}$. Then the residual and the innovation follow $e_k = H\varepsilon_k + v_k$, and

$$\varepsilon_{k+1} = \underbrace{(F - FKH)}_{\bar{F}} \varepsilon_k + \underbrace{[I_{n_x}, -FK]}_{G} \underbrace{\begin{bmatrix} w_k \\ v_k \end{bmatrix}}_{\bar{w}_k}. \tag{6}$$

In the steady state, the Kalman gain converges to a constant value, which can be computed offline. The steady-state residual covariance matrix, denoted by P_ε, can be solved via the following Lyapunov equation:

$$P_\varepsilon = \bar{F} P_\varepsilon \bar{F}^{\mathrm{T}} + G \Sigma G^{\mathrm{T}} \tag{7}$$

where $\Sigma = \begin{bmatrix} Q & 0_{n_x \times n_z} \\ 0_{n_z \times n_x} & R \end{bmatrix}$.

Then the auto-covariance of the innovation sequence is:

$$\begin{aligned} C_{e,0} &\triangleq \mathbb{E}[e_k e_k^{\mathrm{T}}] = H P_\varepsilon H^{\mathrm{T}} + R \\ C_{e,j} &\triangleq \mathbb{E}[e_{k+j} e_k^{\mathrm{T}}] = H \bar{F}^j P_\varepsilon H^{\mathrm{T}} - H \bar{F}^{j-1} FKR \end{aligned} \tag{8}$$

where $j = 1, 2, \ldots, N-1$, and N is a user-defined parameter defining the window size. Substitute the solution to (7) into (8), and the noise covariance matrix is derived as a linear matrix equation by using the least-squares method:

$$\mathcal{A}\theta = b \tag{9}$$

where the matrix $\mathcal{A}$ is defined as

$$\begin{aligned} \mathcal{A} &= [D, D(FK \otimes FK) + (I_{n_z} \otimes \Gamma)] \\ D &= (H \otimes \mathcal{O})(I_{n_x^2} - \bar{F} \otimes \bar{F})^{-1} \\ \mathcal{O} &= [H^{\mathrm{T}}, (H\bar{F})^{\mathrm{T}}, \ldots, (H\bar{F}^{N-1})]^{\mathrm{T}} \\ \Gamma &= [I_{n_z}, -(HFK)^{\mathrm{T}}, \ldots, -(H\bar{F}^{N-2}FK)^{\mathrm{T}}]^{\mathrm{T}} \end{aligned} \tag{10}$$

where the symbol $\otimes$ stands for the Kronecker product. Denote the dependent variables as $\theta = [Q_s^{\mathrm{T}}, R_s^{\mathrm{T}}]^{\mathrm{T}}$ and $b = (C_e(N))_s$ with $C_e(N) = [C_{e,0}, C_{e,1}^{\mathrm{T}}, \ldots, C_{e,N-1}^{\mathrm{T}}]^{\mathrm{T}}$. Q_s means the column-wise stacking of the matrix Q into a vector.

The ALS method estimates the covariance matrices by minimizing the difference between the sample and theoretical innovations' autocovariance by minimizing the following cost function:

$$\begin{aligned} \hat{\theta} &= \arg\min_{\theta} \|\mathcal{A}\theta - b\|_2^2 \\ s.t.,& \quad Q, R_i \geq 0. \end{aligned} \tag{11}$$

Then the parameter θ is estimated by solving the semi-definite constrained least squares problem. The optimal solution can be found by setting the derivative of the cost function to zero when the matrix inequality holds:

$$\hat{\theta} = (\mathcal{A}^{\mathrm{T}}\mathcal{A})^{-1}\mathcal{A}^{\mathrm{T}}\hat{b} = \mathcal{A}^{\dagger}\hat{b} \tag{12}$$

where $\hat{b}$ denotes the unbiased estimate of the vector b, and is computed by using the ergodic property of the N-innovations to estimate the auto-covariance matrix $C_e(N)$:

$$\begin{aligned} \hat{b} &= (\hat{C}_e(N))_s \\ \hat{C}_e(N) &= [\hat{C}_{e,0}^{\mathrm{T}}, \hat{C}_{e,1}^{\mathrm{T}}, \ldots, \hat{C}_{e,N-1}^{\mathrm{T}}]^{\mathrm{T}} \\ \hat{C}_{e,j} &= \frac{1}{\tau - j} \sum_{k=1}^{\tau - j} e_{k+j} e_k^{\mathrm{T}}, \quad j = 0, 1, \ldots, N-1 \end{aligned} \tag{13}$$

where τ is denoted as the window size of the auto-covariances.

The steady-state solution P_ε to the Lyapunov equation in (7) exists and the innovation sequence is a stationary process if the prediction error transitional matrix $\bar{F}$ is stable.

$$P_\varepsilon = F(P_\varepsilon - P_\varepsilon H^T (HP_\varepsilon H^T + R_A)^{-1} HP_\varepsilon) + Q_A \tag{14}$$

where Q_A and R_A are positive user-defined matrices.

3. ALS-KF Based Visual Object Tracking Algorithm

The visual object tracking algorithm based on the Kalman filter combined with the autocovariance least square estimation method called the AKF-ALS algorithm is proposed in this paper. It consists of three procedures: object detection via background subtraction, object tracking using the Kalman filter, and noise covariance estimation using the ALS method, as shown in the flowchart in Figure 1.

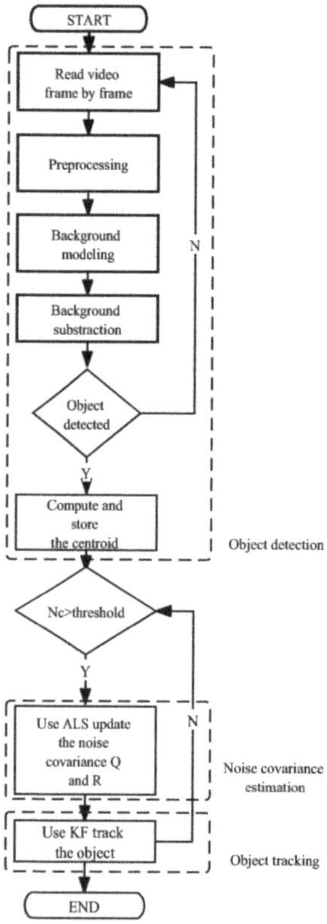

Figure 1. Flowchart of the ALS-KF based VOT algorithm.

3.1. Object Detection Using Background Subtraction

The first step in the tracking process is to identify the object of interest. In our proposed method, we utilize the background subtraction method to detect the moving object. The background is initialized as the first frame of the video. For each subsequent frame, we subtract the background from the current frame. By applying a threshold to the difference image, we convert it into a binary image. The binary image is then processed using morphological operations to reduce noise and fill gaps in the detected objects.

Object tracking in video sequences often requires adaptive thresholding techniques to robustly handle changes in object appearance over time. The process noise covariance $\hat{Q}$

estimated by the Kalman filter provides a principled metric to adapt threshold levels based on the dynamics of the tracked object. The process noise captures uncertainties in modeling the motion of the object, thus larger values of $\hat{Q}$ imply more aggressive maneuvers by the target. This leads to greater shape deformation and scale variation as the object moves rapidly through the scene. By leveraging $\hat{Q}$, the threshold T can be adapted during tracking according to:

$$T_t = \alpha \hat{Q}_t + T_{\min} \tag{15}$$

where α is a scaling parameter and $T_{\min}$ is a minimum threshold level. The key idea is that a larger process noise covariance $\hat{Q}t$ at time t indicates faster motion, which requires relaxing the threshold to accommodate greater changes in appearance. This avoids losing the target due to overly tight thresholds based on stale measurements.

Connected components in the resulting binary image are then identified, and a bounding box is generated for each component. The centroid of each component, which is computed as the geometric center of the bounding box, is then stored as the object's location. The process is repeated for each frame in the video, resulting in a series of centroids representing the object's trajectory.

3.2. Object Tracking Using Kalman Filter

Once the objects are detected via the background subtraction, the next step is to track the detected object using the Kalman filter. Denote the object O_i is represented by its state x_i, which includes its position and velocity. In the visual object tracking task, the Kalman filter is used to predict the state of each object in the next frame, given its current state and the observations (detected positions). The Kalman filter is a recursive algorithm that uses a series of measurements observed over time, containing statistical noise and other inaccuracies, and produces estimates of unknown variables that tend to be more precise than those based on a single measurement alone. It includes the prediction and update steps: The prediction step projects forward the current state and error covariance estimates to obtain the a priori estimates for the next time step; the update step incorporates a new measurement into the a priori estimate to obtain an improved a posteriori estimate.

Given the current state of the object O_i at time t denoted as $x_{i,t}$, and observation z_t, we want to estimate the state $x_{i,t+1}$ at time $t+1$. The observation z_t can be seen as the detected position of the object from the background subtraction. The dynamic model of the system and the measurement model can be represented as:

$$\begin{aligned} x_{k+1} &= F x_k + B u_k + w_k \\ z_k &= H x_k + v_k \end{aligned} \tag{16}$$

where x_k is the state vector, u_k is the control input, w_k is the process noise, z_k is the measurement, and v_k is the measurement noise. F, B, and H are system matrices.

Based on the Kalman filter introduced in (3) and (4); the KF derivation for the object tracking problem is given below.

Prediction:

$$\begin{aligned} \hat{x}_{k|k-1} &= F \hat{x}_{k-1|k-1} + B u_k \\ P_{k|k-1} &= F P_{k-1|k-1} F^T + Q \end{aligned} \tag{17}$$

Correction:

$$\begin{aligned} K_k &= P_{k|k-1} H^T (H P_{k|k-1} H^T + R)^{-1} \\ \hat{x}_{k|k} &= \hat{x}_{k|k-1} + K_k (z_k - H \hat{x}_{k|k-1}) \\ P_{k|k} &= (I - K_k H) P_{k|k-1} \end{aligned} \tag{18}$$

3.3. Noise Covariance Estimation Using ALS

An important part of the Kalman filter is the estimation of the noise covariance matrices Q (process noise covariance) and R (measurement noise covariance). The accuracy of these estimates significantly affects the filter's performance. In this work, we use the ALS method to estimate Q and R. The ALS method is based on the autocovariance of the innovation sequence (also known as the difference between the actual measurement and the predicted measurement based on the predicted state) from the Kalman filter. The ALS method computes the autocovariance of this sequence and formulates a least-squares problem to solve for Q and R. By continuously updating the noise covariance estimates, the ALS method allows the Kalman filter to adapt to changes in the system dynamics and measurement noise, resulting in improved tracking accuracy.

The adaptive Kalman filter (AKF) with the autocovariance least square estimation (ALS) algorithm, termed AKF-ALS algorithm is outlined in Algorithm 1.

Algorithm 1 AKF-ALS

Initialize: $\hat{x}_0, P_0, F, H, Q, R$
for $k = 1, 2, \ldots$ until convergence **do**
  Prediction:
  $\hat{x}_{k|k-1} = F\hat{x}_{k-1|k-1}$
  $P_{k|k-1} = FP_{k-1|k-1}F^{\mathrm{T}} + Q$
  Update:
  $e_k = z_k - H\hat{x}_{k|k-1}$
  $K_k = P_{k|k-1}H^{\mathrm{T}}(HP_{k|k-1}H^{\mathrm{T}} + R)^{-1}$
  $\hat{x}_{k|k} = \hat{x}_{k|k-1} + K_k e_k$
  $P_{k|k} = (I - K_k H)P_{k|k-1}$
  if $k \geq \tau + N$ **then**
    Estimate Q, R via ALS:
    Compute $\hat{C}_{e,0}, \hat{C}_{e,1}, \ldots, \hat{C}_{e,N-1}$ in Equation (13)
    Stack $\hat{C}_{e,0}, \hat{C}_{e,1}, \ldots, \hat{C}_{e,N-1}$ into $\hat{b}$
    Compute $\mathcal{A}$ using Equation (10)
    Solve for $\hat{\theta}$ using Equation (12)
    Unstack $\hat{\theta}$ into $\hat{Q}$ and $\hat{R}$
  end if
end for

The algorithm includes the following procedures.

1. Initialize the state estimate $\hat{x}_0$, the state covariance matrix P_0, the system matrices F and H, the Kalman gain matrix K, the window sizes N and τ, and the noise covariance matrices Q and R.
2. The main loop of the algorithm iterates until the ALS method has converged. At each iteration, The state estimate and covariance matrix are updated recursively using the observed measurement z_k, the measurement model defined by H, and the noise covariance matrix R.
3. After a sufficient number of iterations k exceeds $\tau + N$, the ALS method can be applied to estimate the noise covariance matrices. We compute the autocovariance of the innovations, stack them into a vector $\hat{b}$, and compute the matrix $\mathcal{A}$ as per the provided equations. A least-squares problem is then solved to get an estimate $\hat{\theta}$, which we then unstack into the individual matrices $\hat{Q}$ and $\hat{R}$.

Process model owing to its computational simplicity and efficiency, which is essential for adherence to real-time tracking constraints. The LTI model's predictive strength in capturing linear motion makes it particularly suitable for VOT scenarios, and its compatibility with Kalman filtering allows for optimal tracking in systems characterized by Gaussian noise and linear dynamics. Concurrently, our measurement model is robustly designed to withstand VOT-specific challenges such as occlusions and dynamic appearance changes, employing autocovariance least square estimation (ALS) to dynamically adapt to the non-stationary noise statistics typical of real-world environments. This adaptability is augmented by an adaptive background subtraction pre-processing step that effectively differentiates the target from its backdrop, thus enhancing measurement accuracy and, by extension, the reliability of object tracking.

The computation complexity of the proposed AKF-ALS algorithm is derived to show the real-time performance in Theorem 1.

Theorem 1 (Computation complexity). *The computational complexity of the proposed AKF-ALS algorithm in (1) is $O(N^2 n_x^4 n_z^4)$, where n_x and n_z are the dimensions of the state and measurement, respectively.*

Proof. In (10), the computation of $\bar{F}_{n_x \times n_x}$ matrix involves $O(n_x^2)$ for matrix addition, $O(n_x^3)$ for matrix multiplication and $O(n_x^3)$ for solving the Lyapunov equation in (7). Computing $\bar{F} \otimes \bar{F}$ involves $O(n_x^3)$ for fast Kronecker product computation, then the inverse matrix $\left(I_{n_x^2} - \bar{F} \otimes \bar{F}\right)_{n_x^2 \times n_x^2}$ involves $O(n_x^6)$ using the fastest matrix inversion method. $\mathcal{O}_{Nn_z \times n_x}$ and the fourth term $\Gamma_{Nn_z \times n_z}$ involve concatenating matrices, with a complexity of $O(Nn_x^2 n_z)$ and $O(Nn_x^2 n_z)$, respectively. $\mathcal{A}_{Nn_z^2 \times (n_x^2 + n_z^2)}$ has a complexity of approximately $O(max(n_x^6, Nn_x^4 n_z^2))$. Then, the ALS estimate $\hat{\theta}$ involves $O(N^2 n_z^4 n_x^4)$ for computing $\mathcal{A}_i^{\mathrm{T}} \mathcal{A}_i$ and $O(n_x^6)$ for the matrix inversion $(\mathcal{A}_i^{\mathrm{T}} \mathcal{A}_i)^{-1}$. Given the computation complexity of KF is $O((\max(n_x^3, n_z^3))$, the proposed AKF-ALS algorithm has the computation complexity of $O(\max(n_x^6, N^2 n_x^4 n_z^4, (\tau - j) N n_z^2))$. Since n_x and n_z are near and $(\tau - j)N$ is smaller than $n_x^4 n_z^2$ in most cases, the computation complexity is streamlined into $O(N^2 n_x^4 n_z^4)$. □

As indicated by the above theorem, incorporating the ALS into the KF does not produce a significant rise in the computation complexity when n_x and n_z are small.

4. Experiment Study

4.1. Numerical Analysis of Noise Covariance Estimation

A numerical simulation of noise covariance estimation for the LTI system with the LTI measurement model in (16) is provided to show the performance of the ALS algorithm, where the sampling time, the state, and measurement transitional matrices are set as $T_s = 1$, $F = \begin{bmatrix} 0.7T_s & 0.3 \\ -0.1 & 0.9T_s \end{bmatrix}$ and $H = [0.5, 0.2]$.

In this simulation study, the ALS algorithm was operated with $N_{MC} = 500$ Monte Carlo simulations over $N_{time} = 150$ time steps. The initial state for each simulation was defined as $x_0 = [1.1, 1.1, 1.1, 1.1]^{\mathrm{T}}$, expressed in meters (m). The true noise covariance matrices were initialized with $Q_{true} = \begin{bmatrix} 0.5 & 0 \\ 0 & 1 \end{bmatrix}$ and $R_{true} = 0.8$, where the units are square meters (m^2). The initial estimations for the noise covariance matrices were set to $Q_A = 0.1 I_4$ for the process noise and $R_A = I_2$ for the measurement noise.

The noise covariance estimates $\hat{Q}_{ALS}$, $\hat{R}_{ALS}$ is estimated using the ALS method. Then the steady-state state covariance is computed by solving the Lyapunov equation in (7) as $P_\varepsilon = \begin{bmatrix} 0.32 & 0.14 \\ 0.14 & 0.35 \end{bmatrix}$, the steady-state Kalman gain is $K = [0.17, 0.13]$.

By setting the window size N_w as 5, The empirical estimation of N-innovations autocovariance is computed as $\hat{b} = [2.019, 1.144, 1.177, 0.861, 0.794]^{\mathrm{T}}$. Then according to the Equation (12), the estimated noise covariance matrix with 500 MC initialization is computed and stored as $\hat{Q}_{ALS,MC}$ and $\hat{R}_{ALS,MC}$, and the mean value of $\hat{Q}_{ALS,MC}$ and $\hat{R}_{ALS,MC}$ is $\hat{Q}_{ALS} = \begin{bmatrix} 0.48 & 0.214 \\ 0.214 & 0.879 \end{bmatrix}$ and $\hat{R}_{ALS} = 0.801$.

The effectiveness of the estimated noise covariance matrices, based on 500 Monte Carlo simulations, is depicted in Figures 2 and 3. Figure 2 plots the diagonal elements of the estimated model noise covariance matrix, $\hat{Q}_{11}$ and $\hat{Q}_{22}$, on the x-axis and y-axis, respectively. Figure 3 presents the estimated measurement noise covariance matrix R on the x-axis, with the density of the Monte Carlo simulations displayed on the y-axis.

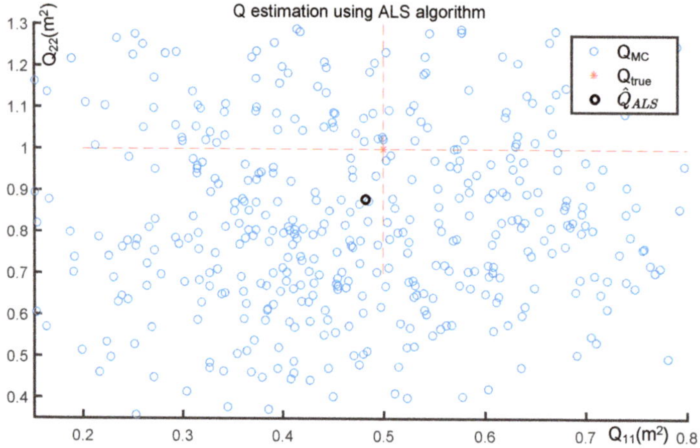

Figure 2. Performance of model noise covariance estimation Q_{ALS} using ALS algorithm over 500 Monte Carlo simulations. The blue dots denote the estimated diagonal-form model noise covariance matrix $Q_{MC} = diag(\hat{Q}_{11}, \hat{Q}_{22})$ over each Monte Carlo simulation, where the x-axis and y-axis represent the first and second diagonal elements respectively. The black bold dot is the model noise covariance estimate $\hat{Q}_{ALS}$ via the ALS algorithm. The intersection of two red dotted lines denotes the true noise covariance Q_{true}.

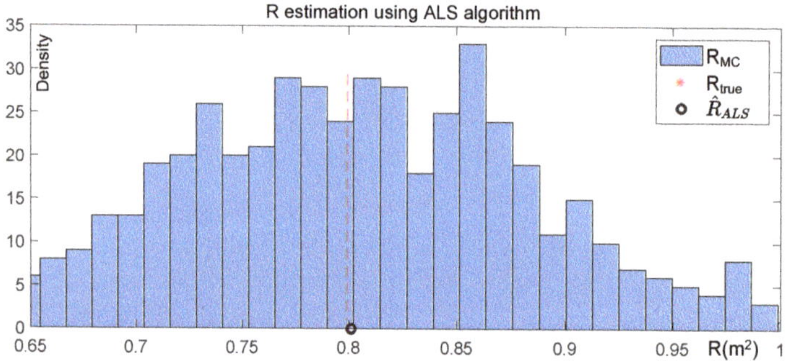

Figure 3. Performance of measurement noise covariance estimation R_{ALS} using ALS algorithm over 500 Monte Carlo simulations. The blue bars denote the estimated measurement noise covariance value R_{MC} over each Monte Carlo simulation, where the x-axis and y-axis respectively represent the measurement noise covariance value and the probabilistic density. The black bold dot is the measurement noise covariance estimate $\hat{Q}_{ALS}$ via the ALS algorithm. The intersection of the red dotted line and x-axis denotes the true noise covariance Q_{true}.

As indicated above, the noise covariance estimates with 500 MC initialization follow the Gaussian distribution where the mean value Q_{ALS} and R_{ALS} are near the true value. The results demonstrate that the ALS method can accurately estimate the noise covariance for the KF, and the ALS-KF framework which combines the Kalman filter with the ALS method is well-suited for real-world visual tracking applications where the underlying noise distributions are unknown.

4.2. Centroid-Based Moving Ball Tracking Subjected to Three Motions

The experiment on the video of a moving ball separately subjected to projectile motion, free-fall bouncing motion, and back-and-forth linear motion across the video is conducted to assess the performance of the proposed AKF-ALS algorithm. The video consists of 100 frames, with a resolution of 480×640 pixels.

- Projectile motion following a parabolic path refers to the motion of an object thrown into the air and subject to downward acceleration due to gravity. The challenge in tracking such motion is the constantly changing speed and direction of the object.
- Free-fall bouncing motion is the motion of an object falling under gravity and then bouncing back upwards. The challenge is the object's velocity changes rapidly at the point of impact, which can be difficult for a tracking algorithm to handle.
- Back-and-forth linear motion refers to an object moving to and fro along a straight line. The challenge is the abrupt change in velocity when the object changes direction.

For visual object tracking (VOT), these different types of motion pose unique challenges. Rapid changes in direction and velocity may affect the ability of a VOT system to accurately track an object. Therefore, the proposed AKF-ALS algorithm is applied to show the performance. In the algorithm, the background subtraction method is first applied to each frame in the video to detect the ball. The background is initialized as the first frame of the video, and an adaptive threshold is applied to the difference image to create a binary image. The connected components in the binary image are identified as potential objects, and the centroid of the largest component is stored as the location of the ball. Then the Kalman filter is used to track the ball's movement, i.e., the centroid of a series of video frames. The ALS method is used to estimate the process and measurement noise covariance. The performance of the proposed method was compared with a standard Kalman filter with fixed noise covariance.

Specifically, we consider the state of the ball as its position and velocity, and the observation as the detected position of the ball from the background subtraction. Based on the kinematic equation, the relation between the position x and velocity $\dot{x}$ can be written as follows:

$$\mathrm{x}_k = \begin{bmatrix} x_k \\ \dot{x}_k \end{bmatrix} = \begin{bmatrix} x_{k-1} + \dot{x}_{k-1}\Delta t + 1/2\ddot{x}_{k-1}\Delta t^2 \\ \dot{x}_{k-1} + \ddot{x}_{k-1}\Delta t \end{bmatrix} \tag{19}$$

where Δt is the sampling time and can be set as 0.2. The 2-D kinetic equation can be shown for state x_k as:

$$\mathrm{x}_k = \begin{bmatrix} x_k \\ y_k \\ \dot{x}_k \\ \dot{y}_k \end{bmatrix} = \begin{bmatrix} x_{k-1} + \dot{x}_{k-1}\Delta t + 1/2\ddot{x}_{k-1}\Delta t^2 \\ y_{k-1} + \dot{y}_{k-1}\Delta t + 1/2\ddot{y}_{k-1}\Delta t^2 \\ \dot{x}_{k-1} + \ddot{x}_{k-1}\Delta t \\ \dot{y}_{k-1} + \ddot{y}_{k-1}\Delta t \end{bmatrix} \tag{20}$$

which can be simplified as:

$$\mathrm{x}_k = \begin{bmatrix} 1 & 0 & \Delta t & 0 \\ 0 & 1 & 0 & \Delta t \\ 0 & 0 & 1 & 0 \\ 0 & 0 & 0 & 1 \end{bmatrix} \mathrm{x}_{k-1} + \begin{bmatrix} 1/2\Delta t^2 & 0 \\ 0 & 1/2\Delta t^2 \\ \Delta t & 0 \\ 0 & \Delta t \end{bmatrix} a_{k-1} \tag{21}$$

where the transitional matrices F, Γ and H are defined as

$$\begin{aligned} \mathrm{F} &= \begin{bmatrix} 1 & 0 & \Delta t & 0 \\ 0 & 1 & 0 & \Delta t \\ 0 & 0 & 1 & 0 \\ 0 & 0 & 0 & 1 \end{bmatrix} \\ \boldsymbol{\Gamma} &= \begin{bmatrix} 1/2\Delta t^2 & 0 \\ 0 & 1/2\Delta t^2 \\ \Delta t & 0 \\ 0 & \Delta t \end{bmatrix} \\ \mathrm{H} &= \begin{bmatrix} 1 & 0 & 0 & 0 \\ 0 & 1 & 0 & 0 \end{bmatrix}. \end{aligned} \tag{22}$$

The performance of the proposed AKF-ALS-based VOT algorithm is compared to that of the traditional KF-based VOT algorithm, as introduced in [64], across three types of motion: projectile motion, free-fall with a bouncing trajectory, and linear back-and-forth motion. The KF-based VOT algorithm uses the same kinematic equation in (19), but the noise covariance is initially determined as a constant value, i.e., $Q = 1$ and $R = 0.5$. The comparative results are illustrated in Figure 4.

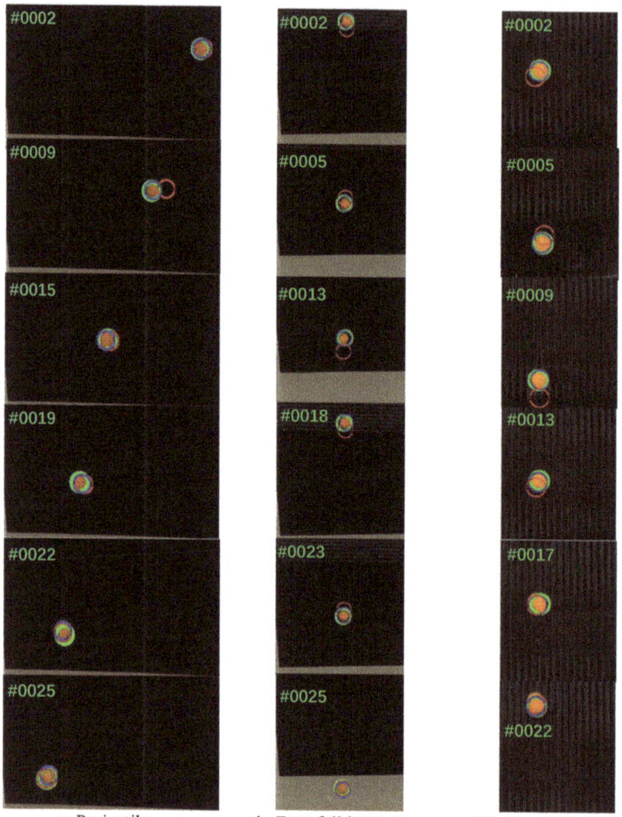

a. Projectile b. Free-fall bouncing c. Back-and-forth linear

Figure 4. VOT using the proposed AKF-ALS algorithm and traditional KF algorithm. The green, blue, and red circles, respectively, represent the circle of the ball with true centroid, with estimated centroid using AKF-ALS algorithm, and with estimated centroid using KF algorithm.

As indicated from the above figure, the proposed AKF-ALS algorithm shows improved robustness in handling constantly changing speed and direction, rapidly changing velocity, and the abrupt change in velocity when the object changes direction, while continuing tracking accurately after the change ends. As the tracking progresses, the performance of KF-based VOT approaches that of AKF-ALS-based VOT because the state covariance matrix approximates the steady-state value. This highlights the advantage of the proposed AKF-ALS-based VOT method in handling challenging conditions in visual object tracking.

The performance of the proposed ALS-KF-based VOT algorithm is evaluated using the root mean square error (RMSE) (in pixels) between the estimated position $\hat{x}_{1:2,t}$ (the first two elements) and the true position $x_{1:2,t}$ of the moving ball subjected to projectile motion, free-fall bouncing motion, and back-and-forth linear motion across the video, as shown in Table 1.

$$RMSE = \sqrt{\frac{1}{N}\sum_{t=1}^{N}(x_{1:2,t} - \hat{x}_{1:2,t})^2} \quad (23)$$

Table 1. Comparisons of RMSE (in pixels) between AKF-ALS-based and KF-based VOT subjected to projectile motion, free-fall bouncing motion, and back-and-forth linear motion.

RMSE (in Pixels)	Projectile	Free-Fall Bouncing	Back-and-Forth Linear
AKF-ALS	1.05	2.76	0.89
KF	4.67	8.35	2.31

The quantitative analysis of the table above reveals that AKF-ALS consistently achieves lower root mean square error (RMSE) values across various motion patterns. The RMSE for AKF-ALS is 1.05 pixels for projectile motion, significantly better than the 4.67 pixels of the standard KF. For free-fall bouncing motion, AKF-ALS records an RMSE of 2.76 pixels, in contrast to 8.35 pixels for the KF. Additionally, in back-and-forth linear motion scenarios, AKF-ALS demonstrates an RMSE of 0.89 pixels, outperforming the KF's 2.31 pixels. The results show that the proposed AKF-ALS algorithm achieves a lower RMSE compared to a standard KF VOT algorithm with fixed noise covariance, demonstrating the proposed method outperformed the standard Kalman filter in terms of RMSE.

4.3. Bounding Box-Based Pedestrian Tracking with Occlusions

The simulation was also conducted on the PETS09-S2L1 dataset [65] within the MOT-15 challenge, a complex scenario with up to eight pedestrians engaging in unusual patterns. The IOU matching utilized the detection box from the previous frame to compute the IOU with the current frame detections. A threshold of $IOU_{\text{Threshold}} = 0.3$ was set for a match to be considered valid. If the IOU matching failed to find a maximum value due to occlusion, the Kalman filter's prediction was used to estimate the target's new position and continue tracking. However, the KF-based VOT algorithm failed to maintain tracking under occlusion conditions.

The AKF-ALS VOT algorithm demonstrated robustness in maintaining the identity of the pedestrian target even when the IOU matching failed due to occlusion due to its dynamic covariance estimation, which allows the filter to adapt its parameters in response to the evolving state of the system. This adaptive mechanism is crucial for handling occlusions and erratic pedestrian movements. The AKF-ALS VOT algorithm improves upon the standard KF VOT by dynamically adjusting the process and measurement noise covariances. The state vector remains consistent with the classic AKF, with the following representation:

$$X = [x, y, h, w, \dot{x}, \dot{y}]^T \tag{24}$$

where x and y denote the horizontal and vertical position of the pedestrian in the video frame (in pixels), respectively, and $\dot{x}$ and $\dot{y}$ denote the horizontal and vertical velocity of the pedestrian (in pixels per frame). h and w denote the height and width of the bounding box around the pedestrian (in pixels), respectively.

The ALS algorithm updates the process noise covariance matrix Q and the observation noise covariance matrix R iteratively. The innovation sequence is used to calculate the autocovariance, which in turn adjusts Q and R to respond to abrupt changes in the tracked object's motion. The trajectory is shown in Figure 5.

As indicated by the figure, the tracking of pedestrians in scenarios obstructed by objects such as a light pole, can present challenges for bounding box detection methods. However, the implementation of the adaptive Kalman filter with adaptive least squares (AKF-ALS) algorithm demonstrates robustness in tracking despite failures in bounding box detection under such conditions.

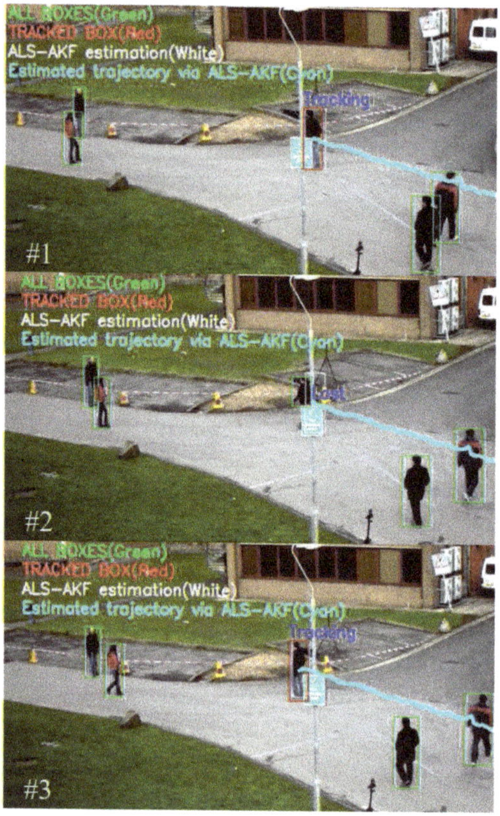

Figure 5. Bounding box-based pedestrian tracking with occlusions using AKF-ALS algorithm.

To demonstrate the comparative performance of the accuracy and robustness, we use the RMSE in (23), the identity switches, and the track fragmentation as the metrics. The identity switches metric measures the number of times the tracking algorithm incorrectly switches the identity of the tracked object. Fewer identity switches indicate a more robust tracking performance. The track fragmentation evaluates the continuity of the tracking throughout the sequence. Lower numbers imply fewer instances where tracks are broken and later reacquired. We compare the proposed ALS-AKF algorithm with the classic adaptive KF algorithm using Otsu's method in [66], as shown in Table 2.

Table 2. Comparisons of RMSE (in pixels), between AKF-ALS-based and KF-based VOT algorithm in pedestrian tracking with occlusions.

Metric	AKF ALS VOT	Classic AKF VOT [66]
RMSE	5.5 pixels	10.7 pixels
Identity switches	1	8
Track fragmentation	2	8

As is indicated from the table above, the AKF-ALS VOT algorithm demonstrated a markedly higher precision with a root mean square error (RMSE) of only 5.5 pixels, compared to the classic AKF VOT's 10.7 pixels. Robustness in maintaining pedestrian identity was also significantly improved in the AKF-ALS VOT, which had a mere single identity switch, whereas the classic AKF VOT recorded eight. Additionally, the AKF-ALS VOT showed greater continuity in tracking with only two instances of track fragmentation, in contrast to the classic AKF VOT's eight, indicating a more stable tracking performance throughout the sequences. These enhancements are likely due to the AKF-ALS's iterative updates to the process and observation noise covariance matrices, enabling it to adeptly adjust to sudden changes in the tracked object's motion.

5. Conclusions

In conclusion, the AKF-ALS algorithm introduced in this paper represents a significant advancement in the field of real-time visual object tracking. By integrating the autocovariance least squares estimation method with an adaptive Kalman filter, we have demonstrated a robust approach to tracking that compensates for system model mismatches and invalid offline calibration, which are critical in dynamic environments. The computational complexity analysis indicates that AKF-ALS is efficient enough for real-time applications, striking an optimal balance between computational load and tracking accuracy. Our experimental results have shown that the AKF-ALS algorithm exhibits superior performance over traditional Kalman filter methods that rely on fixed noise statistics. The algorithm's capability to adapt to changing noise covariances in real-time ensures greater accuracy in tracking the centroid of a moving object under various motion patterns, including projectile motion, free-falling with bounces, and linear back-and-forth motion. Experimental results on the complex PETS09-S2L1 dataset from the MOT-15 challenge demonstrate that the AKF-ALS VOT algorithm is capable of tracking pedestrians with occlusion where traditional methods fail. The algorithm's dynamic adaptation to changing noise covariances in real-time significantly enhances accuracy and reliability in tracking various motion patterns. Overall, the AKF-ALS algorithm's ability to iteratively refine noise covariances and adjust to abrupt motion changes makes it an excellent candidate for autonomous vehicles or mobile robotics.

Future work should focus on expanding the validation of the AKF-ALS algorithm across a broader range of applications, including those with more complex and nonlinear object dynamics. Additionally, further research could explore the integration of this algorithm with deep learning techniques to enhance tracking performance in highly cluttered and dynamic visual scenes.

Author Contributions: Conceptualization, J.L. and X.X.; methodology, J.L. and B.J.; software, J.L. and Z.J.; validation, J.L. and B.J.; writing—original draft preparation, J.L. and X.X. All authors have read and agreed to the published version of the manuscript.

Funding: This research is supported in part by the National Natural Science Foundation of China under grant 61803034, 61973035, R&D Program of Beijing Municipal Education Commission (KM202311417006), and the National Natural Science Foundation of China Basic Science Center Program under Grant 62088101.

Institutional Review Board Statement: Not applicable.

Informed Consent Statement: Not applicable.

Data Availability Statement: The dataset established is available upon requests from the corresponding author. The data are not publicly available due to the data are generated randomly based on constraints and parameter settings.

Conflicts of Interest: The authors declare no conflicts of interest.

References

1. Yadav, S.P. Vision-based detection, tracking, and classification of vehicles. *IEIE Trans. Smart Process. Comput.* **2020**, *9*, 427–434. [CrossRef]
2. Abdulrahim, K.; Salam, R.A. Traffic surveillance: A review of vision based vehicle detection, recognition and tracking. *Int. J. Appl. Eng. Res.* **2016**, *11*, 713–726.
3. Gad, A.; Basmaji, T.; Yaghi, M.; Alheeh, H.; Alkhedher, M.; Ghazal, M. Multiple Object Tracking in Robotic Applications: Trends and Challenges. *Appl. Sci.* **2022**, *12*, 9408. [CrossRef]
4. Gammulle, H.; Ahmedt-Aristizabal, D.; Denman, S.; Tychsen-Smith, L.; Petersson, L.; Fookes, C. Continuous human action recognition for human-machine interaction: A review. *ACM Comput. Surv.* **2023**, *55*, 1–38. [CrossRef]

5. Dong, X.; Shen, J.; Yu, D.; Wang, W.; Liu, J.; Huang, H. Occlusion-aware real-time object tracking. *IEEE Trans. Multimed.* **2016**, *19*, 763–771. [CrossRef]
6. Setitra, I.; Larabi, S. Background subtraction algorithms with post-processing: A review. In Proceedings of the 2014 22nd International Conference on Pattern Recognition, Stockholm, Sweden, 24–28 August 2014; pp. 2436–2441.
7. Shaikh, S.H.; Saeed, K.; Chaki, N.; Shaikh, S.H.; Saeed, K.; Chaki, N. Moving Object Detection Using Background Subtraction. In *Moving Object Detection Using Background Subtraction*; SpringerBriefs in Computer Science; Springer: Cham, Switzerland, 2014. [CrossRef]
8. Bouwmans, T.; Javed, S.; Sultana, M.; Jung, S.K. Deep neural network concepts for background subtraction: A systematic review and comparative evaluation. *Neural Netw.* **2019**, *117*, 8–66. [CrossRef]
9. Kalsotra, R.; Arora, S. Background subtraction for moving object detection: Explorations of recent developments and challenges. *Vis. Comput.* **2022**, *38*, 4151–4178. [CrossRef]
10. Zhang, H.; Wu, K. A vehicle detection algorithm based on three-frame differencing and background subtraction. In Proceedings of the 2012 Fifth International Symposium on Computational Intelligence and Design, Hangzhou, China, 28–29 October 2012; Volume 1, pp. 148–151. [CrossRef]
11. Li, J.; Pan, Z.M.; Zhang, Z.H.; Zhang, H. Dynamic ARMA-based background subtraction for moving objects detection. *IEEE Access* **2019**, *7*, 128659–128668. [CrossRef]
12. Zhang, R.; Gong, W.; Grzeda, V.; Yaworski, A.; Greenspan, M. An adaptive learning rate method for improving adaptability of background models. *IEEE Signal Process. Lett.* **2013**, *20*, 1266–1269. [CrossRef]
13. Shi, P.; Jones, E.G.; Zhu, Q. Median model for background subtraction in intelligent transportation system. In *Image Processing: Algorithms and Systems III*; SPIE: Bellingham, WA, USA, 2004; Volume 5298, pp. 168–176.
14. Li, X.; Ng, M.K.; Yuan, X. Median filtering-based methods for static background extraction from surveillance video. *Numer. Linear Algebra Appl.* **2015**, *22*, 845–865. [CrossRef]
15. Roy, S.M.; Ghosh, A. Real-time adaptive histogram min-max bucket (HMMB) model for background subtraction. *IEEE Trans. Circuits Syst. Video Technol.* **2017**, *28*, 1513–1525. [CrossRef]
16. Stauffer, C.; Grimson, W.E.L. Adaptive background mixture models for real-time tracking. In Proceedings of the 1999 IEEE Computer Society Conference on Computer Vision and Pattern Recognition (Cat. No PR00149), Fort Collins, CO, USA, 23–25 June 1999; Volume 2, pp. 246–252.
17. Zivkovic, Z.; Van Der Heijden, F. Efficient adaptive density estimation per image pixel for the task of background subtraction. *Pattern Recognit. Lett.* **2006**, *27*, 773–780. [CrossRef]
18. Sheikh, Y.; Shah, M. Bayesian modeling of dynamic scenes for object detection. *IEEE Trans. Pattern Anal. Mach. Intell.* **2005**, *27*, 1778–1792. [CrossRef]
19. Han, B.; Comaniciu, D.; Zhu, Y.; Davis, L.S. Sequential kernel density approximation and its application to real-time visual tracking. *IEEE Trans. Pattern Anal. Mach. Intell.* **2008**, *30*, 1186–1197. [PubMed]
20. Zhu, Q.; Shao, L.; Li, Q.; Xie, Y. Recursive kernel density estimation for modeling the background and segmenting moving objects. In Proceedings of the 2013 IEEE International Conference on Acoustics, Speech and Signal Processing, Vancouver, BC, Canada, 26–31 May 2013; pp. 1769–1772.
21. Lin, H.H.; Liu, T.L.; Chuang, J.H. A probabilistic SVM approach for background scene initialization. In Proceedings of the International Conference on Image Processing, Rochester, NY, USA, 22–25 September 2002; Volume 3; pp. 893–896.
22. Cheng, L.; Gong, M.; Schuurmans, D.; Caelli, T. Real-time discriminative background subtraction. *IEEE Trans. Image Process.* **2010**, *20*, 1401–1414. [CrossRef]
23. Bouwmans, T. Subspace learning for background modeling: A survey. *Recent Patents Comput. Sci.* **2009**, *2*, 223–234. [CrossRef]
24. Djerida, A.; Zhao, Z.; Zhao, J. Background subtraction in dynamic scenes using the dynamic principal component analysis. *IET Image Process.* **2020**, *14*, 245–255. [CrossRef]
25. Buccolieri, F.; Distante, C.; Leone, A. Human posture recognition using active contours and radial basis function neural network. In Proceedings of the IEEE Conference on Advanced Video and Signal Based Surveillance, Como, Italy, 15–16 September 2005; pp. 213–218.
26. Maddalena, L.; Petrosino, A. A self-organizing approach to background subtraction for visual surveillance applications. *IEEE Trans. Image Process.* **2008**, *17*, 1168–1177. [CrossRef] [PubMed]
27. Gemignani, G.; Rozza, A. A robust approach for the background subtraction based on multi-layered self-organizing maps. *IEEE Trans. Image Process.* **2016**, *25*, 5239–5251. [CrossRef]
28. Braham, M.; Van Droogenbroeck, M. Deep background subtraction with scene-specific convolutional neural networks. In Proceedings of the 2016 International Conference on Systems, Signals and Image Processing (IWSSIP), Bratislava, Slovakia, 23–25 May 2016; pp. 1–4.
29. Babaee, M.; Dinh, D.T.; Rigoll, G. A deep convolutional neural network for video sequence background subtraction. *Pattern Recognit.* **2018**, *76*, 635–649. [CrossRef]
30. Bakkay, M.C.; Rashwan, H.A.; Salmane, H.; Khoudour, L.; Puig, D.; Ruichek, Y. BSCGAN: Deep background subtraction with conditional generative adversarial networks. In Proceedings of the 2018 25th IEEE International Conference on Image Processing (ICIP), Athens, Greece, 7–10 October 2018; pp. 4018–4022.

31. Sultana, M.; Mahmood, A.; Bouwmans, T.; Jung, S.K. Dynamic background subtraction using least square adversarial learning. In Proceedings of the 2020 IEEE International Conference on Image Processing (ICIP), Abu Dhabi, United Arab Emirates, 25–28 October 2020; pp. 3204–3208.
32. McHugh, J.M.; Konrad, J.; Saligrama, V.; Jodoin, P.M. Foreground-adaptive background subtraction. *IEEE Signal Process. Lett.* **2009**, *16*, 390–393. [CrossRef]
33. Li, X.; Hu, W.; Shen, C.; Zhang, Z.; Dick, A.; Hengel, A.V.D. A survey of appearance models in visual object tracking. *ACM Trans. Intell. Syst. Technol. (TIST)* **2013**, *4*, 1–48. [CrossRef]
34. Fiaz, M.; Mahmood, A.; Javed, S.; Jung, S.K. Handcrafted and deep trackers: Recent visual object tracking approaches and trends. *ACM Comput. Surv. (CSUR)* **2019**, *52*, 1–44. [CrossRef]
35. Javed, S.; Danelljan, M.; Khan, F.S.; Khan, M.H.; Felsberg, M.; Matas, J. Visual object tracking with discriminative filters and siamese networks: A survey and outlook. *IEEE Trans. Pattern Anal. Mach. Intell.* **2022**, *45*, 6552–6574. [CrossRef]
36. Ondrašovič, M.; Tarábek, P. Siamese visual object tracking: A survey. *IEEE Access* **2021**, *9*, 110149–110172. [CrossRef]
37. Chen, F.; Wang, X.; Zhao, Y.; Lv, S.; Niu, X. Visual object tracking: A survey. *Comput. Vis. Image Underst.* **2022**, *222*, 103508. [CrossRef]
38. Zhou, H.; Yuan, Y.; Shi, C. Object tracking using SIFT features and mean shift. *Comput. Vis. Image Underst.* **2009**, *113*, 345–352. [CrossRef]
39. Khan, Z.H.; Gu, I.Y.H.; Backhouse, A.G. Robust visual object tracking using multi-mode anisotropic mean shift and particle filters. *IEEE Trans. Circuits Syst. Video Technol.* **2011**, *21*, 74–87. [CrossRef]
40. Cai, Z.; Gu, Z.; Yu, Z.L.; Liu, H.; Zhang, K. A real-time visual object tracking system based on Kalman filter and MB-LBP feature matching. *Multimed. Tools Appl.* **2016**, *75*, 2393–2409. [CrossRef]
41. Farahi, F.; Yazdi, H.S. Probabilistic Kalman filter for moving object tracking. *Signal Process. Image Commun.* **2020**, *82*, 115751. [CrossRef]
42. Kim, T.; Park, T.H. Extended Kalman filter (EKF) design for vehicle position tracking using reliability function of radar and lidar. *Sensors* **2020**, *20*, 4126. [CrossRef] [PubMed]
43. Zhou, Z.; Wu, D.; Zhu, Z. Object tracking based on Kalman particle filter with LSSVR. *Optik* **2016**, *127*, 613–619. [CrossRef]
44. Iswanto, I.A.; Li, B. Visual object tracking based on mean-shift and particle-Kalman filter. *Procedia Comput. Sci.* **2017**, *116*, 587–595. [CrossRef]
45. Rao, G.M.; Satyanarayana, C. Visual object target tracking using particle filter: A survey. *Int. J. Image Graph. Signal Process.* **2013**, *5*, 1250. [CrossRef]
46. Kalman, R.E. A New Approach to Linear Filtering and Prediction Problems. *J. Basic Eng.* **1960**, *82*, 35–45. [CrossRef]
47. Chen, J.; Li, J.; Yang, S.; Deng, F. Weighted optimization-based distributed Kalman filter for nonlinear target tracking in collaborative sensor networks. *IEEE Trans. Cybern.* **2016**, *47*, 3892–3905. [CrossRef]
48. Zheng, B.; Fu, P.; Li, B.; Yuan, X. A robust adaptive unscented Kalman filter for nonlinear estimation with uncertain noise covariance. *Sensors* **2018**, *18*, 808. [CrossRef] [PubMed]
49. Jeong, J.; Yoon, T.S.; Park, J.B. Mean shift tracker combined with online learning-based detector and Kalman filtering for real-time tracking. *Expert Syst. Appl.* **2017**, *79*, 194–206. [CrossRef]
50. Sun, J.; Xu, X.; Liu, Y.; Zhang, T.; Li, Y. FOG random drift signal denoising based on the improved AR model and modified Sage-Husa adaptive Kalman filter. *Sensors* **2016**, *16*, 1073. [CrossRef]
51. Särkkä, S.; Nummenmaa, A. Recursive Noise Adaptive Kalman Filtering by Variational Bayesian Approximations. *IEEE Trans. Autom. Control* **2009**, *54*, 596–600. [CrossRef]
52. Kashyap, R. Maximum likelihood identification of stochastic linear systems. *IEEE Trans. Autom. Control* **1970**, *15*, 25–34. [CrossRef]
53. Zagrobelny, M.A.; Rawlings, J.B. Identifying the uncertainty structure using maximum likelihood estimation. In Proceedings of the 2015 American Control Conference (ACC), Chicago, IL, USA, 1–3 July 2015; pp. 422–427.
54. Myers, K.; Tapley, B. Adaptive sequential estimation with unknown noise statistics. *IEEE Trans. Autom. Control* **1976**, *21*, 520–523. [CrossRef]
55. Solonen, A.; Hakkarainen, J.; Ilin, A.; Abbas, M.; Bibov, A. Estimating model error covariance matrix parameters in extended Kalman filtering. *Nonlinear Process. Geophys.* **2014**, *21*, 919–927. [CrossRef]
56. Yu, C.; Verhaegen, M. Subspace identification of distributed clusters of homogeneous systems. *IEEE Trans. Autom. Control* **2016**, *62*, 463–468. [CrossRef]
57. Mehra, R.K. On the identification of variances and adaptive Kalman filtering. *IEEE Trans. Autom. Control* **1970**, *15*, 175–184. [CrossRef]
58. Odelson, B.J.; Rajamani, M.R.; Rawlings, J.B. A new autocovariance least-squares method for estimating noise covariances. *Automatica* **2006**, *42*, 303–308. [CrossRef]
59. Duník, J.; Straka, O.; Simandl, M. On Autocovariance Least-Squares Method for Noise Covariance Matrices Estimation. *IEEE Trans. Autom. Control* **2017**, *62*, 967–972. [CrossRef]
60. Duník, J.; Straka, O.; Kost, O.; Havlík, J. Noise covariance matrices in state-space models: A survey and comparison of estimation methods—Part I. *Int. J. Adapt. Control Signal Process.* **2017**, *31*, 1505–1543. [CrossRef]

61. Li, J.; Ma, N.; Deng, F. Distributed noise covariance matrices estimation in sensor networks. In Proceedings of the 2020 59th IEEE Conference on Decision and Control (CDC), Jeju, Republic of Korea, 14–18 December 2020; pp. 1158–1163.
62. Arnold, T.J.; Rawlings, J.B. Tractable Calculation and Estimation of the Optimal Weighting Matrix for ALS Problems. *IEEE Trans. Autom. Control* **2021**, *67*, 6045–6052. [CrossRef]
63. Meng, Y.; Gao, S.; Zhong, Y.; Hu, G.; Subic, A. Covariance matching based adaptive unscented Kalman filter for direct filtering in INS/GNSS integration. *Acta Astronaut.* **2016**, *120*, 171–181. [CrossRef]
64. Fu, Z.; Han, Y. Centroid weighted Kalman filter for visual object tracking. *Measurement* **2012**, *45*, 650–655. [CrossRef]
65. Ferryman, J.; Shahrokni, A. PETS2009: Dataset and challenge. In Proceedings of the 2009 Twelfth IEEE International Workshop on Performance Evaluation of Tracking and Surveillance, Snowbird, UT, USA, 7–9 December 2009; pp. 1–6. [CrossRef]
66. Tripathi, R.P.; Singh, A.K. Moving Object Tracking Using Optimal Adaptive Kalman Filter Based on Otsu's Method. In *Innovations in Electronics and Communication Engineering, Proceedings of the 8th ICIECE 2019, Hyderabad, India, 2–3 August 2019*; Springer: Singapore, 2020; pp. 429–435.

applied sciences

MDPI

Article

STA-Net: A Spatial–Temporal Joint Attention Network for Driver Maneuver Recognition, Based on In-Cabin and Driving Scene Monitoring

Bin He [1], Ningmei Yu [1,*], Zhiyong Wang [2,3] and Xudong Chen [2]

1 School of Automation and Information Engineering, Xi'an University of Technology, Xi'an 710048, China; 1210311008@stu.xaut.edu.cn
2 National Key Laboratory of Human–Machine Hybrid Augmented Intelligence, National Engineering Research Center for Visual Information and Applications, and Institute of Artificial Intelligence and Robotics, Xi'an Jiaotong University, Xi'an 710049, China; zhywang@stu.xjtu.edu.cn (Z.W.); 2223515328@stu.xjtu.edu.cn (X.C.)
3 School of Software Engineering, Xi'an Jiaotong University, Xi'an 710049, China
* Correspondence: yunm@xaut.edu.cn

Abstract: Next-generation advanced driver-assistance systems (ADASs) are a promising direction for intelligent transportation systems. To achieve intelligent security monitoring, it is imperative that vehicles possess the ability to accurately comprehend driver maneuvers amidst diverse driver behaviors and complex driving scenarios. Existing CNN-based and transformer-based driver maneuver recognition methods face challenges in effectively capturing global and local features across temporal and spatial dimensions. This paper proposes a Spatial–Temporal Joint Attention Network (STA-Net) to realize high-efficient temporal and spatial feature extractions in driver maneuver recognition. First, we introduce a two-stream architecture for a concurrent analysis of in-cabin driver behaviors and out-cabin environmental information. Second, we propose a Multi-Scale Transposed Attention (MSTA) module and Multi-Scale Feedforward Network (MSFN) to extract features at multiple scales, addressing receptive field inadequacies and combining high-level and low-level information. Third, to address the information redundancy in multi-scale features, we propose a Cross-Spatial Attention Module (CSAM) and Multi-Scale Cross-Spatial Fusion Module (MCFM) to select essential features. Additionally, we introduce an asymmetric loss function to effectively tackle the issue of sample imbalance across diverse categories of driving maneuvers. The proposed method demonstrates a remarkable accuracy of 90.97% and an F1 score of 89.37% on the Brain4Cars dataset, surpassing the performance of the methods compared. These results substantiate the fact that our approach effectively enhances driver maneuver recognition.

Keywords: driver maneuver recognition; deep learning; multi-scale spatial–temporal attention

1. Introduction

Research on road safety indicates that the majority of traffic accidents stem from improper driver maneuvers. Despite achieving satisfactory performance in specific scenarios, fully autonomous driving remains a long-term objective due to the ongoing necessity for comprehensive legislation, regulations, and infrastructure development [1]. Consequently, human–machine cooperative driving continues to be a crucial research direction in intelligent transportation systems. Understanding the driver's intent is an essential prerequisite for effective human–machine interaction and facilitating autonomous vehicle decision-making that aligns with drivers' preferences in uncertain environments as well as alerting drivers during hazardous situations is necessary [2]. However, predicting human intent poses challenges, which are attributed to factors affecting human drivers, such as distraction, the driver's emotional state, and lack of concentration, thereby leading to potential road hazards.

In recent years, research has focused on exploring new technologies for comprehensive perception in intelligent vehicles, making it possible to predict driver intent over time. Jain et al. [3] introduced the Autoregressive Input–Output HMM (AIO-HMM) model, which processes both internal (2D facial features) and external features, predicting potential driver actions seconds before driving maneuvers. They also provided the Brain4Cars dataset, consisting of 1180 miles of natural highway and city driving, for method evaluation.

The dataset's video frames contain dynamic information on driver maneuver patterns and road traffic conditions. Gebert [4] and Xing [5], among others, conducted statistical analyses on different driving intent expressions, finding a high correlation between driving intent and driver maneuvers. When drivers are about to change their maneuver, they exhibit corresponding actions, such as head posture [3,6–8], specific maneuvers, and eye movements while checking the rearview mirror [9,10], providing crucial evidence for intent inference.

This work is significant, as it lays the foundation for driver assistance systems and proposes a method for predicting driver actions using dynamic visual data inside and outside the vehicle. It addresses the challenge of predicting driver operations seconds in advance, allowing for timely warnings to drivers and contributing to the development of next-generation advanced driver-assistance systems (ADASs), reducing road hazard risks.

Several researchers have proposed improvements to this pipeline. Jain et al. [11] suggested a deep learning architecture based on recurrent neural networks (RNN-LSTMs), which upgrades internal driver features from 2D to 3D facial features, enhancing the accuracy of driver maneuver prediction by fusing information from multiple sensors. Moussaid et al. [12] presented a method using driver facial information to predict lane-changing actions, implementing a model based on CNN-LSTMs for analyzing driver actions before lane changes. Tonutti et al. [8] introduced a method based on Domain-Adversarial Recurrent Neural Networks (DA-RNNs), improving the generalization capability of driving manipulation prediction. Gebert et al. [4] combined 3D-ResNet with an LSTM to predict driver intent by analyzing driver motion and external vehicle video data. Rong et al. [13] proposed a driver intent prediction method based on monitoring internal and external scenes, achieving better prediction performance with fewer parameters.

Analyzing Gebert et al. [4] and Rong et al. [13], with both using two branches for internal and external video processing, reveals their distinct approach. While Gebert et al. [4] computed optical flow from internal videos, Rong et al. [13] calculated it from external videos. These studies differ from previous approaches that use numerical data (e.g., lane numbers, speed) as external features. Instead, they directly extract external features from external videos by using CNN models.

Previously, LSTMs played a crucial role in driver maneuver recognition due to their ability to capture long-distance dependencies. However, LSTMs face challenges in capturing extended dependencies, and they present other challenges, such as high computational complexity, susceptibility to video noise, and interpretability issues. Recent studies favor 3D-CNN models for spatiotemporal feature extraction, addressing LSTMs' limitations. However, learning effective spatiotemporal representations remains a challenge due to local redundancy and global dependence issues.

A combination of 3D convolutional neural networks (3D-CNNs) and spatiotemporal transformers has emerged as a promising solution for better driver intent inference. However, both have limitations. While 3D-CNNs reduce spatiotemporal redundancy, their finite receptive fields make learning long-term dependencies difficult. Spatiotemporal transformers excel at capturing global dependencies but introduce redundancy in shallow layers when encoding local spatiotemporal features.

Additionally, two challenges affect driver intent inference accuracy. First, inadequate utilization of external video information limits the perception and understanding of the surrounding environment. Rong et al. [13] demonstrated that external videos complement internal driver videos, providing essential information. This external information is necessary to avoid misidentification in challenging situations. Second, the imbalance in training data samples, with straight driving maneuvers being more common than turns and lane changes, poses difficulties in machine learning model training. This imbalance may cause the model to favor dominant classes, reducing accuracy in predicting minority classes.

Inspired by transformer models, we propose the Spatial–Temporal Joint Attention Network (STA-Net), combining CNN and transformers in a dual-stream framework. The contributions of this study can be summarized as follows:

(1) We propose a two-stream network to extract in-cabin driver behavior and out-cabin environmental information, addressing spatiotemporal redundancy and insufficient use of driving scene information.
(2) We employ the joint learning of the CNN and the transformer to fuse spatiotemporal information at different levels. CNN focuses on low-level local features to reduce redundancy, while the transformer captures high-level global information to address long-term dependencies.
(3) We introduce an asymmetric loss function to tackle the problem of imbalanced training data, reducing the negative impact of sample imbalance on model optimization.

2. Related Works

2.1. Driver Maneuver Recognition

Driver Maneuver Recognition is primarily about intelligent vehicles monitoring the driver's maneuver in real-time. Using machine learning methods, the vehicle analyzes this maneuver data to determine the current actions of the driver and anticipate their future intentions. This enables the provision of various driver assistance features or timely alerts to the driver, thereby enhancing driving safety and efficiency. Current research includes traditional machine learning methods, combinations of 2D CNNs and RNNs, methods combining the 3D CNN and optical flow estimation, and dual-stream frameworks. Here are some representative works:

Combination of 2D CNNs and RNN Techniques: Xing et al. [14] proposed a method that combines RNN technology with 2D CNN in video processing to handle spatial and temporal information. The 2D CNN is used as an encoder to extract spatial features from the driver's maneuver sequence, and RNN technology serves as a decoder for time modeling to infer the driver's intentions. This method leverages both advantages, improving the efficiency of feature extraction and temporal processing. However, the combined model of 2D CNNs and RNNs is often more complex than using either method alone, leading to challenges in model complexity, increased parameter count, and higher computational resource requirements.

Combining 3D CNN and Optical Flow Estimation: Gebert et al. [4] proposed a vision-based 3D convolutional residual learning method using optical flow images from the driver's cabin to predict driver intentions. While this method has advantages in capturing spatiotemporal information, accurate maneuver recognition, efficiency, and flexibility in video understanding, it suffers from drawbacks such as numerous parameters, high hardware requirements, large data volume requirements, and sensitivity to lighting and occlusion conditions.

Dual-Stream Framework: Rong et al. [13] introduced a ConvLSTM-based autoencoder for extracting vehicle motion information from traffic scenes. They proposed a deep network framework to simultaneously study features from two directions (inside and outside the vehicle) without manual encoding or handcrafted features. This structure achieves advanced driver maneuver prediction performance with fewer parameters than previous works. However, it requires more data to train a finer decoder to interpret long-term motion better, presenting challenges in predicting longer-term motion and computational efficiency due to optical flow estimation. Chen et al. [15] introduce an intelligent vehicle driving intention inference method based on spatiotemporal feature enhancement (STEDII-GRU), aiming to improve the accuracy of driving intention inference. The method first utilizes a pre-trained dual-stream network (SlowFast network) as the backbone for feature extraction. Subsequently, the low frame rate and high frame rate paths are employed to process internal driver behavior video data and external forward traffic scene data. Finally, the joint spatiotemporal features are input into the GRU to obtain the most probable intention. The proposed STEDII-GRU method demonstrates high accuracy in driving intention inference, providing a promising solution for enhancing intelligent vehicles' safety and driving performance. Ma et al. [16] introduce a novel framework for driver intention prediction. The framework, CEMFormer, employs spatial-temporal transformers to unify memory representations for improved prediction accuracy. It integrates data from both in-cabin and external cameras, enhancing the prediction through historical data fusion and a novel context-consistency loss. Bonyani et al. [17] explore a deep neural network framework to anticipate driver maneuvers and enhance takeover readiness in automated driving. Utilizing the Brain4Cars dataset, the model integrates DenseNet, LSTM, attention mechanisms, and FlowNet2 to predict driver intentions up to 4 s in advance. The study assesses driver readiness through in-cabin and out-cabin video data and demonstrates improved prediction accuracy and performance against existing models.

Other latest methods: Zhang et al. [18] present a novel method for recognizing driver lane-changing intention (LCI) in a connected environment. Utilizing a driving simulator, it determines LCI time windows and feature parameters, including yaw rate, vehicle speed, and driver gaze. A new LCI model using phase-space reconstruction and Swin Transformer for classification is proposed, surpassing classical machine learning algorithms in accuracy and addressing vanishing gradient issues in long time-series data. This method is significant for lane-changing assistance and human–machine co-driving systems, enhancing traffic safety and efficiency. Li et al. [19] present a driving behavior prediction model that blends gradient boosting decision tree (GBDT), convolutional neural networks (CNN), and long short-term memory networks (LSTM) in a wide-deep framework. This model aims to extract comprehensive driving behavior characteristics and enhance the interpretability of CNN-LSTM models. The integration of GBDT allows for the quantitative analysis of vehicular interactions. Chen et al. [20] introduce a transfer learning-based approach for recognizing various driving behaviors using a convolutional neural network (CNN) model. This method utilizes vehicle kinematic data and drivers' facial expressions, enhancing recognition accuracy for patterns like acceleration, deceleration, turning, lane changing, and lane keeping. The transfer learning technique effectively refined pre-trained models with limited data, significantly improving performance and training cost efficiency. This approach is particularly valuable for challenging data collection scenarios, such as with heavy-duty freight vehicles. Wu et al. [1] present a model for identifying lane-changing maneuvers using the HighD dataset. Focusing on acceleration and velocity as physical data, the research implements a k-nearest neighbor (KNN) classification model. The findings indicate high classification accuracy, suggesting the model's potential utility in advanced driver-assistance systems to improve road safety.

2.2. Video Understanding

The key to understanding driver intent based on visual information lies in accurately interpreting both in-car and out-of-car visual data. In recent years, various deep learning methods for video understanding have been proposed. The purpose of video understanding is to enable computers to interpret and understand video content like humans. Models process videos into sequentially ordered frames, and current video understanding primarily employs recurrent neural networks (RNNs), convolutional neural networks (CNNs), and transformer-based architectures to extract and analyze spatiotemporal features from image sequences, ultimately making predictions based on these data.

Recurrent Neural Networks (RNNs): Sun et al. [21] proposed a novel dual-stream LSTM architecture called L2STM for human action recognition in videos, addressing non-stationary dynamics in long-term motion. Li et al. [22] introduced the RTPR network architecture for video action detection, utilizing action proposals from previous frames to assist detection in the current frame through a recurrent neural network. In summary, RNNs in video understanding can capture temporal information, handle sequence data, facilitate end-to-end learning, and exhibit high flexibility. However, they also face challenges in training, handling variable-length sequences, dealing with multimodal data, and capturing long-term dependencies.

Convolutional Neural Networks (CNNs): Carreira et al. [23] proposed a novel Two-Stream Inflated 3D ConvNets model (I3D) that extends 2D ConvNets to 3D ConvNets, addressing spatiotemporal modeling issues in video understanding. Feichtenhofer et al. [24] introduced a new video recognition model, the SlowFast network, which incorporates two different paths (Slow Path and Fast Path) to handle slow and fast video information separately. CNNs in video understanding can effectively capture local patterns and structural information, automatically learn features, and perform parallel computation on large-scale data. However, they also have limitations in capturing global information, dealing with variable-length sequences, having many parameters, and facing imbalances in spatial and temporal information.

Transformer-based Architectures: Xu et al. [25] presented an algorithm called Long Short-Term Transformer (LSTR) for online action detection, addressing the effective modeling of long-time sequence data and online action detection in videos. Li et al. [26] proposed a multi-scale visual transformer (MViT) for video and image recognition, connecting the basic idea of a multi-scale feature hierarchy with transformer models. Bertasius et al. [27] introduced a video classification model, TimeSformer, based on self-attention mechanisms, incorporating Divided Space-Time Attention (T + S) to calculate time and space self-attention scores separately. Arnab et al. [28] presented a pure transformer-based video classification model, ViViT, introducing a method called Tubelet Embedding to capture spatiotemporal information in videos effectively.

While transformer-based models in video understanding have advantages such as capturing long-term dependencies, strong parallel computation capabilities, applicability to multiple tasks, and end-to-end learning, they also come with disadvantages, including high computational resource requirements, limitations on sequence length, high training time, and cost, and a high number of parameters.

3. Methods

The spatiotemporal joint reasoning process of driving intention is a solution to the sequence image classification problem. In our work, we propose a novel driving intention inference framework, Spatial–Temporal Joint Attention Network (STA-NET), which simultaneously utilizes two input sources: internal and external videos, as shown in Figure 1. One branch learns spatial semantic features in traffic videos. In contrast, the other branch learns spatial semantic features in driver videos, thereby addressing the deficiency of relying solely on in-vehicle driver spatiotemporal features for driving intention inference. As illustrated in Figure 1, the backbone network is a parallel dual-branch network. The basic structure of the backbone network is mainly composed of Spatial–Temporal Joint Attention Block (STA Block) and Cross-Spatial Attention Module (CSAM), where the STA Block consists of Multi-Scale Transposed Attention (MSTA) and Multi-Scale Feedforward Network (MSFN). The STA Block adopts a joint CNN and transformer approach to simultaneously extract driver maneuver features and spatiotemporal features of the traffic scene. MSTA and MSFN alleviate the insufficient receptive field at different levels and enhance the richness of spatiotemporal feature information. At each stage, features extracted by the STA Block at the same scale are aggregated through CSAM. The MCFM (Multi-CSAM Fusion Module) aggregates different scales of in-vehicle driver features and traffic scene features at different stages for driving intention inference.

More specifically, we hierarchically stack STA Block units to construct our network for spatiotemporal learning. As shown in Figure 1, our network comprises four stages with channel numbers 64, 128, 256, and 512, respectively. We build the backbone network of the STA framework based on the quantities of STA Block units in each stage, which are [5,7,8,20]. We employ MSTA (Equation (1)) at each STA Block to reduce spatiotemporal redundancy. We normalize the data using LN [29]. Before the first stage, we apply a $3 \times 4 \times 4$ convolution with a stride of $2 \times 4 \times 4$, meaning both spatial and temporal dimensions are downsampled. Before the other stages, we use a $1 \times 1 \times 2$ convolution with a stride of $1 \times 1 \times 2$. Finally, the spatiotemporal average pooling and fully connected layers are employed for the ultimate prediction. In this way, our STA-Net, with an insightful unified framework, addresses video redundancy and dependencies. Each module is detailed as follows.

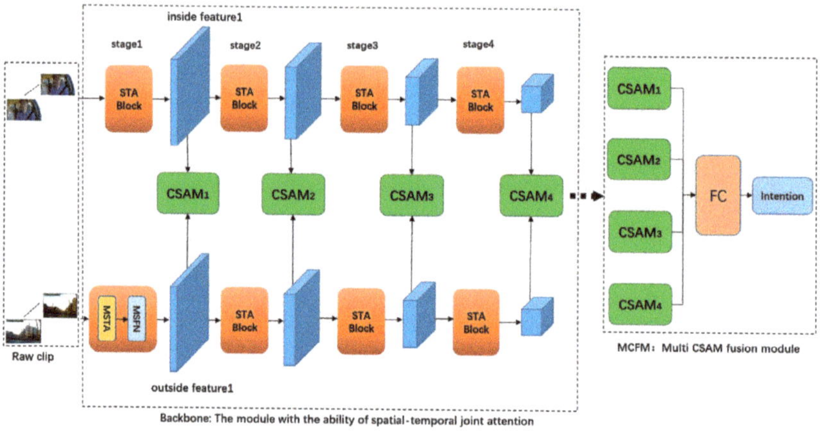

Figure 1. Spatial–Temporal Joint Attention Network (STA-NET).

3.1. *Framework for Spatiotemporal Feature Extraction Based on Dual-Stream Networks*

As previously mentioned, the innovative driving intent inference framework, STA-NET, simultaneously handles video data inside and outside the vehicle. One branch learns spatial semantic information from traffic videos, while the other learns from driver videos. The specific structures are described below.

3.1.1. STA-Block

To overcome spatiotemporal redundancy and dependency issues, we propose a novel module called Spatial–Temporal Joint Attention Block (STA-Block), as illustrated in Figure 1. We leverage the fundamental Transformer architecture [30] and tailor it specifically for efficient and effective spatiotemporal representation learning. Specifically, the STA-Block comprises two key modules: the Multi-Scale Transposed Attention (MSTA) and the Multi-Scale Feedforward Network (MSFN). Our MSTA adeptly addresses local video redundancy and global video dependencies by extracting features at different scales in both shallow and deep layers. Finally, we introduce a Feedforward Network (FFN) with two linear layers to enhance each token pointwise.

a. Multi-Scale Transposed Attention

As mentioned above, we aim to address two main challenges: significant local redundancy and intricate global dependencies, aiming for efficient and effective spatiotemporal representation learning. However, existing methods, such as popular 3D CNNs and spatiotemporal transformers, often focus solely on one of these challenges. Therefore, we introduce a novel approach called Multi-Scale Transpose Attention (MSTA). Designed in a concise transformer format, MSTA seamlessly unifies 3D convolution and spatiotemporal self-attention, adeptly tackling video redundancy and dependencies at different levels in both shallow and deep layers.

Due to the substantial computational overhead of transformers, primarily from the self-attention layer, applying the traditional self-attention mechanism (SA) [30,31] becomes impractical for most video-understanding tasks. In the conventional self-attention mechanism, the time and memory complexity of key-query dot-product interactions grows quadratically with the spatial resolution of the input. To address this issue, we propose

Multi-Scale Transpose Attention (MSTA), which exhibits linear complexity, as depicted in Figure 2. The key distinction lies in MSTA applying self-attention across channels, calculating cross-covariance across channels to generate an attention map implicitly encoding global context. As another integral component of MSTA, we introduce depthwise convolution to emphasize 3D local context, performing this operation before computing feature covariance to generate a global attention map.

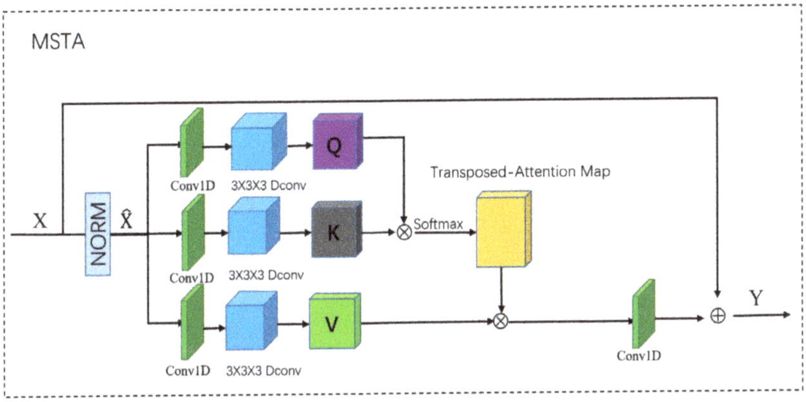

Figure 2. Multi-Scale Transpose Attention (MSTA) module.

From the Layer Normalization tensor $\hat{\mathbf{X}} \in \mathbb{R}^{H \times W \times C \times T}$, our Multi-Scale Transpose Attention (MSTA) initially generates query ($\mathbf{Q}$, key ($\mathbf{K}$), and value ($\mathbf{V}$) projections, enriching local contexts. This is achieved by applying a $1 \times 1 \times 1$ convolution to aggregate spatiotemporal cross-channel context, followed by a $3 \times 3 \times 3$ depthwise convolution to encode channel-level spatiotemporal context, resulting in $\mathbf{Q} = \mathbf{W}_d^Q \mathbf{W}_p^Q \hat{\mathbf{X}}$, $\mathbf{K} = \mathbf{W}_d^K \mathbf{W}_p^K \hat{\mathbf{X}}$, and $\mathbf{V} = \mathbf{W}_d^V \mathbf{W}_p^V \hat{\mathbf{X}}$. Here, $\mathbf{W}_p^{(\cdot)}$ represents a $1 \times 1 \times 1$ pointwise convolution, and $\mathbf{W}_d^{(\cdot)}$ represents a $3 \times 3 \times 3$ depthwise convolution. In summary, the MSTA process is defined as follows:

$$\begin{aligned} \mathbf{Y} &= \mathbf{W}_p \text{Attention}(\mathbf{Q}, \mathbf{K}, \mathbf{V}) + \hat{\mathbf{X}}, \\ \text{Attention}(\mathbf{Q}, \mathbf{K}, \mathbf{V}) &= \mathbf{V} \cdot \text{Soft max}(\mathbf{K} \cdot \mathbf{Q} / \alpha), \end{aligned} \tag{1}$$

Here, $\hat{\mathbf{X}}$ and $\mathbf{Y}$ are the input and output feature maps, respectively. In this context, α is a learnable scaling parameter used to control the magnitude of the dot product between $\mathbf{K}$ and $\mathbf{Q}$ before applying the SoftMax function. Similar to traditional multi-head self-attention [31], we divide the number of channels into "heads" and independently learn attention maps in parallel.

b. Multi-Scale Feedforward Network

A conventional Feedforward Network (FN) [30,31] performs the same operation at each spatiotemporal position to transform features. This network utilizes two $1 \times 1 \times 1$ convolutions, where the first convolution layer is employed to expand feature channels (typically expanded by a factor of $\gamma = 4$), and the second convolution layer reduces the channel count back to the original input dimensions.

The relationship between these two operations lies in that the former aims to increase the spatial dimension of features to capture advanced features of spatiotemporal characteristics more effectively. The latter's task is to map these advanced features back to the original input dimensions to fuse them with the outputs of other layers. Although the goals of these two convolutional layers differ, they operate in the same space, namely the dimensions of the original input. Thus, they can perform dot product calculations in the same space, effectively merging the outputs of the two convolutional layers. This design allows the network to operate in different feature spaces and merge these feature spaces when necessary.

In this work, we made two fundamental modifications to the Feedforward Network (FN) to enhance representation learning: (1) introducing a gating mechanism and (2) adopting depthwise convolutions. The Multi-Scale Feedforward Network (MSFN) structure we designed is shown in Figure 3.

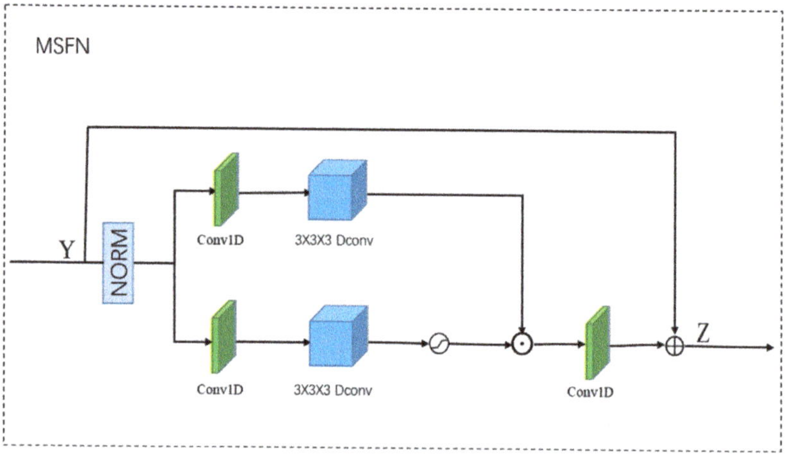

Figure 3. Multi-Scale Feedforward Network (MSFN).

The gating mechanism helps regulate the flow of information in the network hierarchy, enabling each layer to focus on finer image attributes. This mechanism is achieved through the element-wise product of the two parallel paths of the linear transformation layer, with one path passing through the GELU non-linearity [32] activation. Similar to Multi-Scale Transpose Attention (MSTA), we also introduced depth-wise convolutions in Multi-Scale Feedforward Network (MSFN) to encode information from spatially adjacent positions, allowing the model to capture channel-specific information better. This helps improve the model's ability to distinguish between different features, enabling it to learn more discriminative feature representations more effectively. Given that the input tensor $\mathbf{Y} \in \mathbb{R}^{H \times W \times C \times T}$, the representation of MSFN is formulated as:

$$\begin{aligned} \mathbf{Z} &= \mathbf{W}_p^0 \text{Gating}(\mathbf{Y}) + \mathbf{Y}, \\ \text{Gating}(\mathbf{Y}) &= \phi(\mathbf{W}_d^1 \mathbf{W}_p^1 (\text{LN}(\mathbf{Y}))) \odot \mathbf{W}_d^2 \mathbf{W}_p^2 (\text{LN}(\mathbf{Y})), \end{aligned} \tag{2}$$

where $\odot$ represents element-wise multiplication, ϕ represents the GELU non-linearity, and LN is Layer Normalization [29]. Overall, MSFN controls the flow of information at each hierarchical level in our pipeline, allowing each level to focus on complementary fine details with other levels. In other words, compared to MSTA, MSFN provides different roles, focused on enriching features with contextual information). In summary, MSFN can further blend token context at each spatiotemporal position to improve classification accuracy.

3.1.2. Cross-Spatial Attention Module

To address the challenge of integrating the features of in-car driver maneuvers and the traffic motion scene features at the same scale, we propose a novel module called the Cross-Spatial Attention Module (CSAM), as illustrated in Figure 4.

As illustrated in Figure 4, within each of the four stages of the STA backbone, the features of the in-car driver maneuver and the traffic motion scene are aggregated separately at the same scale. Specifically, we initially aggregate the in-car driver maneuver sequence feature (inside feature1) and the traffic motion scene sequence feature (outside feature1) using 3D-CNN. This allows the simultaneous consideration of different aspects of the input data in the spatiotemporal dimensions. As the motion features inside and outside the car often contain information at different levels and types, aggregating these features comprehensively captures the spatiotemporal relationships in the input data. By leveraging their complementarity, the model gains a more comprehensive understanding of the data's characteristics, enhancing its expressive power. Furthermore, feature fusion enables the model to better adapt to various input variations and noise, contributing to improved robustness across different in-car driver environments and external traffic conditions.

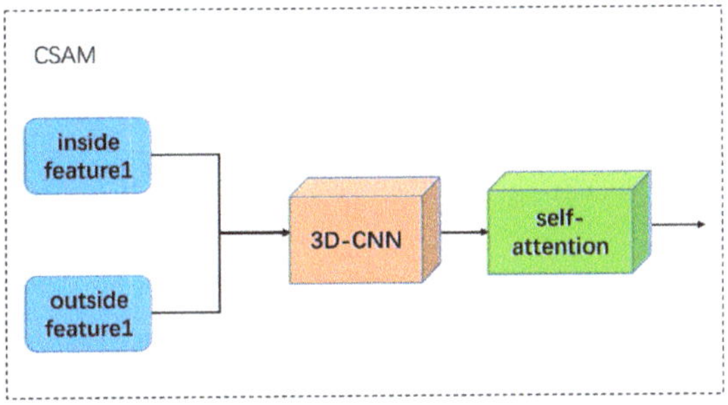

Figure 4. Cross-Spatial Attention Module (CSAM).

To further explore and aggregate the extracted features after the fusion by 3D-CNN, we introduce a self-attention mechanism in CSAM. This allows more effective capture of long-distance dependencies between different positions in spatiotemporal data, enhancing the model's understanding of the global structure. By combating positional biases and dynamically focusing on the importance of different spatiotemporal points, the model's performance and expressive capabilities are improved when dealing with sequential or volumetric data.

3.1.3. Multi-CSAM Fusion Module

To comprehensively capture information, enhance expressiveness, improve robustness, and mitigate overfitting risks, we performed a secondary fusion of features extracted from different stages and scales, as depicted in the Multi-CSAM Fusion Module (MCFM) in Figure 1.

Specifically, we started by downsampling the CSAM1, CSAM2, and CSAM3, fused by the CSAM module at different stages, to match the scale of CSAM4. Subsequently, we applied the same spatial dimension reduction operation to each scale of features, using spatiotemporal average pooling. The dimension-reduced features from all scales were then concatenated to form a single feature vector. A dropout layer was also introduced after the final average pooling layer to prevent overfitting. Finally, the concatenated feature vector was input into fully connected layers to output the ultimate prediction for driver intent recognition.

This process of making different-scale features consistent facilitates parameter sharing, improving computational efficiency, ensuring dimension consistency, and avoiding information loss. It simplifies the model's learning task, reduces the risk of overfitting, and enhances the model's generalization ability and performance across various tasks.

By fusing features from different scales and stages for driver intent recognition, the model comprehensively captures different levels and details of input data, improving its global understanding of data features. Fusing features from different scales enhances the model's expressive power, enabling it to better learn and represent complex data patterns. Furthermore, the fusion of features from multiple scales helps improve the model's robustness to scale and structural variations, making it more resilient in different environments. Multi-scale feature fusion also aids in reducing the model's parameter count, lowering the risk of overfitting, and enhancing its generalization performance. In summary, the multi-scale feature fusion followed by classification through fully connected layers is well-suited for driver intent recognition.

3.2. Asymmetric Loss

According to statistics on driving paths, the straight driving maneuver is more common compared to turning and lane-changing, resulting in the issue of class imbalance across various samples. This imbalance challenges machine learning models during training, as the dominant class samples are more abundant in the training set. The model may tend to learn features and patterns of the dominant class more strongly, leading to insufficient learning for minority classes and increased difficulty in training the model.

We introduce a weighted cross-entropy loss function to address the problem of imbalanced training data in driver intent inference and to ensure that the model adapts better to the sample distribution of different classes while ensuring a more balanced contribution of the loss function to each class. In this context, the loss for each class is multiplied by weight, and we adjust the loss for each sample by setting the weight to the reciprocal of the number of samples in the training set, followed by calculating the average loss. This method helps prevent overfitting to classes with a more significant number of samples, ensuring that each class appropriately impacts the overall loss for more effective model training. Specifically, for the 5-class classification of driver intent inference, the weighted cross-entropy loss function can be defined as follows:

$$L = -\frac{1}{N}\sum_{i=1}^{N} \sum_{j=1}^{5} \omega_j \cdot y_{ij} \cdot \log(\hat{y}_{ij}) \tag{3}$$

where N is the total number of training samples. ω_j is the weight for class j, set as the reciprocal of the number of samples for that class in the training set. y_{ij} is whether the i-th sample in the actual labels belongs to class j. $\hat{y}_{ij}$ is the model's predicted output, representing the probability that sample i belongs to class j.

Setting the weights in this manner helps address the issue of imbalanced samples and improves the prediction accuracy for each class. The specific weight adjustments should be tuned and experimented with based on the dataset's characteristics and the task of finding the optimal weight configuration.

Our experiments showed that identifying certain maneuvers, such as left and right lane changes or turns, can be particularly challenging. This difficulty arises from the dataset's distribution and the inherent characteristics of these maneuvers. Therefore, we adjusted the weights in our model to allocate more significance to these harder-to-recognize categories. Our approach involves increasing the weights for categories based on their difficulty level, with the current coefficients refined through continuous hyperparameter tuning.

Here, we provide an example weight set for the cross-entropy loss function. Based on the distribution of driving actions in our training dataset, as well as finer adjustments made according to the difficulty level of model recognition for different categories, we arrived at these parameter weights as follows (assuming a hypothetical distribution for illustration purposes):

Go Straight: weight = 0.3
Left Lane Change: weight = 1.5
Left Turn: weight = 1.2
Right Lane Change: weight = 1.0
Right Turn: weight = 1.0

We found that applying these weights enhanced the model's prediction accuracy for minority classes and those more challenging to recognize, without significantly impacting overall performance.

4. Experiments and Analysis

4.1. Datasets and Evaluation Metrics

This section delineates the datasets utilized in the experiments and the evaluation criteria employed.

We evaluate the performance of our proposed maneuver prediction method using the publicly available Brain4Cars dataset [33]. This dataset comprises 594 video segments, The

Brain4Cars [3] dataset includes driver observation videos (1088 px × 1920 px, 25 fps) and videos of the outside scenes (480 px × 720 px, 30 fps) recorded simultaneously. The dataset consists of five driving maneuver categories: go straight, left lane change, left turn, right lane change, and right turn. Moreover, samples with no simultaneous recordings of the inside and outside view are considered invalid and not further used in our study.

We use a 5-fold cross-validation for all the experiments in this work, aligning with previous works using the Brain4Cars dataset [3,4,6–8]. The final evaluation metrics include average accuracy and F1 score, along with their standard deviations.

In this research, we utilize accuracy, F1 score, and confusion matrix to evaluate the driver intent recognition performance of both the proposed model and other models. Accuracy (Acc) and F1 score are computed using Equations (4) and (7) [34,35], outlined as follows:

$$\mathrm{Acc} = \frac{\mathrm{TP}+\mathrm{TN}}{\mathrm{TP}+\mathrm{FP}+\mathrm{TN}+\mathrm{FN}} \tag{4}$$

$$\mathrm{Pr} = \frac{\mathrm{TP}}{\mathrm{TP}+\mathrm{FP}} \tag{5}$$

$$\mathrm{Re} = \frac{\mathrm{TP}}{\mathrm{TP}+\mathrm{FN}} \tag{6}$$

$$\mathrm{F1} = \frac{2\mathrm{Pr}\cdot\mathrm{Re}}{\mathrm{Pr}+\mathrm{Re}} \tag{7}$$

In the equations, TP stands for true positive, indicating cases where both the true label and the predicted label are positive; TN stands for true negative, signifying instances where both the true label and the predicted label are negative; FP corresponds to false positive, representing situations where the true label is negative, but the predicted label is positive; and FN denotes false negative, indicating scenarios where the true label is positive, but the predicted label is negative. Pr and Re refer to precision and recall, computed using Equations (5) and (6), respectively, while the F1 score is the harmonic mean.

4.2. Experiment Environment

The training process of STA-NET adopts a transfer learning strategy. The backbone feature extraction network is initialized with weights from Kinetics-400 [36] (a human action dataset). Subsequently, the entire framework is trained. The proposed method, implemented using the PyTorch deep learning framework, The version of PyTorch is 2.1.0, employs AdamW [37] as the optimizer with a weight decay of 0.05. A cosine learning rate scheduler [38] is utilized, setting the base learning rate to 1×10^{-5}. The resolutions of in-cabin and external vehicle camera streams are both set to 224 × 224. The model is trained for 200 epochs on an NVIDIA RTX 3090Ti GPU with 24 GB of memory. The Brain4Cars dataset [3] records video sequences of driving maneuvers. A total of 625 samples are collected, comprising 234 forward samples, 124 left lane change samples, 58 left turn samples, 123 right lane change samples, and 55 right turn samples. Among these, 80% of the video sequences are used for training, and the remaining 20% for testing.

4.3. Pre-Processing and Data Augmentation

To pre-process the data, we first extracted frames from the videos and resized all inputs to a uniform resolution of 224 by 224 pixels. Subsequently, we applied several data augmentation techniques to enhance the dataset and increased the robustness of our model. These techniques include translating the images by 6 pixels in each direction, applying a flip-left-to-right (flipLR), which necessitates a corresponding label change (e.g., a 'turning left' label becomes 'turning right', or the driver's behavior on the left side becomes that on the right side, which covers a broader range of usage scenarios.), and implementing cutout [39]—a method that randomly masks out square regions of the image. Additionally, we employed Augmix [40], which combines various augmentations such as auto contrast, equalization, posterization, and solarization to create a diverse set of training examples. This comprehensive augmentation approach not only amplified our dataset but also ensured broader scenario coverage. Further details of these techniques are discussed in the referenced literature.

4.4. Ablation Experiments

To thoroughly evaluate the practical impact of the modules in improving the performance of the baseline method in real-world scenarios, this study conducted ablation experiments.

Two ablation experiments were performed: (1) Comparison between the improved cross-entropy loss function (ICELF) and the regular cross-entropy loss function, assessing the accuracy of driver intent recognition results under these two scenarios. (2) Comparison of the accuracy of driver intent recognition results obtained by adding the CSAM and MCFM against directly using CSAM4. as shown in Table 1.

Table 1. The results of the ablation experiments on the Brain4Cars dataset.

Module		Accuracy ± SD (%)	F1 ± SD (%)
CSAM and MCFM	ICELF		
		81.35 ± 0.52	80.39 ± 1.63
	√	82.16 ± 5.18	83.73 ± 2.96
√		88.67 ± 0.63	88.32 ± 3.72
√	√	**90.97 ± 0.72**	**89.37 ± 1.56**

√ indicates that this module was added, the bold indicates the final result of this work after adding all modules.

We systematically tested the impacts of CSAM and MCFM and the improved cross-entropy loss function on the accuracy of driver intent recognition by modifying one component at a time while keeping the other unchanged. The experiments utilized both in-cabin and external camera views. According to the results in Table 1, CSAM and MCFM increased the ACC score by 7.32% compared to the baseline model. Simultaneously, the improved cross-entropy loss function raised the ACC score by 0.81%. These results indicate that these two components complement each other in terms of performance and variance, playing crucial roles in enhancing the accuracy of driver intent recognition. When used together, their accuracy is 90.97%, with an F1 score of 89.37%.

4.5. Comparative Experiments of Existing Methods

In Table 2, we summarized and compared our work with other relevant studies, utilizing the most commonly used metrics: precision, recall, and time-to-maneuver (TTM). Time-to-maneuver is defined as the time interval between the time of the model's prediction with the greatest confidence and the actual start of the maneuver. It is observed that our model's precision and recall are on par with the average levels.

Table 2. The summary of the performance of related works on driver intention prediction.

Paper	Data Source	Method	Precision	Recall	TTM (s)
Jain et al. [3]	in-cabin and external	AIO-HMM	77.4%	71.2%	3.53
Jain et al. [11]	in-cabin and external	RNN-LSTM	84.5%	77.1%	3.58
Tonutti et al. [8]	in-cabin and external	DA-RNN	92.3%	90.8%	3.98
Zhou et al. [7]	in-cabin and external	LSTM	91.7%	90.7%	3.30
Rekabdar et al. [41]	in-cabin and external	Dilated CNN	91.8%	92.5%	3.76
Ours	in-cabin and external	CNN + Trans	90.8%	91.1%	0

Table 3 presents the comparative results of the Brain4Cars test set. We compared STA-NET with three widely used end-to-end studies [4,13] because they have demonstrated better performance in driver intent recognition tasks than traditional machine learning methods. In [4,13], researchers applied video action recognition methods to driver intent prediction, such as 3DResNet and ConvLSTM + 3DResNet. All models were pretrained using the Kinetics-400 dataset [36], and all video data with a duration of 5 s were used as input. The table shows results for three different inputs, including only in-cabin driver maneuver videos, only external traffic scene videos, and both in-cabin and external videos. Average accuracy and F1 scores based on five-fold cross-validation are used to illustrate the performance of different methods, where "SD" denotes standard deviation.

Table 3. The comparison of different algorithms with different data inputs.

Algorithms	Data Source	Param. (M)	Accuracy ± SD (%)	F1 ± SD (%)
3DResNet [4]	in-cabin only	240.26	83.10 ± 2.5	81.7 ± 2.6
	external only	240.26	53.20 ± 0.5	43.4 ± 0.9
	in-cabin and external	480.52	75.50 ± 2.4	73.2 ± 2.2
ConvLSTM + 3DResNet [13]	in-cabin only	46.22	77.40 ± 0.02	75.49 ± 0.02
	external only	160.41	60.87 ± 0.01	66.38 ± 0.03
	in-cabin and external	212.92	83.98 ± 0.01	84.3 ± 0.01
STA-Net (Ours)	in-cabin only	**60.56**	**87.31 ± 0.56**	**85.32 ± 1.71**
	external only	**60.56**	**71.12 ± 0.58**	**73.10 ± 0.63**
	in-cabin and external	**137.25**	**90.97 ± 0.72**	**89.37 ± 1.56**

The bold indicates the final result of this work.

Based on Table 3, when exclusively using in-cabin driver maneuver data, all algorithms achieved satisfactory results, with accuracy ranging from 77 to 84%, making further improvements challenging. STA-NET achieved the highest accuracy of 90.97% and an F1 score of 89.37% when provided with dual perspectives—both inside and outside the vehicle. The experiment indicates that the accuracy of intent recognition increased by approximately 3.66% when the STA-NET model used both in-cabin driver maneuver and external traffic scene as inputs. This result strongly suggests that external traffic scene features and in-cabin driver maneuver features contain complementary information for driver intent recognition. Moreover, our model significantly reduces the number of parameters compared to previous methods. Given the potential computational costs associated with model complexity, models with low resource requirements are preferred for automotive applications. Our model achieves the extraction of valuable features with fewer parameters, facilitating easier deployment in resource-constrained environments such as onboard vehicle systems.

The confusion matrix for the proposed method is illustrated in Figure 5, showcasing the classification performance for the five intents. The results indicate that lane-keeping and right lane change intent recognition achieved the most accurate results, with accuracy reaching 91.3% and 90.9%, respectively. The ability to recognize the intent to change lanes to the left is relatively lower, with an accuracy of approximately 70.8%, often confused with the intent to go straight. The accuracy for recognizing the intent to turn left is around 83.3%, and it is also prone to confusion with the intent to go straight. The accuracy for recognizing the intent to turn right is approximately 87.5%, with potential confusion with the intent to go straight and change lanes to the right.

Through an analysis of misclassified samples, three main reasons are proposed. Firstly, for some lane-keeping maneuvers, drivers may be more inclined to perform left-check actions to ensure safe driving, making it easy to infer them as left lane change or left turn. Secondly, some right lane change intents are similar to maneuvers during lane-keeping, suggesting that drivers might occasionally perform right lane changes while maintaining their lane, although infrequently, potentially confusing the model. Thirdly, some right turn intents are very similar to maneuvers during right lane changes, which can confuse the model. These inferences suggest that in more complex traffic scenarios, drivers' maneuvers may adjust frequently based on the traffic conditions on both sides, posing a challenge for driver intent recognition.

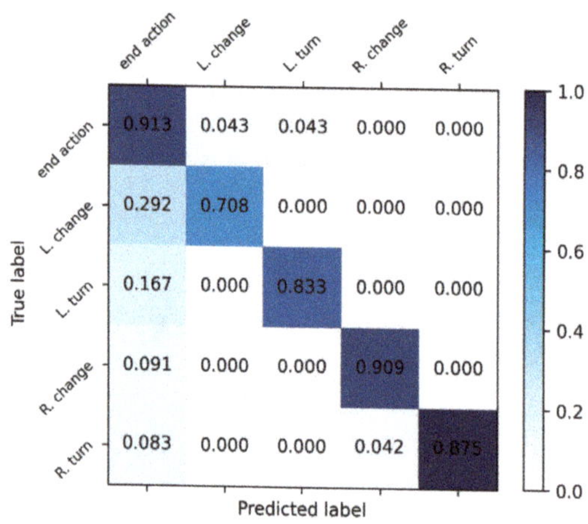

Figure 5. The confusion matrices for five driver intentions are based on the proposed models. The abscissa and the ordinate represent the probability of predicted and true labels.

Upon comparing our STA-NET's performance with the method described in [13], we acknowledge the variation in classification accuracy across different maneuvers. Specifically, the model in [13] exhibits superior performance in identifying left lane changes, left turns, and right turns, whereas STA-NET struggles with right lane change predictions. This discrepancy raises a pertinent discussion on the feasibility and potential advantages of employing a multi-method approach for driver maneuver prediction.

Rationale for Multi-Method Approach:

The diversity in driving behaviors and the complexity of road scenarios necessitate a nuanced approach to maneuver recognition. A single model may not optimally capture the intricacies associated with various maneuvers due to differences in visual cues, driver intentions, and environmental contexts. Therefore, leveraging the strengths of different predictive models based on the predicted maneuver could enhance overall performance and reliability.

Methodological Considerations:

To explore this possibility, we propose a framework where the predictive model dynamically selects between STA-NET and alternative methods, like the one presented in [13], based on the specific maneuver scenario. This selection could be informed by pre-defined criteria, such as the maneuver type, confidence levels of the predictions, or contextual factors like traffic density and road type.

Potential Benefits:

(1) Enhanced Accuracy: By aligning model selection with the maneuver's characteristics, we anticipate improvements in prediction accuracy, particularly for maneuvers where STA-NET's performance is currently lacking.
(2) Reduced False Positives/Negatives: A more tailored approach allows for finer discrimination between maneuvers, potentially reducing misclassifications and enhancing the system's reliability.
(3) Adaptability: This strategy introduces a layer of adaptability, enabling the system to evolve and incorporate new methods or findings from ongoing research.

5. Conclusions

In this study, we proposed an end-to-end driver intent inference framework named STA-NET based on joint spatiotemporal analysis. Firstly, a dual-branch network with spatiotemporal joint extraction modules was employed to extract features related to in-cabin driver maneuvers and external traffic scenes separately. Subsequently, the features were aggregated using fusion at different scales. Finally, the aggregated spatiotemporal features were utilized to obtain the most probable driver maneuver intent. The model was validated on Brain4Cars, a natural dataset containing highway and urban road driving information. Results demonstrate that our model achieved favorable performance compared to other action recognition algorithms and classical methods, with an overall accuracy of 90.97% for driver intent recognition. In conclusion, our work contributes to ongoing efforts to enhance traffic safety and paves the way for further advancements in driver intent prediction and personalized driving assistance systems.

Author Contributions: Conceptualization, B.H.; methodology, B.H.; validation, Z.W. and X.C.; investigation, Z.W. and X.C.; writing—original draft preparation, B.H.; writing—review and editing, B.H.; supervision, N.Y.; project administration, N.Y.; funding acquisition, N.Y. All authors have read and agreed to the published version of the manuscript.

Funding: This work was supported by the National Natural Science Foundation of China under grant 62271388.

Institutional Review Board Statement: Not applicable.

Informed Consent Statement: Not applicable.

Data Availability Statement: The Brain4Cars dataset presented in this study is openly available in [3]. The kinetics-400 dataset presented in this study is openly available in [18,31].

Conflicts of Interest: The authors declare no conflicts of interest. The funders had no role in the design of the study; in the collection, analyses, or interpretation of data; in the writing of the manuscript; or in the decision to publish the results.

References

1. Wu, Y.; Zhang, L.; Lou, R.; Li, X. Recognition of Lane Changing Maneuvers for Vehicle Driving Safety. *Electronics* **2023**, *12*, 1456. [CrossRef]
2. David, R.; Rothe, S.; Söffker, D. State Machine Approach for Lane Changing Driving Behavior Recognition. *Automation* **2020**, *1*, 68–79. [CrossRef]
3. Jain, A.; Koppula, H.S.; Raghavan, B.; Soh, S.; Saxena, A. Car that knows before you do: Anticipating maneuvers via learning temporal driving models. In Proceedings of the IEEE International Conference on Computer Vision (ICCV), Santiago, Chile, 7–13 December 2015; pp. 3182–3190.
4. Gebert, P.; Roitberg, A.; Haurilet, M.; Stiefelhagen, R. End-to-end Prediction of Driver Intention using 3D Convolutional Neural Networks. In Proceedings of the IEEE Intelligent Vehicles Symposium (IV), Paris, France, 9–12 June 2019; pp. 969–974.
5. Xing, Y.; Lv, C.; Wang, H.; Cao, D.; Velenis, B. An ensemble deep learning approach for driver lane change intention inference. *Transp. Res. Part C Emerg. Technol.* **2020**, *115*, 102615. [CrossRef]
6. Jain, A.; Soh, S.; Raghavan, B.; Singh, A.; Koppula, H.S.; Saxena, A. Brain4Cars: Sensory-Fusion Recurrent Neural Models for Driver Activity Anticipation. Available online: http://brain4cars.com/pdfs/baylearn.pdf (accessed on 29 November 2023).
7. Zhou, D.; Ma, H.; Dong, Y. Driving maneuvers prediction based on cognition-driven and data-driven method. In Proceedings of the 2018 IEEE Visual Communications and Image Processing (VCIP), Taichung, Taiwan, 9–12 December 2018; IEEE: Piscataway, NJ, USA, 2018; pp. 1–4.
8. Tonutti, M.; Ruffaldi, E.; Cattaneo, A.; Avizzano, C.A. Robust and subject-independent driving manoeuvre anticipation through Domain-Adversarial Recurrent Neural Networks. *Robot. Auton. Syst.* **2019**, *115*, 162–173. [CrossRef]
9. Braunagel, C.; Kasneci, E.; Stolzmann, W.; Rosenstiel, W. Driver-activity recognition in the context of conditionally autonomous driving. In Proceedings of the 2015 IEEE 18th International Conference on Intelligent Transportation Systems, Gran Canaria, Spain, 15–18 September 2015; pp. 1652–1657.
10. Braunagel, C.; Geisler, D.; Rosenstiel, W.; Kasneci, E. Online recognition of driver-activity based on visual scanpath classification. *IEEE Intell. Transp. Syst. Mag.* **2017**, *9*, 23–36. [CrossRef]
11. Jain, A.; Singh, A.; Koppula, H.S.; Soh, S.; Saxena, A. Recurrent neural networks for driver activity anticipation via sensory-fusion architecture. In Proceedings of the IEEE International Conference on Robotics and Automation (ICRA), Stockholm, Sweden, 16–21 May 2016; pp. 3118–3125.
12. Moussaid, A.; Berrada, I.; El Kamili, M.; Fardousse, K. Predicting driver lane change maneuvers using driver's face. In Proceedings of the International Conference on Wireless Networks and Mobile Communications, (WINCOM), Fez, Morocco, 29 October–1 November 2019; pp. 1–7.
13. Rong, Y.; Akata, Z.; Kasneci, E. Driver intention anticipation based on in-cabin and driving scene monitoring. In Proceedings of the IEEE 23rd International Conference on Intelligent Transportation Systems (ITSC), Rhodes, Greece, 20–23 September 2020; pp. 1–8.
14. Xing, Y.; Tian, B.; Lv, C.; Cao, D. A Two-Stage Learning Framework for Driver Lane Change Intention Inference. *IFAC PapersOnLine* **2020**, *53*, 638–643. [CrossRef]

15. Chen, H.; Chen, H.; Liu, H.; Feng, X. Spatiotemporal Feature Enhancement Aids the Driving Intention Inference of Intelligent Vehicles. *Int. J. Environ. Res. Public Health* **2022**, *19*, 11819. [CrossRef] [PubMed]
16. Ma, Y.; Ye, W.; Cao, X.; Abdelraouf, A.; Han, K.; Gupta, R.; Wang, Z. CEMFormer: Learning to Predict Driver Intentions from In-Cabin and External Cameras via Spatial-Temporal Transformers. *arXiv* **2023**, arXiv:2305.07840.
17. Bonyani, M.; Rahmanian, M.; Jahangard, S.; Rezaei, M. DIPNet: Driver intention prediction for a safe takeover transition in autonomous vehicles. *IET Intell. Transp. Syst.* **2023**, *17*, 1769–1783. [CrossRef]
18. Zhang, H.; Guo, D.; Guo, Y.; Wu, F.; Gao, S. A Novel Method for the Driver Lane-Changing Intention Recognition. *IEEE Sens. J.* **2023**, *23*, 20437–20451. [CrossRef]
19. Li, R.; Shu, X.; Li, C. Driving Behavior Prediction Based on Combined Neural Network Model. *IEEE Trans. Comput. Soc. Syst.* **2024**. [CrossRef]
20. Chen, S.; Yao, H.; Qiao, F.; Ma, Y.; Wu, Y.; Lu, J. Vehicles driving behavior recognition based on transfer learning. *Expert Syst. Appl.* **2023**, *213*, 119254. [CrossRef]
21. Sun, L.; Jia, K.; Chen, K.; Yeung, D.Y.; Shi, B.E.; Savarese, S. Lattice Long Short-Term Memory for Human Action Recognition. In Proceedings of the IEEE International Conference on Computer Vision, Venice, Italy, 22–29 October 2017; pp. 2147–2156.
22. Li, D.; Qiu, Z.; Dai, Q.; Yao, T.; Mei, T. Recurrent Tubelet Proposal and Recognition Networks for Action Detection. In Proceedings of the European Conference on Computer Vision (ECCV), Munich, Germany, 8–14 September 2018; pp. 303–318.
23. Carreira, J.; Zisserman, A. Quo vadis, action recognition? A new model and the kinetics dataset. In Proceedings of the IEEE Conference on Computer Vision and Pattern Recognition, Honolulu, HI, USA, 21–26 July 2017; pp. 6299–6308.
24. Feichtenhofer, C.; Fan, H.; Malik, J.; He, K. SlowFast Networks for Video Recognition. In Proceedings of the IEEE/CVF International Conference on Computer Vision, Seoul, Republic of Korea, 27 October–2 November 2019; pp. 6202–6211.
25. Xu, M.; Xiong, Y.; Chen, H.; Li, X.; Xia, W.; Tu, Z.; Soatto, S. Long Short-Term Transformer for Online Action Detection. *Adv. Neural Inf. Process. Syst.* **2021**, *34*, 1086–1099.
26. Li, Y.; Wu, C.Y.; Fan, H.; Mangalam, K.; Xiong, B.; Malik, J.; Feichtenhofer, C. Feichtenhofer, MViTv2: Improved Multiscale Vision Transformers for Classification and Detection. In Proceedings of the IEEE/CVF Conference on Computer Vision and Pattern Recognition, New Orleans, LA, USA, 18–24 June 2022; pp. 4804–4814.
27. Bertasius, G.; Wang, H.; Torresani, L. Is Space-Time Attention All You Need for Video Understanding? In Proceedings of the 38th International Conference on Machine Learning, PMLR, Virtual, 18–24 July 2021; Volume 139, pp. 813–824.
28. Arnab, A.; Dehghani, M.; Heigold, G.; Sun, C.; Lučić, M.; Schmid, C. Vivit: A video vision transformer. In Proceedings of the IEEE/CVF International Conference on Computer Vision, Montreal, BC, Canada, 11–17 October 2021.
29. Ba, J.L.; Kiros, J.R.; Hinton, G.E. Layer normalization. *arXiv* **2016**, arXiv:1607.06450.
30. Vaswani, A.; Shazeer, N.; Parmar, N.; Uszkoreit, J.; Jones, L.; Gomez, A.N.; Kaiser, Ł.; Polosukhin, I. Attention is all you need. *Adv. Neural Inf. Process. Syst.* **2017**, *30*, 11.
31. Dosovitskiy, A.; Beyer, L.; Kolesnikov, A.; Weissenborn, D.; Zhai, X.; Unterthiner, T.; Dehghani, M.; Minderer, M.; Heigold, G.; Gelly, S.; et al. An image is worth 16 × 16 words: Transformers for image recognition at scale. *arXiv* **2020**, arXiv:2010.11929.
32. Hendrycks, D.; Gimpel, K. Gaussian error linear units (GELUs). *arXiv* **2016**, arXiv:1606.08415.
33. Jain, A.; Koppula, H.S.; Soh, S.; Raghavan, B.; Singh, A.; Saxena, A. Brain4Cars: Car That Knows Before You Do via Sensory-Fusion Deep Learning Architecture. *arXiv* **2016**, arXiv:1601.00740.
34. Wu, Z.; Liang, K.; Liu, D.; Zhao, Z. Driver Lane Change Intention Recognition Based on Attention Enhanced Residual-MBi-LSTM Network. *IEEE Access* **2022**, *10*, 58050–58061. [CrossRef]
35. Yu, B.; Bao, S.; Zhang, Y.; Sullivan, J.; Flannagan, M. Measurement and prediction of driver trust in automated vehicle technologies: An application of hand position transition probability matrix. *Transp. Res. Part C Emerg. Technol.* **2021**, *124*, 102957. [CrossRef]
36. Kay, W.; Carreira, J.; Simonyan, K.; Zhang, B.; Hillier, C.; Vijayanarasimhan, S.; Viola, F.; Green, T.; Back, T.; Natsev, P.; et al. The kinetics human action video dataset. *arXiv* **2017**, arXiv:1705.06950.
37. Loshchilov, I.; Hutter, F. Decoupled Weight Decay Regularization. In Proceedings of the 7th International Conference on Learning Representations, ICLR 2019, New Orleans, LA, USA, 6–9 May 2019.
38. Loshchilov, I.; Hutter, F. SGDR: Stochastic Gradient Descent with Warm Restarts. In Proceedings of the 5th International Conference on Learning Representations, ICLR 2017, Toulon, France, 24–26 April 2017.
39. De Vries, T.; Taylor, G.W. Improved regularization of convolutional neural networks with cutout. *arXiv* **2017**, arXiv:1708.04552.
40. Hendrycks, D.; Mu, N.; Cubuk, E.D.; Zoph, B.; Gilmer, J.; Lakshminarayanan, B. Augmix: A simple data processing method to improve robustness and uncertainty. *arXiv* **2019**, arXiv:1912.02781.
41. Rekabdar, B.; Mousas, C. Dilated convolutional neural network for predicting driver's activity. In Proceedings of the 2018 21st International Conference on Intelligent Transportation Systems (ITSC), Maui, HI, USA, 4–7 November 2018; IEEE: Piscataway, NJ, USA, 2018; pp. 3245–3250.

applied sciences

MDPI

Article

Exploring High-Order Skeleton Correlations with Physical and Non-Physical Connection for Action Recognition

Cheng Wang [1], Nan Ma [2,3,*] and Zhixuan Wu [4]

1 Beijing Key Laboratory of Information Service Engineering, Beijing Union University, Beijing 100101, China; 15056271781@163.com
2 Faculty of Information Technology, Beijing University of Technology, Beijing 100124, China
3 Engineering Research Center of Intelligence Perception and Autonomous Control, Ministry of Education, Beijing University of Technology, Beijing 100124, China
4 School of computer science, Beijing University of Posts and Telecommunications, Beijing 100876, China; zhixuanwuyslya@163.com
* Correspondence: manan123@bjut.edu.cn; Tel.: +86-139-11155086

Abstract: Hypergraphs have received widespread attention in modeling complex data correlations due to their superior performance. In recent years, some researchers have used hypergraph structures to characterize complex non-pairwise joints in the human skeleton and model higher-order correlations of the human skeleton. However, traditional methods of constructing hypergraphs based on physical connections ignore the dependencies among non-physically connected joints or bones, and it is difficult to model the correlation among joints or bones that are highly correlated in human action but are physically connected at long distances. To address these issues, we propose a skeleton-based action recognition method for hypergraph learning based on skeleton correlation, which explores the effects of physically and non-physically connected skeleton information on accurate action recognition. Specifically, in this paper, spatio-temporal correlation modeling is performed on the natural connections inherent in humans (physical connections) and the joints or bones that are more dependent but not directly connected (non-physical connection) during human actions. In order to better learn the hypergraph structure, we construct a spatio-temporal hypergraph neural network to extract the higher-order correlations of the human skeleton. In addition, we use an attentional mechanism to compute the attentional weights among different hypergraph features, and adaptively fuse the rich feature information in different hypergraphs. Extensive experiments are conducted on two datasets, NTU-RGB+D 60 and Kinetics-Skeleton, and the results show that compared with the state-of-the-art skeleton-based methods, our proposed method can achieve an optimal level of performance with significant advantages, providing a more accurate environmental perception and action analysis for the development of embodied intelligence.

Keywords: action recognition based on skeleton; multi-channel features; spatio-temporal hypergraph neural network; cross-channel attention mechanism; high-order semantic correlation; adaptive fusion

1. Introduction

Action recognition [1], a pivotal area within the realm of computer vision, has been extensively applied across various domains, including human–computer interaction [2], virtual reality [3], video surveillance [4], and the field of autonomous driving [5]. Action recognition refers to analyzing and classifying human motion in specific environments. One important method is action recognition based on skeleton, where the skeleton sequence represents compact information on human motion as high-level information, providing an effective and robust representation form for describing human action. With the continuous development of 3D depth sensor technology, researchers can easily obtain the three-dimensional coordinates of human key joints, and thus obtain human skeleton data. Therefore, in recent years, more attention has been paid to some methods based on skeleton in the field of action recognition [6,7].

In real-life scenarios, apart from pairwise relationships, there also exist multiple types of data that contain numerous non-pairwise relationships. However, these intricate relationships cannot be effectively represented using graph structures. Additionally, due to the need for coordination among multiple joints in human actions, these joints may have significant distances in the graph structure of the human skeleton. The graph neural network methods only focus on the local physical connections between joints, while overlooking the non-physical dependency relationships among them. Furthermore, each action has varying degrees of impact on these joints, and the degree of influence of each joint should be captured during the network training process. Therefore, one of the main challenges in skeleton-based action recognition is how to utilize the graph structure of the skeleton data while fully considering the impact of different joints on human action. As a result, researchers have introduced the concept of hypergraphs [8] to represent complex non-pairwise relationships. Hypergraphs can effectively represent higher-order relationships in the data. Feature learning is performed using hypergraph neural networks [9] to obtain the feature representations of complex data. Compared to traditional graph neural networks, hypergraph neural networks are a more versatile representation learning framework that can effectively handle complex high-order correlations through the use of hypergraph structures, thus enabling the efficient processing of diverse types of complex data. Consequently, the utilization of hypergraph neural networks for achieving action recognition in complex environments has become a widely studied problem in recent years. An increasing number of researchers are attempting to model human skeletons using hypergraphs and employ hypergraph neural networks to obtain the feature representations of human actions, thereby facilitating more efficient human action recognition.

To address the above issues, this paper proposes a framework to explore high-order skeleton correlations with physical and non-physical connections for action recognition. Specifically, the contributions of this paper are as follows:

- Firstly, this paper proposes a spatio-temporal hypergraph modeling method of human skeleton correlation. This method focuses on the inherent physical connections and non-physically connected relationships among human skeletons. For different actions, different weights are assigned to different vertex in the hypergraph, which highlights the regions that have a significant impact on human action.
- Secondly, this paper proposes an adaptive multi-channel spatio-temporal hypergraph neural network (AMC-STHGNN). The network captures the high-order correlation among human skeletons, and the features of different data channels are adaptively fused, making full use of the complementarity and diversity among three types of features.
- Finally, our proposed method is tested on two public datasets and shows a superior performance, effectively enhancing the ability of environmental perception and action analysis.

The rest of this paper is organized as follows. We first introduce related work in Section 2, including action recognition based on skeleton and hypergraph neural networks. Then, in Section 3, we review the basic theory of hypergraphs and hypergraph neural networks. Section 4 provides detailed implementation details of the proposed method. The experimental results and discussions of the proposed method are presented in Section 5. Finally, a summary of the entire paper is given in Section 6.

2. Related Work

2.1. Action Recognition Based on Skeleton

Due to the rich spatial and temporal information present in deep skeleton sequences, many researchers have utilized skeleton data as input for neural networks in human action recognition tasks. Xu et al. [10] proposed an architecture called topology-aware CNN for skeleton-based action recognition. This architecture enhances performance through a novel cross-channel feature enrichment module and SkeletonMix strategy. Ref. [11] introduced a design method based on temporal convolutional neural networks (TCNs) to analyze and recognize human activities. The emergence of graph convolutional neural networks (GCNs) has significantly improved the ability to handle non-Euclidean data. Chen et al. [12] proposed the CTR-GCN method, which dynamically learns joint information in the spatial domain to achieve channel topology modeling. Song et al. [13] presented an effective GCN baseline method with high accuracy and fewer trainable parameters, employing a spatial-temporal separation learning strategy for model configuration. Although significant progress has been made in skeleton-based action recognition methods, traditional methods based on skeleton information cannot meet practical requirements for extracting action features in complex environments.

2.2. Hypergraph Neural Networks

As a pioneering work in hypergraph neural networks, Feng et al. [14] introduced a hypergraph neural network framework (HGNN) that leverages the hypergraph structures for feature learning and effectively extracts high-order correlations in the data. HGNN extends spectral-domain convolutional operations from graph learning to hypergraph learning, performing convolution in the spectral domain using the hypergraph Laplacian operator. Jiang et al. [15] proposed a dynamic hypergraph neural network framework (DHGNN) to address the limitation of fixed hypergraph structures in hypergraph neural networks, which hinders the representation capacity for varying data features. Bai et al. [16] introduced two end-to-end operators, which can be inserted into most graph neural networks when non-pairwise relationships exist in the data for model training. Gao et al. [17] proposed a hypergraph neural network framework called HGNN+, which can be extended to directed hypergraphs. Due to the powerful representation capability of complex relationships, hypergraphs have been widely applied in the field of computer vision [18,19]. It is worth noting that research on action recognition based on hypergraph neural networks [20–22] has achieved certain progress. However, further optimization is needed to improve the ability to extract advanced features in the process of human motion, in order to enhance the accuracy of action recognition.

3. Hypergraph Theory

A hypergraph is characterized by the triple $\mathcal{G} = (\mathcal{V}, \mathcal{E}, \mathbf{W})$, where $\mathcal{V}$ represents the set of vertices, the vertex denoted by $v \in \mathcal{V}$; $\mathcal{E}$ is the set of hyperedges, the hyperedge denoted by $e \in \mathcal{E}$; and $\mathbf{W}$ is the matrix of hyperedge weights, indicating the weight associated with each hyperedge, represented as $\omega(e)$. The incidence between hyperedges and vertices is captured through an incidence matrix $\mathbf{H}$, which has dimensions $|\mathcal{V}| \times |\mathcal{E}|$. In this matrix, if a vertex v is part of a hyperedge e, the corresponding element $h(v,e)$ is set to 1; if not, $h(v,e)$ is set to 0. The elements of the incidence matrix $\mathbf{H}$ are defined such that:

$$h(v,e) = \begin{cases} 1, & v \in e \\ 0, & v \notin e \end{cases}. \tag{1}$$

In addition, $h(v,e)$ can also represent the possibility of assigning vertex v to hyperedge e or the importance of vertex v with respect to hyperedge e, using a range of [0,1] for representation. The relationship between the hypergraph and the incidence matrix is illustrated in Figure 1.

The hyperedge degree represents the number of vertices associated with a hyperedge e, which can be defined as

$$\delta(e) = \sum_{v \in \mathcal{V}} h(v,e) \tag{2}$$

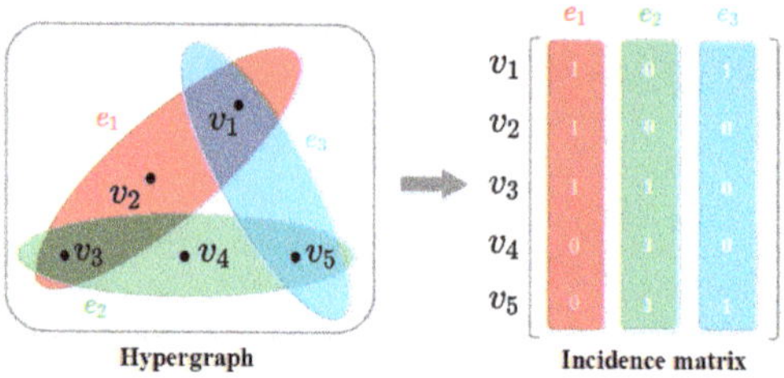

Figure 1. Hypergraph and incidence matrix.

The vertex degree represents the sum of the hyperedge weights associated with a vertex v, which can be defined as

$$d(v) = \sum_{e \in \mathcal{E}} w(e) \times h(v,e) \tag{3}$$

In addition, $\mathbf{D}_e$ and $\mathbf{D}_v$ are defined to denote the diagonal matrices of hyperedge degree and vertex degree, respectively. In [18], the features of the 3D object are used to construct the hypergraph by the k-NN method, where each vertex is connected to its k nearest neighbors. In [23], the l-1 sparse representation of the vertex is learned to construct the hyperedges.

After constructing the hypergraph, hypergraph learning is performed using a deep learning method [24], which is known as a hypergraph neural network. Specifically, according to [25], the Laplacian matrix of the hypergraph is defined as

$$\Delta = \mathbf{I} - \mathbf{D}_v^{-1/2}\mathbf{H}\mathbf{D}_e^{-1}\mathbf{H}^\top\mathbf{D}_v^{-1/2} \tag{4}$$

where $\mathbf{I}$ is the unit matrix. The hypergraph convolution on the spectral domain is parametrized as

$$\mathbf{X}^{t+1} = \mathbf{D}_v^{-1/2}\mathbf{H}\mathbf{D}_e^{-1}\mathbf{H}^\top\mathbf{D}_v^{-1/2}\mathbf{X}^t\mathbf{\Theta} \tag{5}$$

where $\mathbf{\Theta} \in \mathbb{R}^{d_{in} \times d_{out}}$ is the trainable parameter in the hypergraph convolution layer. In [17], inspired by hyperpaths in hypergraphs, the spatial hypergraph-based convolution HGNNConv+ can be written in matrix format as

$$\mathbf{X}^{t+1} = \mathbf{D}_v^{-1}\mathbf{H}\mathbf{D}_e^{-1}\mathbf{H}^\top\mathbf{X}^t\mathbf{\Theta} \tag{6}$$

In the process of hypergraph convolution, vertex features are transformed into hyperedge features, which are then converted into new vertex features as shown in Figure 2. Hypergraph neural networks include spectral domain-based methods [14], where convolution uses a pre-defined hypergraph Laplacian matrix for vertex-to-vertex feature smoothing, and space-based methods [17], where a two-stage message passing method (vertex-edge-vertex) is used, which is more flexible and can be easily extended to more types of higher-order structures, such as directed hypergraphs.

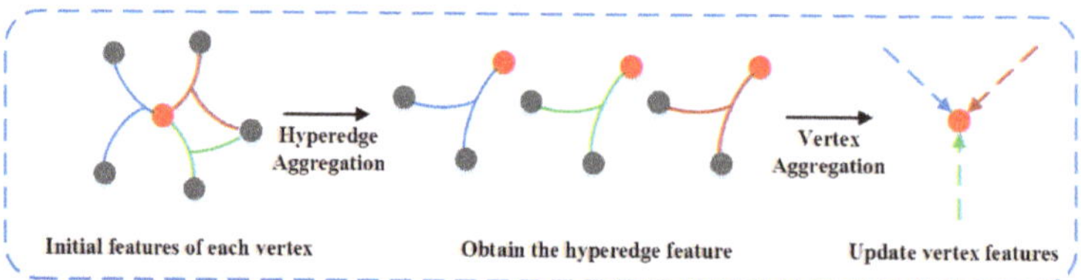

Figure 2. Hypergraph and incidence matrix.

4. The Proposed Method

4.1. Pipeline Overview

In order to explore high-order skeleton correlations with physical and non-physical connections for action recognition, this paper proposes a framework, which is shown in Figure 3. Firstly, this paper uses both human joint and bone to construct a physically connected joint (PCJ) spatio-temporal hypergraph, a non-physically connected joint (N-PCJ) spatio-temporal hypergraph, and a non-physically connected bone (N-PCB) spatio-temporal hypergraph as model inputs to generate three different data channels. Notably, we not only use the inherent physical connectivity associations between joints as a priori knowledge to construct hyperedges, but also use a clustering algorithm to obtain the non-physical connectivity associations between joints to construct hyperedges, where the non-physical connectivity hyperedges are constructed to differentiate the importance of different joints or bones in action recognition and assign different weights to different joints or bones. Then, this paper proposes a network called AMC-STHGNN. In this network, the spatio-temporal information-based hypergraph convolution module is used to more accurately capture the higher-order spatio-temporal feature information during human action; the multi-feature fusion module based on cross-channels attention mechanism makes full use of the complementarity and diversity among different data channels, and obtains the final prediction results by adaptively fusing the rich feature information in the multi-channel data.

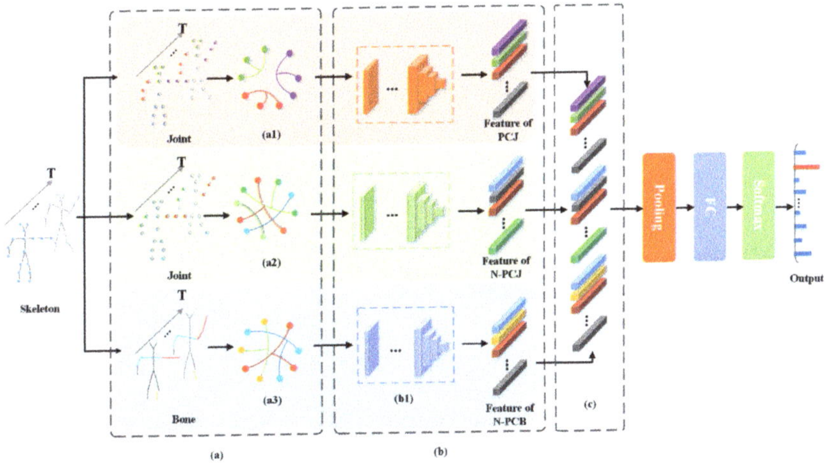

Figure 3. The pipeline of the proposed method: (**a**) spatio-temporal hypergraph modeling of human skeleton correlation; (**b**) adaptive multi-channel spatio-temporal hypergraph neural network; (**a1**) physically connected joint spatio-temporal hypergraph; (**a2**) non-physically connected joint spatio-temporal hypergraph; (**a3**) non-physically connected bone spatio-temporal hypergraph; (**b1**) hypergraph convolution module based on spatio-temporal information; and (**c**) multi-feature fusion module based on cross-channels attention mechanism.

4.2. Spatio-Temporal Hypergraph Modeling of Human Skeleton Correlation

In this paper, a video sequence is used as the model input, which is cropped into T frames containing information about the human skeleton, each frame is noted as $\{I_t\}_{t=1,2,\cdots,T}$. Feature extraction is performed for each frame according to the method in [26], with the initial feature vector of the i-th joint $v_t^{(i)}$ in the t-th frame image noted as $\boldsymbol{v}_i^t \in \mathbb{R}^{C_1}$ and the initial feature vector of the j-th bone $b_t^{(j)}$ in the t-th frame image noted as $\boldsymbol{b}_j^t \in \mathbb{R}^{C_2}$, where C_1 denotes the dimension of the feature vector $\boldsymbol{v}_i^t$ and C_2 denotes the dimension of the feature vector b_j^t, and the hypergraphs are constructed separately using the joints information and the bones information according to different strategies.

4.2.1. Physically Connected Joint Spatio-Temporal Hypergraph

For the physically connected joint (PCJ) spatio-temporal hypergraph, the inherent topology of the human joints is used as a basis to divide them into different regions, including the head, trunk, left hand (*Lhand*), right hand (*Rhand*), left leg (*Lleg*), and right leg (*Rleg*) in six regions $\mathcal{V}_{area}$, $area = \{head, trunk, Lhand, Rhand, Lleg, Rleg\}$. $\mathcal{V}_{area}$ represents the set of all joints belonging to a certain area, and the joints of the human in consecutive frames of the same area are connected by a hyperedge to construct a spatio-temporal hypergraph as a fixed information representation of the human body structure, as shown in Figure 4a.

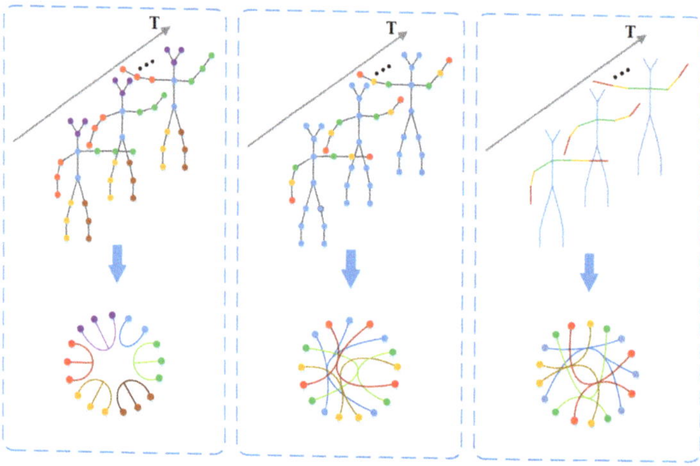

Figure 4. Visualization of different hypergraph constructions: (**a**) the construction of a physically connected joint spatio-temporal hypergraph; (**b**) the construction of a non-physically connected joint spatio-temporal hypergraph; (**c**) the construction of a non-physically connected bone spatio-temporal hypergraph.

The physically connected joint spatio-temporal hypergraph is denoted as $\mathcal{G}_{PCJ} = (\mathcal{V}_{PCJ}, \mathcal{E}_{PCJ}, \mathbf{W}_{PCJ})$, where $\mathcal{V}_{PCJ}$ denotes the set of vertices of this hypergraph, $\mathcal{E}_{PCJ}$ denotes the set of hyperedges of this hypergraph, and $\mathbf{W}_{PCJ}$ denotes the diagonal matrix of hyperedge weights in this hypergraph. Specifically, $\mathcal{V}_{PCJ}$ is the set of all joints $v_t^{(i)}$ (the i-th joint in the image of frame t) in all frames; the elements in $\mathcal{E}_{PCJ}$ are noted as $e_{area} = \{\forall v_t^{(i)} | v_t^{(i)} \in \mathcal{V}_{area}\}$, $\mathcal{E}_{PCJ} = \{e_{area} | area = head, trunk, Lhand, Rhand, Lleg, Rleg\}$. The incidence matrix $\mathbf{H}_{PCJ}$ of the spatio-temporal hypergraph of physically connected joints is constructed based on the association between $\mathcal{V}_{PCJ}$ and $\mathcal{E}_{PCJ}$, and the hyperedge degree $\delta(e_{area})$ and vertex degree $d(v_t^{(i)})$ are calculated using Equations (2) and (3), following which the diagonal matrices $\mathbf{D}_{e_{area}}$ and $\mathbf{D}_{v_t^{(i)}}$ of the hyperedge degree and vertex degree are obtained. In order to optimize the network using higher-order information, the incidence matrix $\mathbf{H}_{PCJ}$ is transformed by Laplace to generate the Laplace matrix $\mathbf{G}_{PCJ}$, which is calculated as

$$\mathbf{G}_{PCJ} = \mathbf{D}_{v_t^{(i)}}^{-1/2} \mathbf{H}_{PCJ} \mathbf{W}_{PCJ} \mathbf{D}_{e_{area}}^{-1} (\mathbf{H}_{PCJ}^{\top}) \mathbf{D}_{v_t^{(i)}}^{-1/2} \tag{7}$$

4.2.2. Non-Physically Connected Joint Spatio-Temporal Hypergraph

For example, in a traffic police gesture, the arm of the command 'go straight' changes more, while the torso and legs change less, as shown in Figure 5. Therefore, we propose to add an attention mechanism to obtain the key areas of the action, so as to highlight the influence of the key areas on human action recognition. This paper proposes an attention-based representation of non-physically connected skeletal information, aiming to capture the key regions that have a greater impact on action recognition, and thus increase the efficiency of action recognition.

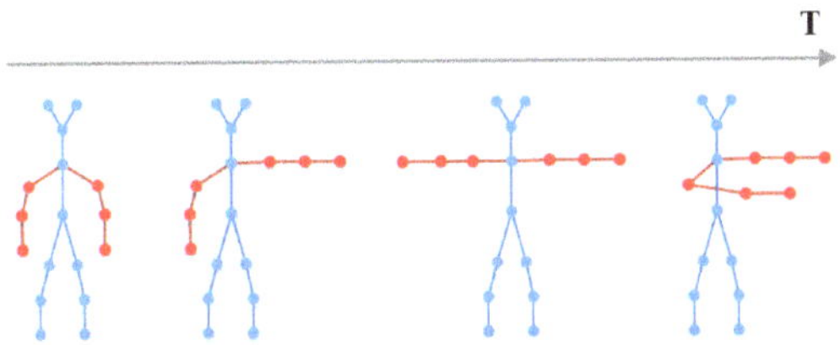

Figure 5. Illustration of the skeletal changes in traffic police gesture of 'go straight'. Red changes more, which indicates the key area, and blue changes less.

For non-physical connections, this paper introduces the concept of magnitude of change values to calculate the magnitude of change that occurs in the temporal sequence for each joint. Each joint has a magnitude of change value, which indicates the degree of influence that a particular joint has on human action recognition. Specifically, a joint is taken and its magnitude of change between the first and second frames is calculated using two-parametric numbers, where the joint is represented using xy-coordinates, such as (x_0, y_0), representing the two-dimensional coordinates of the joint. The magnitude of change of the joint between two consecutive frames is calculated as

$$d(v_i^t, v_i^{t+1}) = \sqrt{(v_i^{t+1} - v_i^t)^\top (v_i^{t+1} - v_i^t)} \tag{8}$$

where v_i^t denotes the coordinates of the i-th joint at time t and v_i^{t+1} denotes the coordinates of the i-th joint at time $t+1$. The value of the magnitude of change of the joint v_i over the whole time series is

$$d_{sum}^{v_i} = \sum_{t=1}^{T-1} d(v_i^t, v_i^{t+1}) \tag{9}$$

where T denotes the number of frames of the video data segmentation. After obtaining the change amplitude value of each joint over the whole time series, it is normalized and the normalization process of the change amplitude value of the joint over the whole time series is as follows:

$$d(v_i) = \frac{d_{sum}^{v_i}}{\sum_{i=1}^{I} d_{sum}^{v_i}} \tag{10}$$

where I indicates the number of human joints in a frame.

The association is modeled for joints that are more dependent but not directly physically connected in the human action, using a clustering algorithm that clusters joints according to the value of the magnitude of change of each joint. Joints of the same class in consecutive frames are connected by a hyperedge, and the joints in the same hyperedge have a similar degree of influence on the action. In general, for a given action, there is a dependency among the joints connected by the hyperedge, and this is used to construct a spatio-temporal hypergraph representing the association information among the non-physically connected joints in the different actions of the human body. Assuming that the clustering algorithm divides the joints into a total of Q classes, denoted by the set $\{\mathcal{V}_q\}_{q=1,2,\cdots,Q}$, the spatio-temporal hypergraph of the non-physically connected joint (N-PCJ) is denoted by $\mathcal{G}_{N-PCJ} = (\mathcal{V}_{N-PCJ}, \mathcal{E}_{N-PCJ}, \mathbf{W}_{N-PCJ})$, where $\mathcal{V}_{N-PCJ}$ denotes the set of vertices of this hypergraph, $\mathcal{E}_{N-PCJ}$ denotes the set of hyperedges of this hypergraph, and $\mathbf{W}_{N-PCJ}$ denotes the diagonal matrix of hyperedge weights in this hypergraph. Specifically, $\mathcal{V}_{N-PCJ}$ is the set of all joints $v_t^{(i)}$ in all frames; the elements in $\mathcal{E}_{N-PCJ}$ are denoted by $e_q = \{v_t^{(i)} | v_t^{(i)} \in \mathcal{V}_q\}$ and $\mathcal{E}_{N-PCJ} = \{e_q | q = 1, 2, \cdots, Q\}$. According to the steps in Section 4.2.1, the incidence matrix $\mathbf{H}_{N-PCJ}$ of the non-physically connected joint spatio-temporal hypergraph is obtained. After calculating the hyperedges and vertex degrees of this hypergraph, the Laplace matrix $\mathbf{G}_{N-PCJ}$ is generated by the Laplace transform. The process of constructing the spatio-temporal hypergraph of non-physically connected joint is shown in Figure 4b.

It is worth noting that, in the constructed non-physically connected joint hypergraph, the incidence matrix of the hypergraph is not a simple 0, 1 matrix; the value of each element in the matrix corresponds to the value of the magnitude of change (normalized value) of each joint or bone, and the elements in the incidence matrix $\mathbf{H}_{N-PCJ}$ of the non-physically connected joint spatio-temporal hypergraph are represented as follows:

$$h(v_t^{(i)}, e_q) = \begin{cases} d(v_i), & v_t^{(i)} \in e_q \\ 0, & v_t^{(i)} \notin e_q \end{cases}. \tag{11}$$

The normalized magnitude of change is used as the weight of a hypergraph vertex to represent the degree of influence of an articulation point or bone on human movement; the sum of the weights of all the vertices in the hypergraph is used as the weight of the hypergraph edge to differentiate the influence of different parts on action recognition.

4.2.3. Non-Physically Connected Bone Spatio-Temporal Hypergraph

For the non-physically connected bone (N-PCB) spatio-temporal hypergraph, the construction process is similar to that for the non-physically connected joint spatio-temporal hypergraph, as shown in Figure 4c. In our study, the bones of the human body are used as the vertices of the hypergraph, and bones with similar values of change in successive frames are connected by a single hyperedge, which characterizes both the spatial and temporal information of the human bones.

In our method, bones are represented using a vector representation, pointing from the source joint to the target joint, such as $(x_1, y_1) \rightarrow (x_2, y_2)$, and the magnitude of change of the bones between two consecutive frames is calculated as

$$d(\boldsymbol{b}_j^t, \boldsymbol{b}_j^{t+1}) = \sqrt{(\boldsymbol{b}_j^{t+1} - \boldsymbol{b}_j^t)^\top (\boldsymbol{b}_j^{t+1} - \boldsymbol{b}_j^t)} \tag{12}$$

where $\boldsymbol{b}_j^t$ denotes the coordinate representation of the j-th bone at moment t and $\boldsymbol{b}_j^{t+1}$ denotes the coordinate representation of the j-th bone at moment $t+1$. The value of the magnitude of change of the bone b_j over the whole time series is

$$d_{sum}^{b_j} = \sum_{t=1}^{T-1} d(\boldsymbol{b}_j^t, \boldsymbol{b}_j^{t+1}) \tag{13}$$

The process of normalizing the magnitude of change values of bones over the entire time series is as follows:

$$d(b_j) = \frac{d_{sum}^{b_j}}{\sum_{j=1}^{J} d_{sum}^{b_j}} \tag{14}$$

where J denotes the number of human bones in a frame.

The non-physically connected bone spatio-temporal hypergraph is denoted by $\mathcal{G}_{N-PCB} = (\mathcal{V}_{N-PCB}, \mathcal{E}_{N-PCB}, \mathbf{W}_{N-PCB})$, where $\mathcal{V}_{N-PCB}$ denotes the set of vertices of this hypergraph, $\mathcal{E}_{N-PCB}$ denotes the set of hyperedges of this hypergraph, and $\mathbf{W}_{N-PCB}$ denotes the diagonal matrix of hyperedge weights in this hypergraph. Similarly to the process of constructing a spatio-temporal hypergraph of non-physically connected joint, the incidence matrix $\mathbf{H}_{N-PCB}$ of this hypergraph is constructed and the Laplacian matrix $\mathbf{G}_{N-PCB}$ of this hypergraph is calculated.

The above process of constructing a hypergraph correlates the temporal and spatial relationships of the target objects. For spatial information, the hyperedge connects joints or bones with similar values of change in the same frame, characterizing the dependency among the joints or bones in the action; for temporal information, the hyperedge connects the same joints or bones in different frames, characterizing the temporal information of the joints or bones in the change of the action. In addition, the joints and bones of the human action are also considered, with the joints as the basic information of the human action and the bones as the complementary information, the modeling process fully reflects the completeness of the data. For the action recognition in complex scenes, the use of both joint and bone can obtain more effective information from the human skeleton than a single joint, which is beneficial to improve the accuracy of human action recognition. In addition, the modeling process characterizes the association among physical connections and the dependency among non-physical connections in the human skeleton, taking into account both the fixed human topology and the dependency among different joint and bone in the changing human action, and more accurately solving the problem of the inaccurate recognition of human action in complex scenes.

4.3. Adaptive Multi-Channel Spatio-Temporal Hypergraph Neural Network

After constructing the hypergraph, this paper proposes an adaptive multi-channel spatio-temporal hypergraph neural network for hypergraph learning, which consists of a hypergraph convolution module based on spatio-temporal information and a multi-feature fusion module based on cross-channel attention mechanism. Specifically, the spatio-temporal information-based hypergraph convolution module is used to iteratively update the features of each vertex to fully learn the spatio-temporal associations and the higher-order semantics of human joints and bones during motion; the multi-feature fusion module based on cross-channel attention mechanism is used to adaptively fuse features from different channels to maximize the use of rich information in each channel and to weaken conflicts and interferences among different channels. The module can also be used to improve the accuracy and robustness of action recognition.

4.3.1. Hypergraph Convolution Module Based on Spatio-Temporal Information

The hypergraph convolution module based on spatio-temporal information consists of a spatial convolution block, a temporal convolution block, and a global pooling layer, with a spatial convolution block and a temporal convolution block representing one layer in series, for a total of L layers, as shown in Figure 6. Among them, the spatial convolutional block is used to aggregate the semantic features of the hypergraph in the spatial dimension, and the temporal convolutional block is used to learn the temporal information of the hypergraph in the temporal dimension. Finally, a global pooling layer is connected to output the updated feature vector.

In order to better learn the temporal information of the human skeleton in the process of movement, the incidence matrix of the constructed hypergraph is split according to the temporal dimension in this paper, and the incidence matrix of the spatio-temporal hypergraph is dimensionally transformed, and the incidence matrix of the spatio-temporal hypergraph of each channel after the transformation is expressed as

$$\tilde{\mathbf{H}}_* = reshape(\mathbf{H}_*) = \{\mathbf{H}_*^{(t)} | t = 1, 2, \cdots, T\} \tag{15}$$

where $\tilde{\mathbf{H}}_*$ can be viewed as the subincidence matrix of $\mathbf{H}_*$, and $*$ denotes the different strategies for constructing the hypergraph (including PCJ, N-PCJ, and N-PCB). The Laplacian matrix $\mathbf{G}_*^{(t)}$ of each subincidence matrix is calculated according to the transformation process of Laplacian matrices in Section 4.2, and stitched according to the temporal dimension to obtain the 3D matrix $\tilde{\mathbf{G}}_*$, denoted by

$$\tilde{\mathbf{G}}_* = \{\mathbf{G}_*^{(t)} | t = 1, 2, \cdots, T\} \quad (16)$$

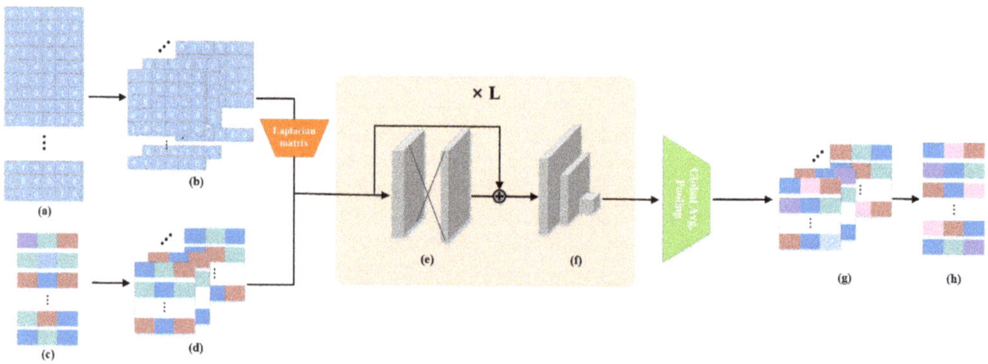

Figure 6. Hypergraph convolution module based on spatio-temporal information: (**a**) incidence matrix; (**b**) subincidence matrix; (**c**) initial features of hypergraph vertices; (**d**) initial features of hypergraph vertices after splitting by time dimension; (**e**) spatial convolution block; (**f**) temporal convolution block; (**g**) hypergraph vertex features after splitting by temporal dimension; and (**h**) trained hypergraph vertex features.

In addition, the initial joint features and bone features are also represented according to the temporal dimension. Specifically, the initial feature matrix of the joints in frame t is denoted by $\mathbf{V}_t = \{v_1^t, v_2^t, \cdots, v_I^t\}$, and the initial feature matrix of the joints in all frames is denoted by $\mathbf{V} = \{\mathbf{V}_t | t = 1, 2, \cdots, T\}$; the initial feature matrix of the bones in frame t is denoted by $\mathbf{B}_t = \{b_1^t, b_2^t, \cdots, b_J^t\}$, and the initial feature matrix of the bones in all frames is denoted by $\mathbf{B} = \{\mathbf{B}_t | t = 1, 2, \cdots, T\}$.

Taking the physically connected joint spatio-temporal hypergraph as an example, $\tilde{\mathbf{G}}_{PCJ}$ and $\mathbf{V}$ are first input to the spatial convolution block for spatial feature learning, and the vertex features after Laplacian matrix learning are weighted and fused with the original vertex features to form the input feature matrix for the l-th temporal convolution block:

$$\tilde{\mathbf{V}}^{(l)} = \sigma(\mathbf{V}^{(l)} + \tilde{\mathbf{G}}_{PCJ} \odot \mathbf{V}^{(l)} \mathbf{\Theta}^{(l)}) \quad (17)$$

where σ is the activation function, $\odot$ denotes the element multiplication, $\mathbf{V}^{(l)}$ denotes the input features of the l-th spatial convolution block, and $\mathbf{\Theta}^{(l)}$ denotes the learnable parameters. $\tilde{\mathbf{V}}^{(l)}$ is then fed into the temporal convolution block for training, inspired by [27], to obtain the new vertex feature matrix as follows:

$$\mathbf{V}^{(l+1)} = tanh(\Gamma_1^{(l)} \circledast \tilde{\mathbf{V}}^{(l)}) \odot \Phi(\Gamma_2^{(l)} \circledast \tilde{\mathbf{V}}^{(l)}) \quad (18)$$

where $\circledast$ is the convolution operation, $\Gamma_1^{(l)}$ and $\Gamma_2^{(l)}$ are two different learnable parameters in the l-th temporal convolution block, and $\Phi(\cdot)$ is the gating unit to control the utilization of historical information. After L iterations of updating, the vertex feature matrix $\mathbf{V}^{(L)}$ is obtained, and the final output feature matrix $\tilde{\mathbf{X}}_{PCJ}$ for this channel is obtained after a global pooling layer. Similarly, the output feature matrices $\tilde{\mathbf{X}}_{N-PCJ}$ and $\tilde{\mathbf{X}}_{N-PCB}$ for the other two channels are obtained.

4.3.2. Multi-Feature Fusion Module Based on Cross-Channel Attention Mechanism

After obtaining the spatio-temporal features of each channel, the features $\tilde{\mathbf{X}}_{PCJ}$, $\tilde{\mathbf{X}}_{N-PCJ}$ and $\tilde{\mathbf{X}}_{N-PCB}$ of the three channels are adaptively fused by a multi-feature fusion module based on the cross-channel attention mechanism, which fully considers the complementarity among the data features of different channels. The structure of the module is shown in Figure 7.

The module takes the spatio-temporal features of the respective channels, the incidence matrix of the spatio-temporal hypergraph, and the corresponding Laplacian matrix as input, and learns the spatio-temporal features of each channel once on the entire spatio-temporal hypergraph separately, using the Laplacian matrix while supplementing the original information in the incidence matrix to make full use of the incidence matrix and Laplacian matrix of the hypergraph to update the spatio-temporal features of each channel. The updated data features of each channel are represented as follows:

$$\mathbf{F}_{PCJ} = (\mathbf{H}_{PCJ}\mathbf{H}_{PCJ}^{\top} + \beta_1\mathbf{G}_{PCJ})\tilde{\mathbf{X}}_{PCJ} \tag{19}$$

$$\mathbf{F}_{N-PCJ} = (\mathbf{H}_{N-PCJ}\mathbf{H}_{N-PCJ}^{\top} + \beta_2\mathbf{G}_{N-PCJ})\tilde{\mathbf{X}}_{N-PCJ} \tag{20}$$

$$\mathbf{F}_{N-PCB} = (\mathbf{H}_{N-PCB}\mathbf{H}_{N-PCB}^{\top} + \beta_3\mathbf{G}_{N-PCB})\tilde{\mathbf{X}}_{N-PCB} \tag{21}$$

where $\mathbf{H}_*^{\top}$ denotes the transpose of the correlation matrix $\mathbf{H}_*$ and β_1, β_2, β_3 are learnable scalars.

Considering that different channel features have different degrees of influence on action recognition, in order to make full use of the correlation and complementarity among different channel features, this paper uses a cross-channel attention mechanism to calculate the attention score of each channel feature in human action, uses this score as the weight of each channel feature, and performs a weighted fusion of the features of each channel to adaptively obtain the final fused features. Specifically, first, the features $\mathbf{F}_*$ of each channel are compressed into a one-dimensional vector f_* for attention weighting using a one-dimensional convolutional layer, and the process is represented as

$$f_* = Conv1d(\mathbf{F}_*) \tag{22}$$

where $Conv1d$ denotes a one-dimensional convolutional layer. Next, the attention matrix between each of the two channel features is calculated separately, as follows:

$$\mathbf{A}_{PCJ,N-PCJ} = Softmax\left(\frac{(\mathbf{W}_{q_1}f_{PCJ})(\mathbf{W}_{k_1}f_{N-PCJ})^{\top}}{\sqrt{d}}\right) \tag{23}$$

$$\mathbf{A}_{N-PCJ,N-PCB} = Softmax\left(\frac{(\mathbf{W}_{q_2}f_{N-PCJ})(\mathbf{W}_{k_2}f_{N-PCB})^{\top}}{\sqrt{d}}\right) \tag{24}$$

$$\mathbf{A}_{PCJ,N-PCB} = Softmax\left(\frac{(\mathbf{W}_{q_3}f_{PCJ})(\mathbf{W}_{k_3}f_{N-PCB})^{\top}}{\sqrt{d}}\right) \tag{25}$$

where $\mathbf{W}_q$ and $\mathbf{W}_k$ are trainable parameter matrices, respectively, and d denotes the dimensionality of f_*. Then, the attention score of each channel feature is calculated separately according to the attention matrix, and the calculation process is represented as follows:

$$\alpha_{PCJ} = \sigma(\mathbf{A}_{PCJ,N-PCJ}\mathbf{W}_{v_1}f_{N-PCJ} + \mathbf{A}_{PCJ,N-PCB}\mathbf{W}_{v_1}f_{N-PCB}) \tag{26}$$

$$\alpha_{N-PCJ} = \sigma(\mathbf{A}_{N-PCJ,N-PCB}\mathbf{W}_{v_2}f_{N-PCB} + \mathbf{A}_{PCJ,N-PCJ}\mathbf{W}_{v_2}f_{PCJ}) \tag{27}$$

$$\alpha_{N-PCB} = \sigma(\mathbf{A}_{N-PCJ,N-PCB}\mathbf{W}_{v_3}f_{N-PCJ} + \mathbf{A}_{PCJ,N-PCB}\mathbf{W}_{v_3}f_{PCJ}) \tag{28}$$

where $\mathbf{W}_v$ is the trainable parameter matrix and σ is the activation function. It is worth noting that the attention score is used to adjust the contribution of different channel features to action recognition, thereby highlighting the features that have a greater impact on human action recognition. The attention score is then used to weight and sum the features of the three channels, and the adaptive fusion of the multi-channel features is represented as

$$\mathbf{F}_{fusion} = \alpha_{PCJ} \odot \mathbf{F}_{PCJ} + \alpha_{N-PCJ} \odot \mathbf{F}_{N-PCJ} + \alpha_{N-PCB} \odot \mathbf{F}_{N-PCB} \tag{29}$$

Finally, the fused features are globally pooled and passed through the fully connected layer before $Softmax$ is used to calculate the action prediction probability values, with the predicted category being the action category z with the highest probability value:

$$z = Softmax\left(FCL(GAP(\mathbf{F}_{fusion}))\right) \tag{30}$$

where GAP denotes the global average pooling operation and FCL denotes full connection layer.

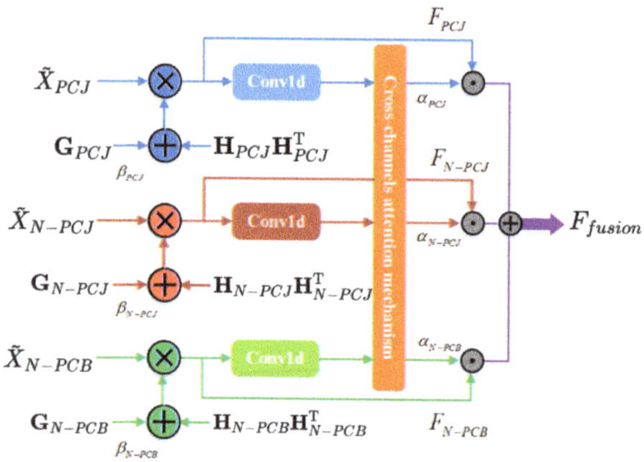

Figure 7. Multi-feature fusion module based on cross-channel attention mechanism.

The multi-feature fusion module based on cross-channel attention mechanism achieves the effective fusion of information features of each channel by learning the features of the three channels, reasonably using the diversity and complementarity of physically connected joints, non-physically connected joint, and non-physically connected bone in human action, weakening the interference of noise and effectively improving the accuracy of action recognition.

5. Experimental Results and Analysis

In this paper, we propose a framework for action recognition, which uses an adaptive multi-channel spatio-temporal hypergraph neural network (AM-STGNN) to learn hypergraphs based on human skeleton information to obtain higher-order semantic features of human actions for accurate action recognition. Then, we conduct experiments using two publicly available datasets, the NTU RGB+D 60 dataset [28] and the Kinetics-Skeleton dataset [29], respectively, and extensive experimental results show that our method achieves significant performance improvements in action recognition on both publicly available datasets. We will then present the datasets and relevant experimental details, compare the action recognition results of our proposed method with recent state-of-the-art methods, and perform ablation experiments and visualization analysis.

5.1. Dataset

NTU RGB+D 60: The dataset has 60 categories of actions, such as drinking, sitting down and standing up, with a total of 56,880 samples. These samples were performed by 40 volunteers and captured simultaneously by 3 Microsoft Kinect v2 cameras from different viewpoints. Each sample contains one action and is guaranteed to have a maximum of two objects. The dataset uses two benchmarks: (1) cross-subject (CS): training data from

20 volunteers and test data from 20 other volunteers; and (2) cross-view (CV): training data from camera views 2 and 3, test data from camera view 1.

Kinetics-skeleton: The dataset is a large-scale human action dataset containing 300,000 video clips in 400 categories. The video clips are derived from YouTube videos and are highly diverse. The dataset only provides raw video clips and does not include skeleton data. [30] used the publicly available OpenPose toolbox [31] to estimate the positions of 18 articulation points on each clip frame. Two individuals were selected for multiplayer clip cropping based on average joint point confidence. We used their published data to evaluate our model. The dataset was divided into a training set (240,000 fragments) and a validation set (20,000 fragments). The model was trained on the training set and top-1 and top-5 accuracies were reported on the validation set, following the evaluation method described in [30].

5.2. Experimental Settings

All experiments were performed on two 2080Ti GPUs with a batch size of 64, trained with the SGD optimization algorithm (momentum of 0.9), with weight decay set to 0.0001, using a cross-entropy loss function to back-propagate the gradient, and the hyperedge weight of the physically connected joint hypergraph set to 1 by default. In addition, considering that the number of bones is one less than the number of articulation points, an empty bone is added to make the number of bones consistent with the number of articulation points.

For the NTU RGB+D 60 dataset, the maximum number of frames in each sample is 300. For samples with less than 300 frames, we repeated these samples until 300 frames were reached. The learning rate was set to 0.1 and divided by 10 at rounds 30 and 40, with the training process ending at round 50. For the Kinetics-Skeleton dataset, the size of the Kinetics input tensor was the same as in [30], and 150 frames were randomly selected from the input skeleton sequence and processed by slightly perturbing the joint coordinates as well as rotation and translation operations. The learning rate was also set to 0.1 and divided by 10 at rounds 45 and 55. the training process ended at round 65.

5.3. Comparison with State-of-the-Art Methods

In order to verify the effectiveness of the proposed method, this paper compares it with other state-of-the-art methods. For the NTU RGB+D 60 dataset, experiments are conducted in both CS (cross-subject) protocols and CV (cross-view) protocols, and the experimental results are shown in Table 1. The methods used for comparison include handcraft-feature-based methods [32], RNN-based methods [33–35], CNN-based methods [10,36–38], GCN-based methods [30,39–41], and the HGNN-based method [42–45]. The experimental results show that the accuracy of our proposed method achieves 91.2% and 96.7% for action recognition in CS (cross-subject) protocols and CV (cross-view) protocols, respectively, and the accuracy of human action recognition is significantly improved, even compared with other hypergraph neural network-based methods. For the Kinetics-Skeleton dataset, the experimental results are shown in Table 1. The methods used for comparison include handcraft-feature-based methods [46], RNN-based methods [28], CNN-based methods [36], GCN-based methods [30,39,40,47] and HGNN-based methods [42–45]. The experimental results show that the accuracy of our proposed method is 39.1% and 61.3% in Top-1 and Top-5, respectively. Compared with other methods based on skeleton, our model achieves the state-of-the-art. From the above experimental results, it is easy to see that our proposed method can achieve a better performance by constructing hypergraphs from skeleton information in human actions. Thus, capturing higher-order correlations among skeletons can extract richer information, which can further improve the performance of action recognition.

Table 1. Comparison of classification accuracy with state-of-the-art methods on the NTU RGB+D 60 dataset and the Kinetics-Skeleton dataset.

Type	Method	NTU RGB+D 60		Kinetics-Skeleton	
		CS(%)	CV(%)	TOP-1(%)	TOP-5(%)
Handcraft feature based	Lie Group [32]	50.1	82.5	-	-
	Feature Enc [46]	-	-	14.9	25.8
RNN based	ST-LSTM [33]	69.2	77.7	-	-
	VA-LSTM [34]	79.4	87.6	-	-
	AGC-LSTM [35]	89.2	95.0	-	-
	Deep LSTM [28]	-	-	16.4	35.3
CNN based	TCN [36]	74.3	83.1	20.3	40.0
	HCN [37]	86.5	91.1	-	-
	VA-CNN [38]	88.7	94.3	-	-
	Ta-CNN+ [10]	90.7	95.1	-	-
GCN based	ST-GCN [39]	81.5	88.3	30.7	52.8
	2S-AGCN [30]	88.5	95.1	36.1	58.7
	MS-AAGCN [40]	90.0	96.2	37.8	61.0
	Shift-GCN [41]	90.7	96.5	-	-
	Sym-GNN [47]	-	-	37.2	58.1
HGNN based	Hyper-GCN(3S) [42]	89.5	95.7	37.1	60.0
	DHGCN [43]	90.7	96.0	37.7	60.6
	Selective-HCN [44]	90.8	96.6	38.0	61.1
	SD-HGCN [45]	90.9	96.7	37.4	60.5
	Ours	91.2	96.7	39.1	61.3

5.4. Ablation Experiments

In order to verify the validity of the different modules in the proposed network model in this paper, without loss of generality, we conducted a series of ablation experiments on the CS (cross-subject) protocols and CV (cross-view) protocols of the NTU RGB+D 60 dataset. In this subsection, we will analyze whether the modules in the proposed model affect the performance of action recognition in order to verify the plausibility of the network model.

5.4.1. Different Channel Data Channels

The influence of data about joints and bones on final experimental results. On the NTU RGB+D 60 dataset, we compared using only the non-physically connected joint hypergraph as model input, using only the non-physically connected bone hypergraph as model input using both the non-physically connected joint hypergraph and the non-physically connected bone hypergraph as model input, and the experimental results are shown in Table 2. The results show that the action recognition accuracy of using both the non-physically connected joint hypergraph and the non-physically connected bone hypergraph as model input was 90.3% and 95.2% for the CS (cross-subject) protocols and CV (cross-view) protocols, respectively, which were better than that of only using the joint or bone. This shows that the use of multi-channel data containing both joints and bones can provide richer feature information than single-channel data only containing joints or bones, thus enhancing the perceptual capability of the model and achieving a complete representation of the data, thus improving the accuracy of action recognition.

The influence of different hypergraph construction methods on experimental results. In the NTU RGB+D 60 dataset, where we compared only using physically connected joint hypergraphs as model input, only using non-physically connected joint hypergraphs as model input, and using both non-physically connected joint hypergraphs and physically connected joint hypergraphs as model input. The experimental results for this are shown in

Table 2. The experimental results show that the action recognition accuracy of using both non-physically connected joint hypergraphs and physically connected joint hypergraphs as model inputs was 90.7% and 95.4% for CS (cross-subject) protocols and CV (cross-view) protocols, respectively, both of which were better than that of using only non-physically connected or physically connected joints. This shows that both the physical connection inherent in the human skeleton and the non-physical connection that are more relevant during movement can provide effective information for action recognition, and their simultaneous use can improve the learning ability of skeletal features in different movements.

Table 2. Comparison of action recognition accuracy of different hypergraph construction methods and different input data information.

Hypergraph Construction Method			Input Data		NTU RGB+D 60	
PCJ	N-PCJ	N-PCB	Joint	Bone	CS(%)	CV(%)
√			√		88.6	94.0
	√		√		88.7	93.8
		√		√	89.2	94.2
√	√		√		90.7	95.4
	√	√	√	√	90.3	95.2
√	√	√	√	√	91.7	96.7

5.4.2. Different Neural Network Structures

To verify the learning effect of the spatio-temporal hypergraph neural network, we conducted experiments on different hypergraph neural networks on CS (cross-subject) protocols and CV (cross-view) protocols of the NTU RGB+D 60 dataset, respectively. As shown in Table 3, for CS (cross-subject) protocols, the accuracy of action recognition was 81.2% when only using spatial convolutional blocks in the hypergraph neural network; 82.1% when using only temporal hypergraph convolutional blocks in the hypergraph neural network; and 91.2% when using both temporal and spatial hypergraph convolutional blocks. Compared to the former, the accuracy of action recognition was improved by about 10% and 9.1%, respectively. For CV (cross-view) protocols, the accuracy of action recognition was 86.3% when only using spatial convolutional blocks in the hypergraph neural network, 87.7% when only using temporal hypergraph convolutional blocks in the hypergraph neural network, and 96.7% when using both temporal and spatial hypergraph convolutional blocks, with an improvement of about 10.4% and 9%, respectively. This shows that the spatio-temporal hypergraph neural network can further improve the mining ability of temporal and spatial information in human actions, and achieve the purpose of improving the accuracy and model generalization ability.

Table 3. Comparison of the action recognition accuracy of different structural neural networks.

Method	NTU RGB+D 60	
	CS(%)	CV(%)
Only temporal convolution blocks	81.2	86.3
Only spatial convolution blocks	82.1	87.7
Temporal convolution blocks + spatial convolution blocks	91.2	96.7

5.4.3. Different Multi-Channel Fusion Modules

To verify the effect of multi-feature fusion module based on the cross-channel attention mechanism, we compared the module with the traditional natural splicing approach on the NTU RGB+D 60 dataset, and the experimental results are shown in Table 4. The experimental results show that, when a multi-feature fusion module based on a cross-channel attention mechanism is not used and the simple coequal fusion method is used to fuse multi-channel features, the accuracy of action recognition on CS (cross-subject) protocols and CV (cross-view) protocols is 88.4% and 93.2%, respectively; when the multi-feature fusion module based on cross-channel attention mechanism is used, the accuracy of action recognition on CS (cross-subject) protocols and CV (cross-view) protocols is 91.2% and 96.7%, respectively, an improvement of 2.8% and 3.5% compared to the former. This shows that although the introduction of higher-order correlations can improve the action recognition performance, the traditional coequal fusion method does not yield the best combination of multi-channel feature performance, suggesting that different higher-order correlations have different effects on representation learning. Therefore, this paper correlates each channel feature with an attention score, thus adaptively identifying the importance of different channel features to the overall action recognition and making better use of the complementary representations of multi-channel features, which helps improve the accuracy of action recognition.

Table 4. Comparison of action recognition accuracy of different ways of fusing multi-channel features.

Method	NTU RGB+D 60	
	CS(%)	CV(%)
Coequal fusion method	88.4	93.2
Multi-feature fusion module based on cross-channel attention mechanism	91.2	96.7

5.5. Visualization

In order to better illustrate the need for modeling non-physical connections, this section provides visualizations for some samples on the NTU RGB+D 60 dataset. From Figure 8, it can be seen that the action of 'drinking water' is mainly an upper limb movement, and the upper limb should be given more attention in the hypergraph. Meanwhile, it can also be observed that when drinking water, the arm and the head have a strong correlation, but they are not directly physically connected. In conclusion, it is shown that constructing a non-physically connected hypergraph can capture more complex and higher-order dependence among the joints and further improve the recognition accuracy of the model. In addition, the proposed method performs well in both CS (cross-subject) protocols and CV (cross-view) protocols, indicating that the proposed method has excellent robustness.

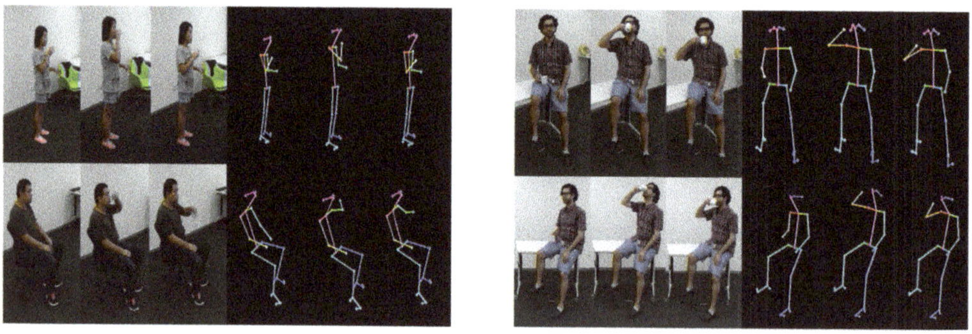

(a) (b)

Figure 8. Visual examples on the NTU RGB + D dataset: (**a**) drink water in CS (cross-subject) protocols; and (**b**) drink water in CV (cross-view) protocols.

The NTU RGB+D 60 dataset contains common human actions such as drinking water, eating a meal, etc., and we analyze the accuracy of the action recognition for each category on this dataset. In order to visualize the improvement in the accuracy on different actions and to analyze the causes of misclassified actions, we plot bar charts based on the accuracy of 2S-AGCN and our proposed method on the NTU RGB+D 60 dataset for 60 action categories. As shown in Figure 9, the blue bars indicate the performance of 2S-AGCN, and the orange bars and the numbers on them indicate the performance improvement of our proposed method compared to using 2S-AGCN. Observation reveals that our proposed method has some performance improvement over 2S-AGCN for the vast majority of action categories. Figure 9a shows the CS (cross-subject) protocols of the NTU RGB+D 60 dataset. We can observe that the recognition accuracy of 'write' and 'read' is lower compared to other actions when using 2S-AGCN. For these two actions, the recognition accuracy is improved by about 13.9% and 4.4%, respectively, when using our proposed method. Figure 9b shows the CV (cross-view) protocols of the NTU RGB+D 60 dataset. We found that the recognition accuracy of some action, such as 'writing', 'reading', 'playing with the phone', and 'typing on keyboard' is lower when using 2S-AGCN. The accuracy of action recognition is improved in all cases after using our proposed method. From the above experimental results, it can be seen that, compared to 2S-AGC, there is a large performance improvement for actions such as 'writing', 'reading', etc., which only use part of the human body region. This is due to the fact that the proposed method can focus on regions with large variation and focus on the effective information of the action.

In addition, in order to comprehensively evaluate the model, it is necessary to consider not only the accuracy but also the convergence speed. To better demonstrate the comparative performance with other methods, we conducted accuracy tests on the NTU RGB+D 60 dataset using the CS (cross-subject) and CV (cross-view) protocols, and generated corresponding charts. As shown in Figure 10a,b, it can be observed that, as the number of epochs increases, all curves tend to reach their maximum values and stabilize. Our proposed method exhibits the fastest convergence speed, while Hyper-GNN, 2S-AGCN, AS-GCN, and ST-GCN show slower convergence rates. This indicates that our proposed network can achieve optimal results at the fastest speed compared to other methods.

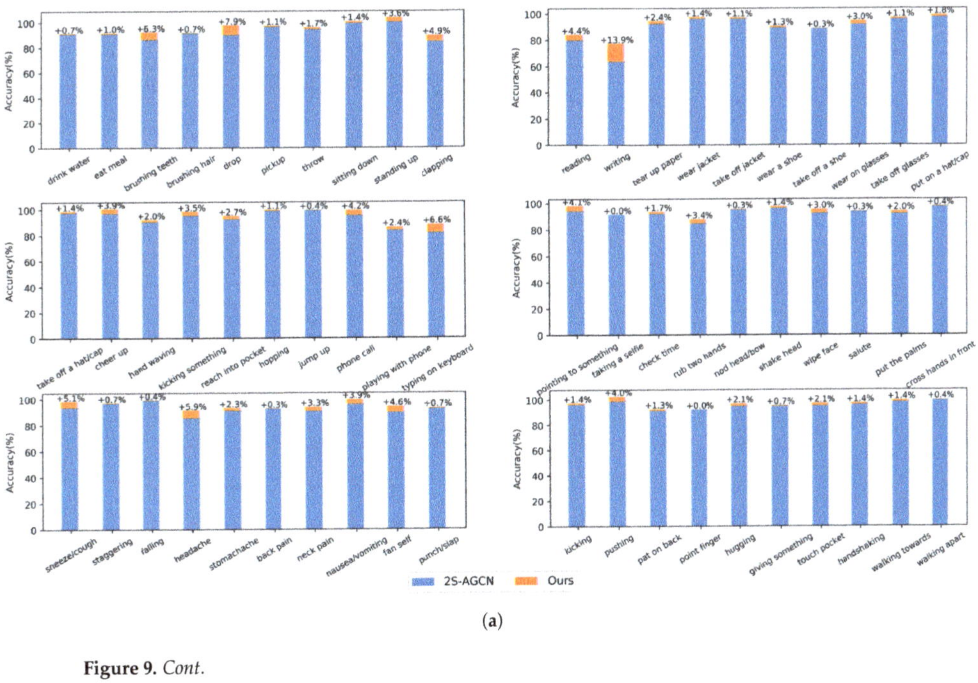

(a)

Figure 9. *Cont.*

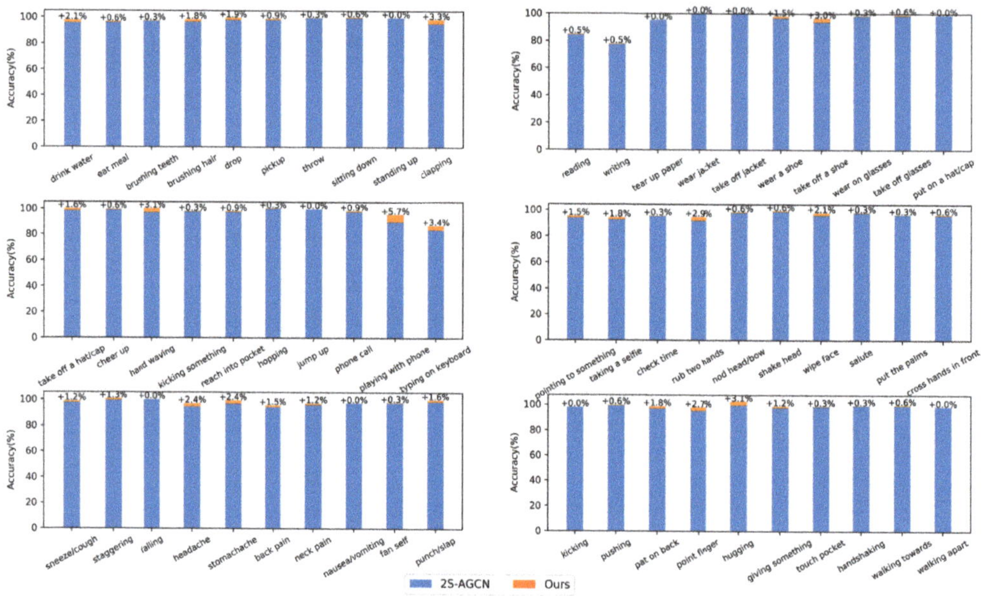

(b)

Figure 9. Accuracy in each action category of our method and 2S-AGCN on the NTU RGB+D 60 dataset. The horizontal coordinates indicate different action categories and the vertical coordinates indicate the accuracies of different actions: (**a**) CS (cross-subject) protocols; (**b**) CV (cross-view) protocols.

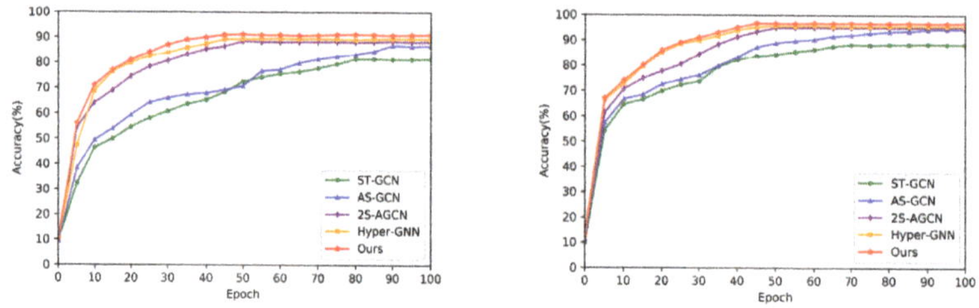

(c) (d)

Figure 10. Classification accuracy of different methods on the NTU RGB+D 60 dataset. The x axis represents Epoch in (**a**,**b**), and the y axis represents accuracy.

6. Conclusions

In this paper, we explore the higher-order correlations of skeleton-based action recognition in physical and non-physical connections. For human actions, the higher-order correlations of skeleton sequences is characterized using hypergraphs to model the physical and non-physical connections of the skeleton before forming different data. Then, AMC-STHGNN effectively fuses the features of each channel, reasonably using the diversity and complementarity of physically connected joints, non-physically connected joints, and non-physically connected bone in human action, weakening the interference of noise and effectively improving the accuracy of action recognition. Experimental results using the NTU RGB+D 60 dataset and the Kinetics-Skeleton dataset demonstrate the superior accuracy and robustness of the proposed model compared to mainstream human action recognition methods.

Future research directions include exploring lightweight models to improve the network computation and real-time performance. The optimization of the hypergraph structure and reduction in noise impact on hypergraph modeling are also important areas of focus. Moreover, the application of hypergraph neural network-based action recognition methods in unmanned fields, such as self-driving vehicles and interactive wheeled robots, shows promising potential [48,49].

Author Contributions: Conceptualization, C.W.; methodology, C.W. and Z.W.; validation, C.W. and Z.W.; investigation, C.W.; data curation, C.W.; writing—original draft preparation, C.W., N.M. and Z.W.; writing—review and editing, C.W. and N.M.; visualization, Z.W.; supervision, N.M.; project administration, N.M.; funding acquisition, N.M. All authors have read and agreed to the published version of the manuscript.

Funding: This work was supported by the National Key R&D Program of China (No.2023YFF0615800), the Beijing Natural Science Foundation (No. 4222025), the National Natural Science Foundation of China (No. 61931012), and QIYUAN LAB Innovation Foundation (Innovation Research) Project (No. S20210201107).

Institutional Review Board Statement: Not applicable.

Informed Consent Statement: Not applicable.

Data Availability Statement: NTU RGB+D 60 dataset was analyzed in this study. This data can be found here: [https://rose1.ntu.edu.sg/dataset/actionRecognition/]. Kinetics-skeleton was analyzed in this study. This data can be found here: https://github.com/activitynet/ActivityNet/tree/master/Crawler/Kinetics.

Conflicts of Interest: The authors declare no conflicts of interest.

References

1. Ma, N.; Wu, Z.; Cheung, Y.m.; Guo, Y.; Gao, Y.; Li, J.; Jiang, B. A survey of human action recognition and posture prediction. *Tsinghua Sci. Technol.* **2022**, *27*, 973–1001. [CrossRef]
2. Zhang, R.; Jiang, C.; Wu, S.; Zhou, Q.; Jing, X.; Mu, J. Wi-Fi sensing for joint gesture recognition and human identification from few samples in human-computer interaction. *IEEE J. Sel. Areas Commun.* **2022**, *40*, 2193–2205. [CrossRef]
3. Dallel, M.; Havard, V.; Dupuis, Y.; Baudry, D. Digital twin of an industrial workstation: A novel method of an auto-labeled data generator using virtual reality for human action recognition in the context of human–robot collaboration. *Eng. Appl. Artif. Intell.* **2023**, *118*, 105655. [CrossRef]
4. Mabrouk, A.B.; Zagrouba, E. Abnormal behavior recognition for intelligent video surveillance systems: A review. *Expert Syst. Appl.* **2018**, *91*, 480–491. [CrossRef]
5. Wu, Z.; Ma, N.; Gao, Y.; Li, J.; Xu, X.; Yao, Y.; Chen, L. Attention Mechanism Based on Improved Spatial-Temporal Convolutional Neural Networks for Traffic Police Gesture Recognition. *Int. J. Pattern Recognit. Artif.* **2022**, *36*, 2256001. [CrossRef]
6. Xu, B.; Shu, X.; Zhang, J.; Dai, G.; Song, Y. Spatiotemporal Decouple-and-Squeeze Contrastive Learning for Semisupervised Skeleton-Based Action Recognition. *IEEE Trans. Neural Netw. Learn. Syst.* **2023**. [CrossRef]
7. Dai, M.; Sun, Z.; Wang, T.; Feng, J.; Jia, K. Global spatio-temporal synergistic topology learning for skeleton-based action recognition. *Pattern Recognit.* **2023**, *140*, 109540. [CrossRef]
8. Gao, Y.; Zhang, Z.; Lin, H.; Zhao, X.; Du, S.; Zou, C. Hypergraph Learning: Methods and Practices. *IEEE Trans. Pattern Anal. Mach. Intell.* **2022**, *44*, 2548–2566. [CrossRef] [PubMed]
9. Wang, C.; Ma, N.; Wu, Z.; Zhang, J.; Yao, Y. Survey of Hypergraph Neural Networks and Its Application to Action Recognition. In Proceedings of the CAAI International Conference on Artificial Intelligence, Beijing, China, 27–28 August 2022; pp. 387–398.
10. Xu, K.; Ye, F.; Zhong, Q.; Xie, D. Topology-aware convolutional neural network for efficient skeleton-based action recognition. In Proceedings of the AAAI Conference on Artificial Intelligence, Vancouver, BC, Canada, 28 February–1 March 2022; Volume 36, pp. 2866–2874.
11. Andrade-Ambriz, Y.A.; Ledesma, S.; Ibarra-Manzano, M.A.; Oros-Flores, M.I.; Almanza-Ojeda, D.L. Human activity recognition using temporal convolutional neural network architecture. *Expert Syst. Appl.* **2022**, *191*, 116287. [CrossRef]
12. Chen, Y.; Zhang, Z.; Yuan, C.; Li, B.; Deng, Y.; Hu, W. Channel-wise topology refinement graph convolution for skeleton-based action recognition. In Proceedings of the IEEE/CVF International Conference on Computer Vision, Virtual, 11–17 October 2021; pp. 13359–13368.
13. Song, Y.F.; Zhang, Z.; Shan, C.; Wang, L. Constructing Stronger and Faster Baselines for Skeleton-Based Action Recognition. *IEEE Trans. Pattern Anal. Mach. Intell.* **2023**, *45*, 1474–1488. [CrossRef] [PubMed]
14. Feng, Y.; You, H.; Zhang, Z.; Ji, R.; Gao, Y. Hypergraph neural networks. In Proceedings of the AAAI Conference on Artificial Intelligence, Honolulu, HA, USA, 29–31 January 2019; Volume 33, pp. 3558–3565.

15. Jiang, J.; Wei, Y.; Feng, Y.; Cao, J.; Gao, Y. Dynamic Hypergraph Neural Networks. In Proceedings of the IJCAI, Macao, China, 10–16 August 2019; pp. 2635–2641.
16. Bai, S.; Zhang, F.; Torr, P.H. Hypergraph convolution and hypergraph attention. *Pattern Recognit.* **2021**, *110*, 107637. [CrossRef]
17. Gao, Y.; Feng, Y.; Ji, S.; Ji, R. HGNN+: General Hypergraph Neural Networks. *IEEE Trans. Pattern Anal. Mach. Intell.* **2023**, *45*, 3181–3199. [CrossRef] [PubMed]
18. Zhang, Z.; Lin, H.; Zhao, X.; Ji, R.; Gao, Y. Inductive Multi-Hypergraph Learning and Its Application on View-Based 3D Object Classification. *IEEE Trans. Image Process.* **2018**, *27*, 5957–5968. [CrossRef] [PubMed]
19. Shi, H.; Zhang, Y.; Zhang, Z.; Ma, N.; Zhao, X.; Gao, Y.; Sun, J. Hypergraph-induced convolutional networks for visual classification. *IEEE Trans. Neural Netw. Learn. Syst.* **2018**, *30*, 2963–2972. [CrossRef] [PubMed]
20. Chen, Y.; Li, Y.; Zhang, C.; Zhou, H.; Luo, Y.; Hu, C. Informed Patch Enhanced HyperGCN for skeleton-based action recognition. *Inf. Process. Manag.* **2022**, *59*, 102950. [CrossRef]
21. Wang, S.; Zhang, Y.; Qi, H.; Zhao, M.; Jiang, Y. Dynamic Spatial-temporal Hypergraph Convolutional Network for Skeleton-based Action Recognition. *arXiv* **2023**, arXiv:2302.08689.
22. Zhou, Y.; Li, C.; Cheng, Z.Q.; Geng, Y.; Xie, X.; Keuper, M. Hypergraph transformer for skeleton-based action recognition. *arXiv* **2022**, arXiv:2211.09590.
23. Wang, M.; Liu, X.; Wu, X. Visual classification by ℓ_1-hypergraph modeling. *IEEE Trans. Knowl. Data Eng.* **2015**, *27*, 2564–2574. [CrossRef]
24. Zhang, Z.; Feng, Y.; Ying, S.; Gao, Y. Deep Hypergraph Structure Learning. *arXiv* **2022**, arXiv:2208.12547.
25. Zhou, D.; Huang, J.; Schölkopf, B. Learning with hypergraphs: Clustering, classification, and embedding. *Adv. Neural Inf. Process. Syst.* **2006**, *19*, 1601–1608.
26. Lu, J.; Wan, H.; Li, P.; Zhao, X.; Ma, N.; Gao, Y. Exploring High-Order Spatio–Temporal Correlations From Skeleton for Person Re-Identification. *IEEE Trans. Image Process.* **2023**, *32*, 949–963. [CrossRef]
27. He, B.; Guan, Y.; Dai, R. Convolutional gated recurrent units for medical relation classification. In Proceedings of the 2018 IEEE International Conference on Bioinformatics and Biomedicine (BIBM), Madrid, Spain , 3–6 December 2018; pp. 646–650. [CrossRef]
28. Shahroudy, A.; Liu, J.; Ng, T.T.; Wang, G. Ntu rgb+ d: A large scale dataset for 3d human activity analysis. In Proceedings of the IEEE Conference on Computer Vision and Pattern Recognition, Las Vegas, NV, USA, 27–30 June 2016; pp. 1010–1019.
29. Kay, W.; Carreira, J.; Simonyan, K.; Zhang, B.; Hillier, C.; Vijayanarasimhan, S.; Viola, F.; Green, T.; Back, T.; Natsev, P.; et al. The kinetics human action video dataset. *arXiv* **2017**, arXiv:1705.06950.
30. Shi, L.; Zhang, Y.; Cheng, J.; Lu, H. Two-stream adaptive graph convolutional networks for skeleton-based action recognition. In Proceedings of the IEEE/CVF Conference on Computer Vision and Pattern Recognition, Long Beach, CA, USA, 16–20 June 2019; pp. 12026–12035.
31. Cao, Z.; Simon, T.; Wei, S.E.; Sheikh, Y. Realtime multi-person 2d pose estimation using part affinity fields. In Proceedings of the IEEE Conference on Computer Vision and Pattern Recognition, Honolulu, HA, USA, 21–26 July 2017; pp. 7291–7299.
32. Vemulapalli, R.; Arrate, F.; Chellappa, R. Human action recognition by representing 3d skeletons as points in a lie group. In Proceedings of the IEEE Conference on Computer Vision and Pattern Recognition, Columbus, OH, USA, 23–28 June 2014; pp. 588–595.
33. Liu, J.; Shahroudy, A.; Xu, D.; Wang, G. Spatio-temporal lstm with trust gates for 3d human action recognition. In Proceedings of the Computer Vision–ECCV 2016: 14th European Conference, Amsterdam, The Netherlands, 11–14 October 2016; pp. 816–833.
34. Zhang, P.; Lan, C.; Xing, J.; Zeng, W.; Xue, J.; Zheng, N. View adaptive recurrent neural networks for high performance human action recognition from skeleton data. In Proceedings of the IEEE International Conference on Computer Vision, Venice, Italy, 22–29 October 2017; pp. 2117–2126.
35. Si, C.; Chen, W.; Wang, W.; Wang, L.; Tan, T. An attention enhanced graph convolutional lstm network for skeleton-based action recognition. In Proceedings of the IEEE/CVF Conference on Computer Vision and Pattern Recognition, Long Beach, CA, USA, 15–20 June 2019; pp. 1227–1236.
36. Soo Kim, T.; Reiter, A. Interpretable 3d human action analysis with temporal convolutional networks. In Proceedings of the IEEE Conference on Computer Vision and Pattern Recognition Workshops, Honolulu, HI, USA, 21–26 July 2017; pp. 20–28.
37. Li, C.; Zhong, Q.; Xie, D.; Pu, S. Co-occurrence feature learning from skeleton data for action recognition and detection with hierarchical aggregation. *arXiv* **2018**, arXiv:1804.06055.
38. Zhang, P.; Lan, C.; Xing, J.; Zeng, W.; Xue, J.; Zheng, N. View adaptive neural networks for high performance skeleton-based human action recognition. *IEEE Trans. Pattern Anal. Mach. Intell.* **2019**, *41*, 1963–1978. [CrossRef]
39. Yan, S.; Xiong, Y.; Lin, D. Spatial temporal graph convolutional networks for skeleton-based action recognition. In Proceedings of the AAAI Conference on Artificial Intelligence, New Orleans, LA, USA, 2–7 February 2018; Volume 32.
40. Shi, L.; Zhang, Y.; Cheng, J.; Lu, H. Skeleton-based action recognition with multi-stream adaptive graph convolutional networks. *IEEE Trans. Image Process.* **2020**, *29*, 9532–9545. [CrossRef]
41. Cheng, K.; Zhang, Y.; He, X.; Chen, W.; Cheng, J.; Lu, H. Skeleton-based action recognition with shift graph convolutional network. In Proceedings of the IEEE/CVF Conference on Computer Vision and Pattern Recognition, Seattle, WA, USA, 14–19 June 2020; pp. 183–192.
42. Hao, X.; Li, J.; Guo, Y.; Jiang, T.; Yu, M. Hypergraph neural network for skeleton-based action recognition. *IEEE Trans. Image Process.* **2021**, *30*, 2263–2275. [CrossRef] [PubMed]

43. Wei, J.; Wang, Y.; Guo, M.; Lv, P.; Yang, X.; Xu, M. Dynamic hypergraph convolutional networks for skeleton-based action recognition. *arXiv* **2021**, arXiv:2112.10570.
44. Zhu, Y.; Huang, G.; Xu, X.; Ji, Y.; Shen, F. Selective hypergraph convolutional networks for skeleton-based action recognition. In Proceedings of the 2022 International Conference on Multimedia Retrieval, Newark, NJ, USA, 27–30 June 2022; pp. 518–526.
45. He, C.; Xiao, C.; Liu, S.; Qin, X.; Zhao, Y.; Zhang, X. Single-skeleton and dual-skeleton hypergraph convolution neural networks for skeleton-based action recognition. In Proceedings of the Neural Information Processing: 28th International Conference, ICONIP 2021, Sanur, Bali, Indonesia, 8–12 December 2021; pp. 15–27.
46. Fernando, B.; Gavves, E.; Oramas, J.M.; Ghodrati, A.; Tuytelaars, T. Modeling video evolution for action recognition. In Proceedings of the IEEE Conference on Computer Vision and Pattern Recognition, Boston, MA, USA, 7–12 June 2015; pp. 5378–5387.
47. Li, M.; Chen, S.; Chen, X.; Zhang, Y.; Wang, Y.; Tian, Q. Symbiotic graph neural networks for 3d skeleton-based human action recognition and motion prediction. *IEEE Trans. Pattern Anal. Mach. Intell.* **2021**, *44*, 3316–3333. [CrossRef] [PubMed]
48. Ma, N.; Li, D.; He, W.; Deng, Y.; Li, J.; Gao, Y.; Bao, H.; Zhang, H.; Xu, X.; Liu, Y.; et al. Future vehicles: Interactive wheeled robots. *Sci. China Inf. Sci.* **2021**, *64*, 156101. [CrossRef]
49. Li, D.; Ma, N.; Gao, Y. Future vehicles: Learnable wheeled robots. *Sci. China Inf. Sci.* **2020**, *63*, 193201. [CrossRef]

applied sciences

MDPI

Article

RL-Based Sim2Real Enhancements for Autonomous Beach-Cleaning Agents

Francisco Quiroga [1], Gabriel Hermosilla [1,*], German Varas [2], Francisco Alonso [1] and Karla Schröder [1]

[1] Escuela de Ingeniería Eléctrica, Pontificia Universidad Católica de Valparaíso (PUCV), Valparaíso 2340025, Chile; francisco.quiroga.a@mail.pucv.cl (F.Q.); francisco.alonso@pucv.cl (F.A.); karla.schroder@pucv.cl (K.S.)

[2] Instituto de Física, Pontificia Universidad Católica de Valparaíso (PUCV), Valparaíso 2340025, Chile; german.varas@pucv.cl

* Correspondence: gabriel.hermosilla@pucv.cl

Abstract: This paper explores the application of Deep Reinforcement Learning (DRL) and Sim2Real strategies to enhance the autonomy of beach-cleaning robots. Experiments demonstrate that DRL agents, initially refined in simulations, effectively transfer their navigation skills to real-world scenarios, achieving precise and efficient operation in complex natural environments. This method provides a scalable and effective solution for beach conservation, establishing a significant precedent for the use of autonomous robots in environmental management. The key advancements include the ability of robots to adhere to predefined routes and dynamically avoid obstacles. Additionally, a newly developed platform validates the Sim2Real strategy, proving its capability to bridge the gap between simulated training and practical application, thus offering a robust methodology for addressing real-life environmental challenges.

Keywords: deep reinforcement learning; mobile robotics; position control; obstacle avoidance; simulated environment; real laboratory validation; beach cleaning; Sim2Real

1. Introduction

The degradation of coastal ecosystems due to pollution is a pressing issue, particularly on beaches where waste accumulation adversely impacts both human and marine life [1,2]. In Chile, the severity of this issue is compounded by inadequate waste management practices, leading to significant shoreline pollution [3]. This paper introduces the deployment of autonomous beach-cleaning robots, equipped with advanced artificial intelligence, as a novel solution to mitigate these effects and contribute significantly to marine conservation.

The motivation for this work stems from the urgent need to preserve marine biodiversity and improve the health of coastal ecosystems. Previous efforts to tackle beach pollution have primarily focused on manual cleanup operations, which are often inefficient and unsustainable. Moreover, existing robotic technologies have not yet fully adapted to the dynamic and complex nature of beach environments, where fluctuating conditions present substantial operational challenges.

Building upon existing knowledge in the fields of robotics and artificial intelligence, this paper employs DRL, an advanced AI technique that integrates deep learning with reinforcement learning principles, to develop effective control systems for autonomous robots [4,5]. Through extensive training in simulated environments, these robots master essential skills for navigating complex terrains, avoiding obstacles, and maintaining precise position control, crucial for their effective operation in real-world scenarios [6].

Utilizing the Khepera IV robot within the CoppeliaSim environment, this paper pioneers a systematic approach to controller design. Robots are trained in a Gym-simulated environment, carefully designed to mirror real-world conditions. This methodology ensures a seamless transition from simulation to actual deployment, with the robots demonstrating robustness and adaptability in their behaviors.

The contributions of this work are twofold: it establishes a robust framework for developing autonomous systems capable of efficiently operating in unpredictable beach environments, and it sets a benchmark in the methodology for transitioning from simulated to real-world applications, enhancing the practical applicability of reinforcement learning in environmental conservation efforts.

In conclusion, by implementing this innovative technology the issues of beach pollution are addressed effectively, establishing a precedent for the application of advanced AI in environmental protection and setting the stage for future initiatives in autonomous robotic systems for ecological restoration.

2. State of Art

The issue of beach cleaning represents a significant environmental challenge, where traditional solutions have demonstrated clear limitations [7,8]. Despite the prevalence of conventional methods such as the use of shovels and tractorized systems, their effectiveness is compromised in the face of increasingly complex challenges, such as escalating pollution and the urgent need to adopt more sustainable and efficient approaches [9].

The application of autonomous robotics in beach cleaning represents a developing area of research, still in its early stages but showing significant potential for advancement [10,11]. Traditional beach-cleaning robots, which are predominantly large-scale [12], often require human supervision and/or control, limiting their autonomy and operational efficiency [13–15]. However, recent advancements in DRL and simulation technologies have introduced new opportunities for improving these systems. Studies employing DRL agents, such as DDPG and DQN, have been conducted solely in simulation [16], demonstrating DRL's capability to train agents that can navigate and operate in complex [17,18] and dynamic [6,19] environments, typical of coastal settings. The success of DRL in autonomous vehicles [20,21] and drones [22–24] highlights its potential for enhancing autonomous navigation and real-time decision-making, which are crucial for efficient beach cleaning. Integrating Imitation Learning (IL) with RL could offer a robust strategy, where specific routes and cleaning tasks are predefined [25].

Simulating to real-world transfer (Sim2Real) has become a cornerstone in robotics, proving invaluable for training agents in simulated settings before their real-world application. This method's efficacy in translating simulated learning to practical scenarios is comprehensively analyzed in references [26,27], with specific applications to mobile robotics discussed in [28,29]. Extending beyond traditional robotics, Sim2Real has also facilitated advancements in diverse areas. For instance, deep reinforcement learning has been applied to optimize wind turbine energy output, demonstrating the versatility of Sim2Real techniques in energy sectors, as explored in [30]. Similarly, convolutional proximal policy optimization has been utilized for mapless navigation [31] and deep deterministic policy gradient methods have been employed for precise target tracking in [32].

In comparison to traditional AI or robotics control techniques, Sim2Real offers a distinctive advantage, especially in handling unpredictable and dynamic environments like beaches. Traditional methods often struggle to account for the complexities inherent in coastal landscapes, lacking the capability to incorporate factors such as variability in weather conditions, the intermittent presence of objects, and variations in terrain characteristics. Simulations also provides a safe and controllable space to test and refine algorithms before implementation in the field, as discussed in [33–35], where efforts are made to incorporate safety factors within simulated environments and the policies that the agents learn.

The Sim2Real technique presents significant challenges in robotics and other fields where real-world training is impractical or risky. This process involves training an RL model in a simulated environment and then deploying it in a real-world setting. However, this transition is not without complexities. One major hurdle is the reality gap [36], which refers to the substantial differences between simulated and real environments. These disparities, including unmodeled dynamics and variabilities in the real world, can undermine the effectiveness of the learned policy. Additionally, ensuring the generalization and robustness of the model poses a challenge. Models trained in deterministic simulations may struggle to adapt to the complexities and variability of real-world environments, potentially leading to over-specialization in simulation-specific features. Moreover, the presence of uncertainty and variability in real-world scenarios further complicates matters. The model may encounter novel situations during deployment that were not present during training, highlighting the need for the model to handle uncertainties adeptly. Addressing these challenges is crucial for the successful application of Sim2Real techniques in real-world scenarios.

Despite progress in DRL and Sim2Real technologies, their application in beach cleaning is minimally addressed in the existing literature, highlighting a significant research and development opportunity. This paper addresses this gap by applying advanced DRL and Sim2Real techniques to develop an autonomous robotic system for beach cleaning. The research primarily focuses on adapting position control skills from simulated environments to real-world operations, thereby advancing the development of effective and sustainable robotic solutions for preserving coastal ecosystems.

3. Reinforcement Learning

Reinforcement learning is a computational paradigm where an agent interacts with its environment to learn optimal behaviors through trial and error, guided by a reward system. This learning process involves the agent making decisions, executing actions, and receiving feedback in the form of rewards aimed at maximizing long-term benefits.

The interaction between the agent and the environment occurs at each discrete time step, t. As depicted in Figure 1, the agent receives an observation or state from the environment (S_t), through which it chooses an action (A_t) from the set allowed by the environment. In return for this action, the agent receives a reward (R_t) and a new state.

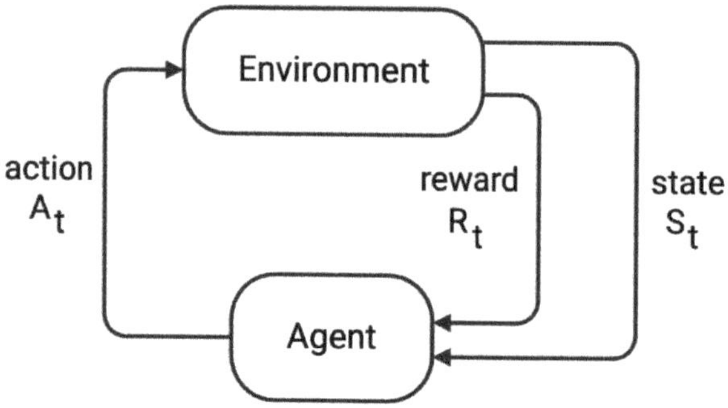

Figure 1. Interaction of the RL agent and environment.

This work utilizes two prominent RL algorithms: Proximal Policy Optimization (PPO) and Deep Deterministic Policy Gradient (DDPG). Both algorithms are implemented using the Stable Baselines3 library, known for its robust and efficient implementation of advanced RL algorithms [37].

The choice of Stable Baselines3 is based on its preference due to its extensive support and efficient implementation of RL algorithms, simplifying the experimentation and comparison of strategies. This library choice is crucial to ensure a robust and standardized foundation in the implementation of RL algorithms. The use of PPO and DDPG is motivated by the intention to explore and compare different approaches in solving specific tasks, such as position control and obstacle avoidance, leveraging the capabilities and facilities offered by Stable Baselines3.

3.1. Deep Deterministic Policy Gradient (DDPG)

DDPG is a model-free, off-policy actor-critic algorithm optimized for continuous action spaces, making it ideal for applications such as mobile robotics. It integrates the strengths of policy-based and value-based approaches, facilitating the development of complex control strategies that involve precise movements and real-time decision-making. The actor network proposes actions based on the current state, while the critic evaluates these actions by computing the value function as follows:

$$Q(s,a) \leftarrow Q(s,a) + \alpha\left(r + \gamma \max_{a'} Q(s',a') - Q(s,a)\right), \tag{1}$$

where α represents the learning rate and γ is the discount factor. The actor's policy is updated using the sampled policy gradient:

$$\nabla_{\theta^{\mu}} J \approx \frac{1}{N} \sum_{i} \nabla_{a} Q(s,a|\theta^{Q})|_{s=s_i,a=\mu(s_i)} \nabla_{\theta^{\mu}} \mu(s|\theta^{\mu})|_{s_i}. \tag{2}$$

For a more detailed understanding of the DDPG agent, readers are referred to the foundational paper [38], which provides comprehensive insights into the algorithm's architecture and performance metrics in various environments.

3.2. Proximal Policy Optimization (PPO)

PPO, known for its stability and efficiency, particularly in environments with high variability, implements a clipped surrogate objective function to manage policy updates [39]. This method prevents drastic policy changes that could destabilize the learning process. The primary equation of PPO reflects the limitation on policy updates:

$$L^{CLIP}(\theta) = \hat{\mathbb{E}}_t\left[\min(r_t(\theta)\hat{A}_t, \text{clip}(r_t(\theta), 1-\epsilon, 1+\epsilon)\hat{A}_t)\right], \tag{3}$$

where $r_t(\theta)$ denotes the ratio of the new policy probability to the old policy probability and $(\hat{A}_t)$ represents the advantage estimate at time t.

For a comprehensive examination of the Proximal Policy Optimization algorithm and its application across diverse environments, readers are encouraged to refer to [40]. This paper provides an in-depth exploration of PPO's innovative clipped surrogate objective function, which is crucial for maintaining stability in policy updates by preventing excessive deviations in policy behavior.

Both DDPG and PPO have been effectively applied in various robotic tasks, providing a solid foundation for addressing the unique challenges of autonomous beach cleaning. The selected algorithms allow for the development of deterministic policies and flexible, safe policy adjustments, critical for autonomous systems operating in unpredictable environments.

4. Simulation and Mobile Robot

The simulation of autonomous robotic systems is crucial for evaluating advanced RL agents, particularly those requiring sophisticated position control and obstacle avoidance capabilities. CoppeliaSim (V4.6.0), a versatile 3D robotic simulation software, has been employed to model the Khepera IV robot meticulously, enabling detailed examinations of autonomous navigation within a controlled environment.

The Khepera IV robot, developed by K-Team and renowned for its modular design and independent wheel motorization, is equipped with a comprehensive suite of sensors. These include eight infrared sensors for obstacle detection, additional sensors for fall prevention and line following, and ultrasonic sensors for long-range object detection, complemented by an accelerometer, gyroscope, encoders, and a color camera. This array facilitates the sophisticated perception and navigation capabilities crucial for RL applications in dynamic environments like beach cleaning [41].

CoppeliaSim supports this project by allowing precise replication of the Khepera IV's physical and sensory attributes through its KH4VREP library [42], which models the robot's rigid structure and non-deformable wheels, essential for accurate simulation outcomes. The robot model operates within CoppeliaSim under control via a remote Python API, fostering efficient programming and execution of RL algorithms. This setup ensures effective bidirectional communication between the simulation environment and the RL agents, facilitating the development and testing of algorithms in a risk-free setting [43].

Figure 2 showcases the Khepera IV robot in both simulated and real environments, highlighting the fidelity of the simulation in replicating real-world conditions.

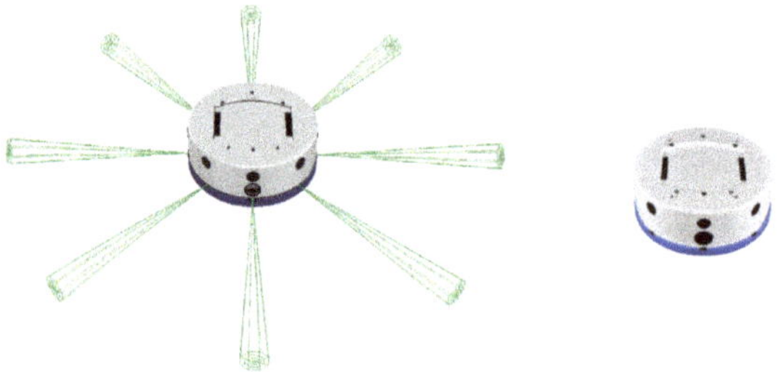

(a) (b)

Figure 2. Comparison of Simulated Khepera IV with IR Sensors (**a**) and Real Khepera IV [44] (**b**).

Utilizing CoppeliaSim in this project offers significant advantages, including the capability to accurately simulate the physical and sensory attributes of the Khepera IV robot. This provides a reliable platform for developing and validating reinforcement learning algorithms in tasks such as navigation and obstacle avoidance before their real-world application, thereby reducing potential deployment risks and costs.

5. Environment Project Setup

This section details the development and integration of the simulated and real testing environments used for the RL training of the Khepera IV robot, employing the OpenAI Gym library to facilitate effective communication between the RL agents and these environments.

5.1. Gym Environments

Two environments have been developed to bridge the gap between theoretical models and practical application. The first, a simulated environment crafted within CoppeliaSim, mirrors the intricate dynamics of robot–environment interactions with high fidelity. The second, its physical counterpart, is established in a laboratory setting, ensuring that RL agents receive consistent training experiences across both settings.

The primary task in these environments involves guiding the Khepera IV robot to reach a visually marked Target Point (TP) while navigating through scattered obstacles within a 2 m $\times$ 1 m area. Training sessions initiate with the robot in a central position, concluding upon successful TP navigation, a collision, or when exceeding a predefined step limit (Figure 3).

The observation space of the environments, which provides the current state to the agent, consists of a list that includes the distance between the robot and the target (d), the angular difference between the robot's orientation and the target position (O_c), ranging from π to $-\pi$, and a set of normalized measurements (between 0 and 1) from the robot's eight infrared sensors (as illustrated in Figure 4). These values are either derived from the simulation or calculated through requests made via the simulation API. In actual environments, the absolute position and orientation of the robot are determined through a tracking platform equipped with a camera.

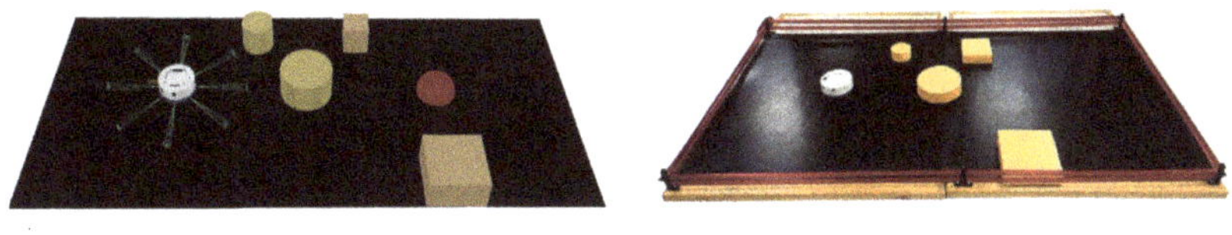

(a) (b)

Figure 3. Visual comparison between the simulated (**a**) and the real environment (**b**), highlighting the consistency in layout and interactive elements.

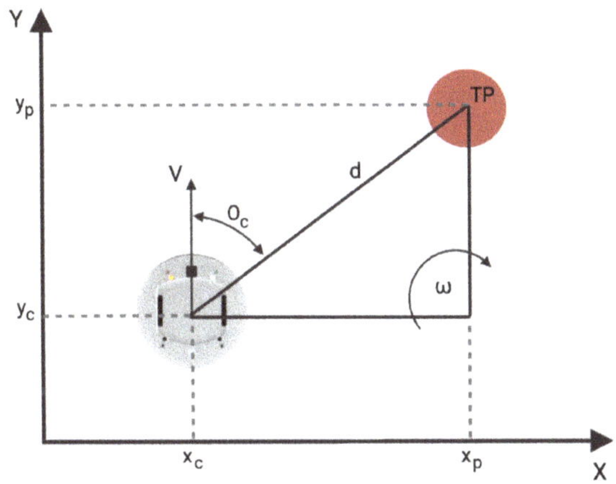

Figure 4. Key variables influencing the robot's perception and action within the environments. It illustrates measurements of distance (d) and angular orientation (O_c) relative to the TP, as well as the robot's linear (V) and angular (ω) velocities.

The action space of the environments is defined by the linear (V) and angular (ω) velocities of the mobile robot, reflecting the fundamental kinematic model of a differential drive robot. This framework allows for precise control over the robot's movements, as supported by the well-established kinematic equations cited in [45]. The linear velocity facilitates forward motion, while the angular velocity governs the rotation around the robot's central axis. The control signals for both linear and angular velocities are normalized to ranges of 0 to 1 and -1 to 1, respectively, ensuring compatibility with the simulation environment.

The reward function, depicted in Equation (4), is structured to reward proximity to the TP (distance d) and penalize collisions, thereby promoting a strategic and cautious approach. The reward increases as the robot nears the TP and adjusts negatively if sensor readings indicate proximity to obstacles. Parameters $R_{collision}$ and $R_{arrival}$ were empirically calibrated to enhance the learning and operational efficacy of the robot, with assigned values of -10 and 10, respectively.

$$Reward = \begin{cases} R_{arrival} & \text{if the robot reaches the TP} \\ R_{collison} & \text{if the robot collides} \\ -d^2 - \sum sensors & \text{in another case} \end{cases} \quad (4)$$

5.2. Laboratory Platform

Control over the Khepera IV within the lab is managed through a socket communication protocol, where the computer serves as a server, running a Python script, and the robot as a client, utilizing a script programmed in the C programming language. This arrangement employs the 'khepera4toolbox' [46] library, enabling efficient interaction with the robot's components. The use of this library facilitates a reliable bidirectional exchange of sensor data and motor control commands between the server and the robot, ensuring precise control over the robot's movements and interactions with its environment, as depicted in Figure 5.

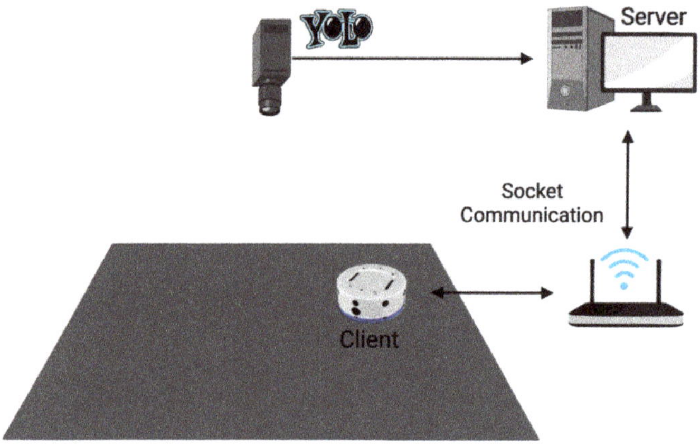

Figure 5. Schematic of the laboratory platform's component connections.

The robot's position within the platform is tracked using a webcam to provide input for a YOLO v8 [47] model trained on a dataset of 2000 manually labeled images of the Khepera on the platform. This model performs pose estimation tasks for the robot. Upon detection within an image, YOLO provides a bounding box from which the robot's (x, y) coordinates and three key points are extracted, as referenced in Figure 6. Through the application of trigonometric formulas, the relative positions of these points enable the calculation of the robot's angular orientation with respect to its surroundings. A black rectangle indicating the robot's front has been added to assist YOLO in determining the robot's orientation.

Figure 6. Keypoints the Khepera. Number 1 indicates the left marker, number 2 indicates the front marker, and number 3 indicates the right marker. These markers are used to calculate the robot's (x, y) coordinates and angular orientation through pose estimation tasks performed by the YOLO v8 model.

The model underwent training for 137 epochs, requiring 3.1 h to converge. It achieved a bounding box mAP50-95 of 0.936 and a pose estimation mAP50-95 of 0.993. These values reflect the model's proficiency in both object localization and pose estimation tasks for the robot within the images, as illustrated in Figure 7.

Given the infrared sensors' imprecisions on the Khepera and the simulation's assumption of linear measurement, detailed sensor calibrations were conducted. The results were plotted to form a sensor measurement curve, as illustrated in Figure 8. To maximize consistency between the simulated environment and the real platform, a process of linear interpolation and normalization was implemented, resulting in an adjusted measurement akin to that of the simulated infrared sensor.

Figure 7. Model predictions for the validation set.

Integration and programming of all these interactions are carried out using the Gym library, identical to that used in the simulated environment. This ensures a seamless and coherent transition between virtual training and physical testing, establishing the necessary synergy for successful policy transfer to the Khepera IV in an operational context.

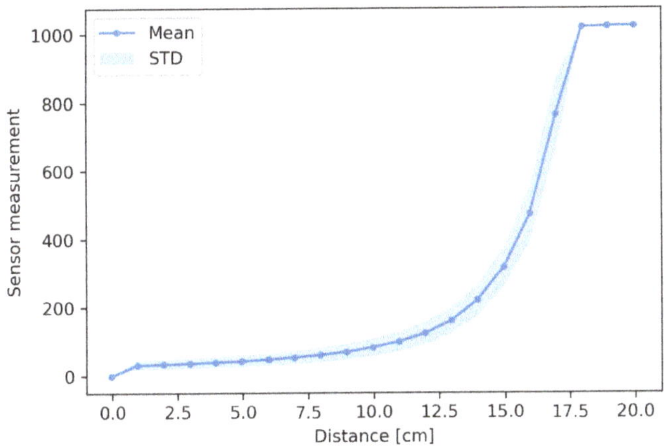

Figure 8. Mean measurement and standard deviation of the Khepera's infrared sensors.

5.3. Sim2Real: Bridging Platforms

The Sim2Real process ensures a seamless transition of control policies from the simulated environment to the physical platform. This approach involves training RL agents within the virtual CoppeliaSim environment, then transferring the learned weights to the real-world Khepera IV robot. By maintaining consistency in software and methodology across both platforms, the transition enables the effective application of sophisticated navigation and obstacle avoidance strategies developed in simulation (Figure 9).

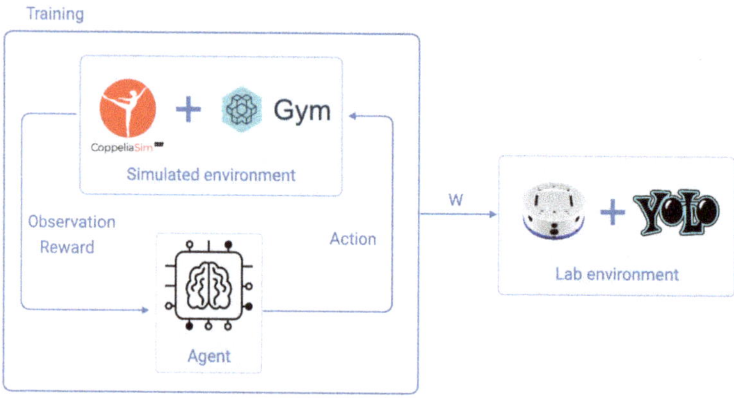

Figure 9. Diagram of the Sim2Real learning transfer process.

Strategies to Improve Sim2Real Transfer

Implementing Sim2Real transfer poses several challenges, including the reality gap between simulated and real environments and the need for robustness in handling uncertainties. To address these challenges and ensure a smoother transition of control policies from simulated environments to real-world applications, several techniques and strategies were implemented during the research. One key technique employed was domain randomization, aimed at bridging the reality gap and improving generalization. This involved systematically varying properties such as physics parameters and obstacle placement within the simulation environment. Random positions and orientations of the robot were introduced for each training episode, exposing the model to a diverse range of simulated scenarios and facilitating the learning of a more generalizable and adaptable policy.

Another crucial strategy utilized was the incorporation of uncertainty into the model. This was achieved through the integration of techniques for handling uncertainty, particularly focusing on model-based reinforcement learning approaches. By embedding these techniques into the model, it became more adept at adapting to uncertainty in the decision-making process, thereby enhancing its resilience to unforeseen situations encountered during deployment.

Furthermore, efforts were made to ensure consistency and compatibility between the simulated and real environments through normalization and standardization. Adjustments were made to the codebase to normalize variables and standardize sensor measurements, ensuring that the model's training data accurately reflected the conditions it would encounter during deployment. These techniques collectively contributed to improving the effectiveness and robustness of Sim2Real transfer, paving the way for more successful applications of reinforcement learning in real-world scenarios.

Applying Sim2Real to a model trained in a simulated environment and using it in a real environment presents significant challenges due to the reality gap and the need for generalization and adaptability. The effectiveness of the transfer will depend on how these challenges are addressed using techniques such as domain randomization, and by incorporating uncertainty into the model. These strategies can help mitigate the effects of real-world environment variability and uncertainty, improving the effectiveness and applicability of the model in real-world applications.

6. Experiments and Results

This section outlines the experiments conducted on position control and obstacle avoidance with the Khepera IV robot and provides a thorough discussion on the training process and subsequent performance analysis of the RL agents.

6.1. Simulated Robot Training Results

Two distinct RL models, PPO and DDPG, were trained over a span of 2 million timesteps, translating to a total training duration of approximately 27.3 h for PPO and 31.3 h for DDPG, respectively. For the DDPG agent, training was conducted using default parameters from the Stable Baselines3 library, with a learning rate of 0.001, buffer size of 1,000,000, learning initiation at the 100th iteration, batch size of 256, τ set to 0.005, and a γ value of 0.99. These parameters were optimized to balance exploration and exploitation, ensuring stable and efficient learning. Similarly, the PPO agent was configured with a learning rate of 0.0003, 2048 steps per update, batch size of 64, and training across 10 epochs. Additional settings included a γ of 0.99, Generalized Advantage Estimation (GAE) λ of 0.95, and a policy clipping range of 0.2, also tailored to foster both stability and efficiency in learning. The performance of each agent is graphically depicted in Figure 10, where the training graph illustrates the mean episode reward over the course of the training period. The primary objective of the training in simulation is to facilitate transfer learning (Sim2Real) for seamless deployment and adaptation in the real-world platform.

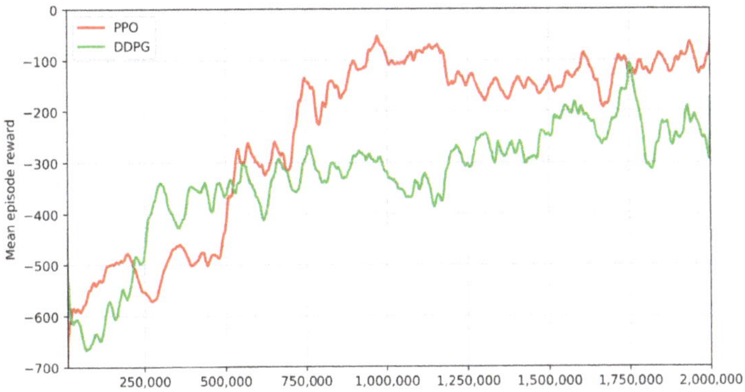

Figure 10. Training progression—mean episode reward graph.

From the graph, it is evident that both agents were capable of effectively learning from the environment, as demonstrated by the gradual increase in the reward over time. Notably, the PPO agent exhibited a particular enhancement in performance, achieving higher rewards in the initial stages of the training process. This suggests that the PPO agent was able to learn more quickly and efficiently compared to the DDPG agent, as can be inferred from the steeper slope of improvement in the corresponding curve.

The graphical analysis indicates a robust learning curve for both models, with the PPO agent demonstrating a consistent and superior performance throughout the training period. The DDPG agent, while showing a steady increase in rewards, lagged slightly behind the PPO, particularly in the early phases. However, both agents ultimately converged adequately, showcasing the efficacy of the RL models in navigating the simulated environment and addressing the complex task of autonomous movement and obstacle negotiation. It is noteworthy that the weights of the models used for the subsequent experiments correspond to those achieving the best rewards and embodying the desired action policy.

6.2. First Experiment: Navigating an Obstacle-Free Environment

In the primary experiment, the capabilities of the RL models, PPO, and DDPG, were evaluated to guide a mobile robot towards a set target within a clear space. This foundational test of positional control was designed to simulate the basic task of beach cleaning, challenging the agents to execute precise navigational maneuvers without the complexity of obstacles. As depicted in Figure 11, the simulation environment places the Khepera IV robot at an opposing start point to the target, set one meter apart, demanding precision in execution for successful task completion, analogous to directing a beach-cleaning robot to specific locations on the sand.

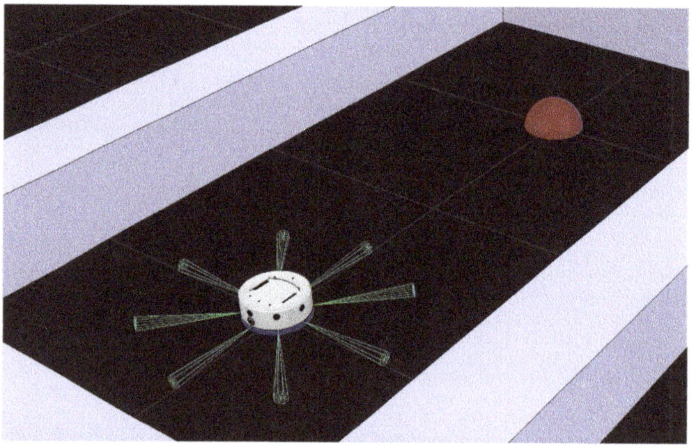

Figure 11. Simulated environment of the first experiment.

Figure 12 illustrates the trajectory paths traversed by the DDPG and PPO agents in simulated and actual environments. The paths showcase the strategic movements from the starting point at the origin (0,0), marked by the robot, to the target at (1,0), highlighted by a red circle. The robot's initial position, oriented away from the target, required a calculated turn, as evidenced by the early curvature of the paths, before proceeding directly towards the endpoint. This maneuvering efficiency is crucial for a beach-cleaning robot that needs to navigate to various locations on a beach while avoiding static and dynamic obstacles.

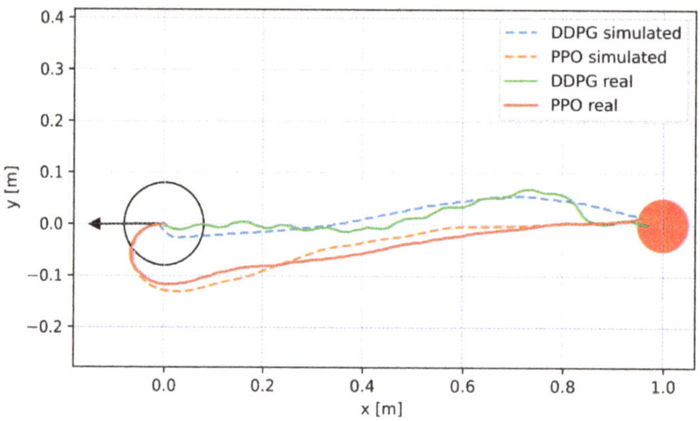

Figure 12. Trajectory plot of the first experiment.

The trajectory graph affirms the success of the Sim2Real transfer process, as both agents reliably achieved the core objective of reaching the target point. This successful transfer from simulation to real-world performance is promising for ongoing research and indicative of the potential applicability of the RL strategies to the domain of autonomous beach-cleaning robots. The ability to maintain consistent performance across both domains suggests that the refined control policies developed through simulation could effectively translate to a beach-cleaning robot navigating in a natural and less predictable environment.

In this work, the evaluation of the algorithms is structured around four specific performance indices, which are analyzed in scenarios both with and without obstacles. These indices include the Integral of Absolute Error (IAE), which measures the total magnitude of error; the Integral of Squared Error (ISE), assessing the cumulative squared error; the Integral of Time Multiplied by Absolute Error (ITAE), reflecting the error magnitude over time; and the Integral of Time Multiplied by Squared Error (ITSE), which combines the error's square with its duration. For these performance indices, the error used is the distance (*d*) from the robot to the TP. These metrics provide a comprehensive framework for quantifying the effectiveness of the algorithms under varying conditions [48].

$$IAE = \int_0^\infty |e(t)|dt \tag{5}$$

$$ISE = \int_0^\infty e^2(t)dt \tag{6}$$

$$ITAE = \int_0^\infty te^2(t)dt \tag{7}$$

$$ITSE = \int_0^\infty t|e(t)|dt \tag{8}$$

Table 1 presents a comparison of performance metrics. The purpose of this comparative table is to establish a metric for evaluating the agent's adaptability to different environments. Specifically, when the performance indices exhibit similarity or equality between the agent in the real and simulated environments, it indicates that the agent's behavior remains consistent across both settings. This alignment in performance metrics signifies a successful Sim2Real transfer, suggesting that the agent is well-adjusted to the complexities of the real-world environment.

Table 1. Performance index comparison of the first experiment.

	Real		Simulated	
Index	**DDPG**	**PPO**	**DDPG**	**PPO**
ISE	**9.48**	11.16	**9.50**	11.00
IAE	**12.90**	14.25	**12.84**	14.15
ITSE	**57.64**	70.84	**59.03**	72.03
ITAE	**102.37**	116.41	**102.91**	118.99

Bold values indicate the best performance (lower index indicates better performance).

The data comparison indicates that both DDPG and PPO agents are competent in directing a robot towards a target, which is fundamental for a beach-cleaning robot's task of reaching specific areas for waste collection. The slightly superior performance of the DDPG agent, as indicated by lower error metrics in both environments, may suggest that this model could be more suitable for the precision required in real-world beach-cleaning operations.

These experimental outcomes validate the robustness of the simulation environment for training RL agents and suggest a promising avenue for deploying such models in the operational context of beach-cleaning robots. The agents' consistent performance across the simulated and actual environments underscores the potential of Sim2Real transfer techniques for future applications in environmental preservation tasks, where autonomous robots can provide a sustainable solution to beach waste challenges.

6.3. Second Experiment: Navigating with Obstacles in Simulated and Real Environments

The second experiment addresses the challenge of the positional control of a mobile robot in an environment enriched with the presence of obstacles. This scenario places a target point one meter away, with the robot oriented in the opposite direction, as depicted in Figure 13, representing the simulated environment for this experimental phase. Unlike the first experiment, which was designed to assess the agents' prowess in precise maneuvers without external complexities, this second scenario tests the agents' abilities to navigate around obstacles while maintaining effective control of their position.

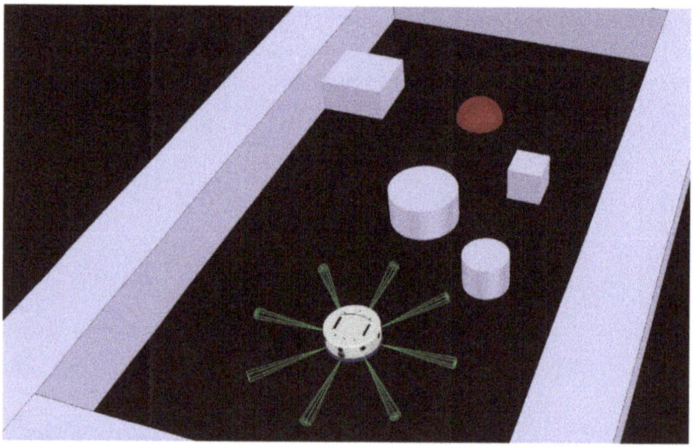

Figure 13. Simulated environment of the second experiment.

Figure 14 eloquently reveals the agents' skill in circumventing obstacles en route to the designated target. Here, a significant adaptation by the agents is observed, as they successfully navigate around the obstacles present in the scenario, a stark contrast to the obstacle-free environment of the first experiment.

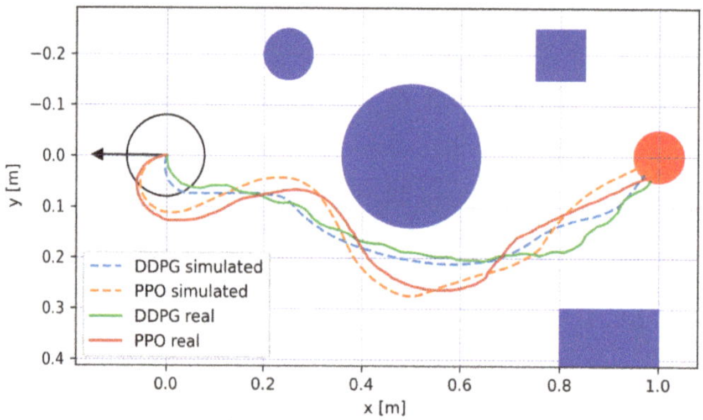

Figure 14. Trajectory graph of the second experiment.

A systematic analysis of the performance indices in Table 2 provides key insights into the agents' competencies in steering a mobile robot in an obstacle-laden environment. The DDPG agent consistently outperforms the PPO agent across all measured indices in both simulated and real-world scenarios. Lower values indicate a higher level of precision and effectiveness in guiding the robot toward the target compared to the PPO. This advantage is particularly notable in the real-world environment, where obstacles add complexity to the navigation task.

The performance indices not only affirm the efficacy of the DDPG algorithm but also highlight the reliability of the simulation environment for training RL agents in obstacle-involving situations. The superior performance of the DDPG agent suggests its potential for precise real-world robotic navigation applications, especially in environments where obstacles present navigational challenges. These findings further underscore the success of the employed Sim2Real transfer methodologies, reaffirming their utility in tackling complex navigational tasks in practical and cluttered environments.

Table 2. Performance index comparison of the second experiment.

	Real		Simulated	
Index	**DDPG**	**PPO**	**DDPG**	**PPO**
ISE	**11.78**	12.45	**9.76**	12.16
IAE	16.71	**16.60**	**13.47**	16.27
ITSE	97.28	**95.04**	**63.66**	92.82
ITAE	178.29	**164.31**	**113.57**	159.50

Bold values indicate the best performance (lower index indicates better performance).

The outcomes of this experiment carry significant implications for the deployment of beach-cleaning robots. Unlike controlled laboratory settings, beach environments are dynamic and unpredictable, with variable elements such as wandering beachgoers, animals, and shifting terrain. The ability of the DDPG agent to adeptly handle these challenges in simulation is promising for the future application in beach-cleaning operations, where autonomous robots must operate effectively in the midst of such uncertainties. The experiment thus sets a precedent for the adaptability required in real-world beach-cleaning applications, illustrating the potential of RL-trained robots to contribute meaningfully to environmental preservation efforts.

6.4. *Third Experiment: Path Following in Dynamic Beach Environments*

The third experiment scrutinizes the robot's ability to precisely follow a pre-established route delineated by a sequence of control points, mimicking a beach-cleaning course. This test scenario is pivotal as it mirrors the practical application of the robot in real-life conditions, where navigational efficiency and precision are paramount.

Figures 15 and 16 visually demonstrate how the robot, under the guidance of reinforcement learning agents, adeptly maneuvers towards the control points, even amidst obstacles. Figure 15 depicts the robot's trajectory in an obstacle-free environment, underscoring the agents' ability to fluidly and accurately adhere to the desired path. In contrast, Figure 16 introduces impediments, simulating the dynamic and ever-changing beach environment where unexpected elements such as beachgoers, animals, or debris must be navigated around.

In the first two experiments, the models' ability to adapt effectively from simulation to reality was assessed. The robust convergence of performance metrics between simulated and real environments adequately supported the models' adaptability. This experiment focused exclusively on the execution of the task of following a predefined route. Given that the primary emphasis was on task execution rather than a detailed performance analysis, reward index tables were omitted to streamline result presentation and concentrate on the task execution capability.

This adaptability and precise navigation are crucial for the goals outlined in this article, which seeks to develop an autonomous robotic system for beach-cleaning tasks. The outcomes of the experiment affirm that the agents can adjust their behavior and make real-time decisions to circumvent obstacles and maintain the set route, promising indicators of their real-world applicability.

Reinforcing the Sim2Real methodology elaborated throughout the article, the experiment confirms that skills transitioned from simulated to actual environments are not only feasible but also effective. This success holds promising implications for scalability and the potential deployment of beach-cleaning robots across various beach environments.

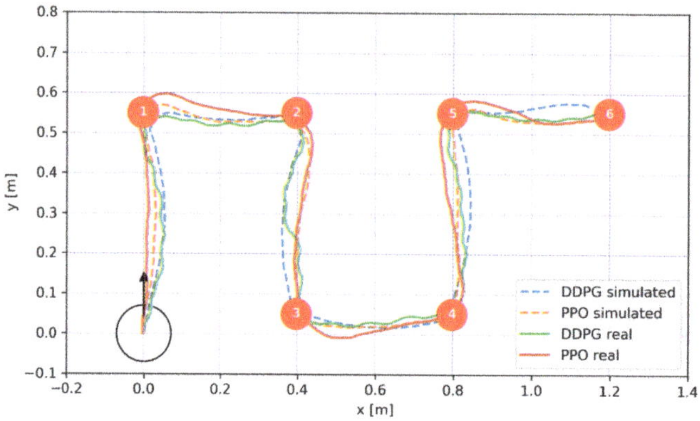

Figure 15. Trajectory graph of the third experiment without obstacles. The numbers 1 to 6 represent the key points along the trajectory followed by the robot, with 1 indicating the starting point and 6 indicating the final point.

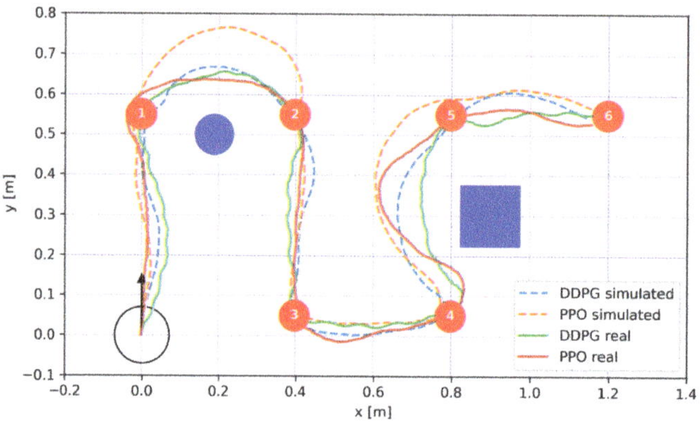

Figure 16. Trajectory graph of the third experiment with obstacles.

The demonstrated effectiveness of transferring learned skills underscores the scalability of the proposed methodology. As the robots showcase adaptability and efficiency in real-world beach landscapes, the potential for deployment in diverse environments becomes evident. The scalability of the approach positions it as a versatile solution with the capability to address the challenges posed by different beach environments, ultimately contributing to more widespread and effective deployment of autonomous beach-cleaning systems.

7. Discussion

The comprehensive experiments conducted using the PPO and DDPG agents with the Khepera IV robot platform have provided insightful data and affirmed the capabilities of these agents in both simulated and real-world settings. During the simulation phase, a commendable learning curve was observed for both agents. The PPO agent displayed an impressive rate of early learning, suggesting potential for rapid adaptation in dynamic environments. Conversely, the DDPG agent, while slower to start, demonstrated high precision in task execution, particularly when transferred to the real-world scenario. This indicates promising applications for tasks where precision is paramount.

In environments devoid of obstacles, the findings showed that both PPO and DDPG agents could navigate efficiently to reach predefined targets. The DDPG agent, however, edged out slightly in terms of precision. This advantage became more pronounced in real-world settings, suggesting that DDPG could be particularly useful in scenarios where navigational accuracy is critical to the task at hand. When obstacles were introduced, the DDPG agent's performance remained consistent, surpassing the PPO in maneuvering through the challenges presented in both simulated and real terrains. This consistency is indicative of DDPG's robustness and makes it an optimal candidate for deployment in environments with variable and unpredictable obstacles.

The third experiment's success in path following, with both agents efficiently navigating a predefined route, solidifies confidence in the RL agents' ability to perform complex navigation tasks. This outcome is particularly pertinent to the ultimate goal of developing autonomous robots for beach cleaning, where precise route adherence is necessary.

8. Conclusions

The outcomes of this research firmly position reinforcement learning as a transformative force in the evolution of autonomous beach-cleaning robots. The rigorous experimental design has enabled the PPO and DDPG agents to learn effectively in a simulated environment and successfully transfer and adapt these skills to real-world scenarios. This seamless transition from simulation to reality, known as Sim2Real, underscores the practicality of the approach and the agents' robust performance under varying conditions. The DDPG agent's consistent outperformance in environments rich with obstacles showcases its potential as an asset in precision-demanding beach-cleaning tasks. It highlights the agent's ability to navigate through complex terrains, reflecting its readiness for real-world applications where unpredictability is a norm.

Looking ahead, the Sim2Real methodology will continue to play a central role in future developments. It facilitates efficient utilization of computational resources during the training phase and mitigates risks associated with real-world testing. By bridging the gap between digital and physical realms, Sim2Real has established itself as a cornerstone of innovation in deploying RL agents for ecological tasks. These advancements encourage further exploration into the intricacies of beach ecosystems, fostering a collaborative synergy between artificial intelligence and environmental conservation. The vision is to harness the power of Sim2Real to deploy autonomous beach-cleaning robots that are not only effective in maintaining coastal cleanliness but are also adept at preserving the natural dynamics of beach environments.

Building on the successful outcomes of this research, the next phase of the investigation will focus on scaling up the results to a larger robot equipped with advanced sensors, designed for real beach-cleaning tasks. This step aims to test and validate the effectiveness and practicality of the learned control policies and Sim2Real strategies in a more challenging and dynamic natural beach environment. This progression marks a significant move towards deploying autonomous robots for environmental conservation and beach maintenance, leveraging the full potential of reinforcement learning and Sim2Real methodologies in real-world applications.

Author Contributions: Conceptualization, F.Q. and G.H.; methodology, G.H.; software, F.Q.; validation, G.H. and F.Q.; formal analysis, F.Q.; investigation, F.Q.; resources, G.V.; data curation, F.Q. and K.S.; writing—original draft preparation, F.Q. and G.H; writing—review and editing, F.A. and G.H.; visualization, F.Q.; supervision, G.H. and G.V.; project administration, G.H.; funding acquisition, G.V. All authors have read and agreed to the published version of the manuscript.

Funding: This work was supported in part by FONDEF under Grant ID23I10249 and FONDECYT under Grant 1240573.

Institutional Review Board Statement: Not applicable.

Informed Consent Statement: Not applicable.

Data Availability Statement: The data supporting the reported results can be found at the following GitHub repository: https://github.com/Fco-Quiroga/gym-kheperaposition.

Conflicts of Interest: The authors declare no conflicts of interest.

References

1. Zielinski, S.; Botero, C.M.; Yanes, A. To clean or not to clean? A critical review of beach cleaning methods and impacts. *Mar. Pollut. Bull.* **2019**, *139*, 390–401. [CrossRef] [PubMed]
2. Deshpande, P.; Milkhe, O.; Kamble, A.; Kudu, N. Beach cleaning robots a comprehensive survey of technologies challenges, and future directions. *Int. Res. J. Mod. Eng. Technol. Sci.* **2023**, *5*, 7182–7188. [CrossRef]
3. Kiessling, T.; Salas, S.; Mutafoglu, K.; Thiel, M. Who cares about dirty beaches? Evaluating environmental awareness and action on coastal litter in Chile. *Ocean. Coast. Manag.* **2017**, *137*, 82–95. [CrossRef]
4. Sutton, R.S.; Barto, A.G. *Reinforcement Learning: An Introduction*; MIT Press: Cambridge, MA, USA, 2018.
5. Mnih, V.; Kavukcuoglu, K.; Silver, D.; Graves, A.; Antonoglou, I.; Wierstra, D.; Riedmiller, M. Playing Atari with Deep Reinforcement Learning. *arXiv* **2013**, arXiv:1312.5602. [CrossRef]
6. Lu, Z.; Huang, R. Autonomous mobile robot navigation in uncertain dynamic environments based on deep reinforcement learning. In Proceedings of the 2021 IEEE International Conference on Real-Time Computing and Robotics (RCAR), Xining, China, 15–19 July 2021; pp. 423–428. [CrossRef]
7. Davenport, J.; Davenport, J.L. The impact of tourism and personal leisure transport on coastal environments: A review. *Estuar. Coast. Shelf Sci.* **2006**, *67*, 280–292. [CrossRef]
8. Vieira, J.V.; Ruiz-Delgado, M.C.; Reyes-Martínez, M.J.; Borzone, C.A.; Asenjo, A.; Sánchez-Moyano, J.E.; García-García, F.J. Assessment the short-term effects of wrack removal on supralittoral arthropods using the M-BACI design on Atlantic sandy beaches of Brazil and Spain. *Mar. Environ. Res.* **2016**, *119*, 222–237. [CrossRef] [PubMed]
9. Naik, A.V.; Raj, E.V.; Chaitra, C.T.; Harshitha, K.S.; Komal. AI Based Robot for Beach Cleaning. In Proceedings of the 2023 International Conference on Applied Intelligence and Sustainable Computing (ICAISC), Dharwad, India, 16–17 June 2023; pp. 1–5. [CrossRef]
10. Schmoeller da Roza, F.; Ghizoni da Silva, V.; Pereira, P.J.; Wildgrube Bertol, D. Modular robot used as a beach cleaner. *Ingeniare Rev. Chil. Ing.* **2016**, *24*, 643–653. [CrossRef]
11. Ichimura, T.; Nakajima, S.I. Development of an autonomous beach cleaning robot "Hirottaro". In Proceedings of the 2016 IEEE International Conference on Mechatronics and Automation, Harbin, China, 7–10 August 2016; pp. 868–872. [CrossRef]
12. Thompson, R. Turtle Friendly Beach Cleaning Device. U.S. Patent 2015/014.4362 A1, 28 May 2015.
13. Praveen, R.; Prabhu, L.; Premjith, P.; Mohan, A.K.; Ajayraj. Design experimental of RF controlled beach cleaner robotic vehicle. *IOP Conf. Ser. Mater. Sci. Eng.* **2020**, *993*, 012030. [CrossRef]
14. Bano, N.; Amin, A.; Boghani, H.; Tariq, H.; Bakhtawar, S.; Waggan, I.; Younas, T. Radio Controlled Beach Cleaning Bot. In Proceedings of the 2019 IEEE 6th International Conference on Engineering Technologies and Applied Sciences (ICETAS), Kuala Lumpur, Malaysia, 20–21 December 2019; pp. 1–6. [CrossRef]
15. Deshpande, P.; Milkhe, O.; Kamble, A. Autonomous beach cleaning robot controlled by mobile application with real-time video feed and object detection. *Int. Res. J. Mod. Eng. Technol. Sci.* **2023**, *5*, 7169–7176. [CrossRef]
16. Quiroga, F.; Hermosilla, G.; Farias, G.; Fabregas, E.; Montenegro, G. Position Control of a Mobile Robot through Deep Reinforcement Learning. *Appl. Sci.* **2022**, *12*, 7194. [CrossRef]
17. Montero, E.E.; Mutahira, H.; Pico, N.; Muhammad, M.S. Dynamic warning zone and a short-distance goal for autonomous robot navigation using deep reinforcement learning. *Complex Intell. Syst.* **2024**, *10*, 1149–1166. [CrossRef]
18. Hu, H.; Wang, Y.; Tong, W.; Zhao, J.; Gu, Y. Path Planning for Autonomous Vehicles in Unknown Dynamic Environment Based on Deep Reinforcement Learning. *Appl. Sci.* **2023**, *13*, 56. [CrossRef]
19. Qiu, X.; Wan, K.; Li, F. Autonomous Robot Navigation in Dynamic Environment Using Deep Reinforcement Learning. In Proceedings of the 2019 IEEE 2nd International Conference on Automation, Electronics and Electrical Engineering (AUTEEE), Shenyang, China, 22–24 November 2019; pp. 338–342. [CrossRef]
20. Tammewar, A.; Chaudhari, N.; Saini, B.; Venkatesh, D.; Dharahas, G.; Vora, D.; Patil, S.; Kotecha, K.; Alfarhood, S. Improving the Performance of Autonomous Driving through Deep Reinforcement Learning. *Sustainability* **2023**, *15*, 3799. [CrossRef]
21. Kiran, B.R.; Sobh, I.; Talpaert, V.; Mannion, P.; Sallab, A.A.A.; Yogamani, S.K.; Pérez, P. Deep Reinforcement Learning for Autonomous Driving: A Survey. *arXiv* **2020**, arXiv:2002.00444. [CrossRef]
22. Çetin, E.; Barrado, C.; Muñoz, G.; Macias, M.; Pastor, E. Drone Navigation and Avoidance of Obstacles Through Deep Reinforcement Learning. In Proceedings of the 2019 IEEE/AIAA 38th Digital Avionics Systems Conference (DASC), San Diego, CA, USA, 8–12 September 2019; pp. 1–7. [CrossRef]
23. Tan, Z.; Karaköse, M. A new approach for drone tracking with drone using Proximal Policy Optimization based distributed deep reinforcement learning. *SoftwareX* **2023**, *23*, 101497. [CrossRef]
24. Azar, A.T.; Koubâa, A.; Mohamed, N.A.; Ibrahim, H.A.; Ibrahim, Z.F.; Kazim, M.; Ammar, A.; Benjdira, B.; Khamis, A.M.; Hameed, I.A.; et al. Drone Deep Reinforcement Learning: A Review. *Electronics* **2021**, *10*, 999. [CrossRef]
25. Fu, Z.; Zhao, T.Z.; Finn, C. Mobile ALOHA: Learning Bimanual Mobile Manipulation with Low-Cost Whole-Body Teleoperation. *arXiv* **2024**, arXiv:2401.02117. [CrossRef]
26. Petrovic, O.; Schäper, L.; Roggendorf, S.; Storms, S.; Brecher, C. Sim2Real Deep Reinforcement Learning of Compliance-based Robotic Assembly Operations. In Proceedings of the 2022 26th International Conference on Methods and Models in Automation and Robotics (MMAR), Międzyzdroje, Poland, 22–25 August 2022; pp. 300–305. [CrossRef]

27. Huang, J.; Zhang, Y.; Giardina, F.; Rosendo, A. Trade-off on Sim2Real Learning: Real-world Learning Faster than Simulations. *arXiv* **2022**, arXiv:2007.10675. [CrossRef]
28. Li, D.; Okhrin, O. A Platform-Agnostic Deep Reinforcement Learning Framework for Effective Sim2Real Transfer in Autonomous Driving. *arXiv* **2023**, arXiv:2304.08235. [CrossRef]
29. Li, D.; Okhrin, O. Vision-based DRL Autonomous Driving Agent with Sim2Real Transfer. *arXiv* **2023**, arXiv:2305.11589. [CrossRef]
30. Bawo, B. Optimizing The Output Energy of a Vertical Axis Wind Turbine Using Deep Deterministic Policy Gradient and Proximal Policy Optimization. Master's Thesis, Sabanci University, Tuzla, Turkey, 2023.
31. Toan, N.D.; Woo, K.G. Mapless Navigation with Deep Reinforcement Learning based on The Convolutional Proximal Policy Optimization Network. In Proceedings of the 2021 IEEE International Conference on Big Data and Smart Computing (BigComp), Jeju Island, Republic of Korea, 17–20 January 2021; pp. 298–301. [CrossRef]
32. You, S.; Diao, M.; Gao, L.; Zhang, F.; Wang, H. Target tracking strategy using deep deterministic policy gradient. *Appl. Soft Comput.* **2020**, *95*, 106490. [CrossRef]
33. Yuan, Z.; Hall, A.W.; Zhou, S.; Brunke, L.; Greeff, M.; Panerati, J.; Schoellig, A.P. Safe-control-gym: A Unified Benchmark Suite for Safe Learning-based Control and Reinforcement Learning in Robotics. *arXiv* **2022**, arXiv:2109.06325. [CrossRef]
34. Gu, S.; Yang, L.; Du, Y.; Chen, G.; Walter, F.; Wang, J.; Yang, Y.; Knoll, A. A Review of Safe Reinforcement Learning: Methods, Theory and Applications. *arXiv* **2023**, arXiv:2205.10330. [CrossRef]
35. Yang, W.C.; Marra, G.; Rens, G.; Raedt, L.D. Safe Reinforcement Learning via Probabilistic Logic Shields. *arXiv* **2023**, arXiv:2303.03226. [CrossRef]
36. Chen, J., Zhang, K., Wang, J., Shen, W. Closing the Simulation-to-Reality Gap for Digital Twin-Assisted Fault Diagnosis: Sim2real Knowledge Transfer with Contrastive Learning. 2024. Available online: https://ssrn.com/abstract=4699149 (accessed on 23 May 2024).
37. Raffin, A.; Hill, A.; Gleave, A.; Kanervisto, A.; Ernestus, M.; Dormann, N. Stable-Baselines3: Reliable Reinforcement Learning Implementations. *J. Mach. Learn. Res.* **2021**, *22*, 1–8.
38. Lillicrap, T.P.; Hunt, J.J.; Pritzel, A.; Heess, N.; Erez, T.; Tassa, Y.; Silver, D.; Wierstra, D. Continuous control with deep reinforcement learning. *arXiv* **2019**, arXiv:1509.02971. [CrossRef]
39. OpenAI Spinning Up. Proximal Policy Optimization. 2020. Available online: https://spinningup.openai.com/en/latest/algorithms/ppo.html (accessed on 1 March 2024).
40. Schulman, J.; Wolski, F.; Dhariwal, P.; Radford, A.; Klimov, O. Proximal Policy Optimization Algorithms. *arXiv* **2017**, arXiv:1707.06347. [CrossRef]
41. Tharin, J.; Lambercy, F.; Carron, T. *Khepera IV User Manual*; K-Team S.A.: Vallorbe, Switzerland, 2019. Available online: https://ftp.k-team.com/KheperaIV/software/Gumstix%20COM%20Y/UserManual/Khepera%20IV%20User%20Manual%204.x.pdf (accessed on 1 March 2024).
42. Farias, G.; Fabregas, E.; Peralta, E.; Torres, E.; Dormido, S. A Khepera IV library for robotic control education using V-REP. *IFAC-PapersOnLine* **2017**, *50*, 9150–9155. [CrossRef]
43. Coppelia Robotics. *CoppeliaSim User Manual*. 2023. Available online: https://manual.coppeliarobotics.com/index.html (accessed on 1 March 2024).
44. K-Team. KHEPERA IV. Available online: https://www.k-team.com/khepera-iv (accessed on 1 March 2024).
45. Peralta, E.; Fabregas, E.; Farias, G.; Vargas, H.; Dormido, S. Development of a Khepera IV Library for the V-REP Simulator. *IFAC-PapersOnLine* **2016**, *49*, 81–86. [CrossRef]
46. Soares, J. Khepera IV Toolbox. 2018. Available online: https://github.com/jsoares/khepera4toolbox (accessed on 1 March 2024).
47. Reis, D.; Kupec, J.; Hong, J.; Daoudi, A. Real-Time Flying Object Detection with YOLOv8. *arXiv* **2023**, arXiv:2305.09972. [CrossRef].
48. Shinners, S.M. *Modern Control System Theory and Design*, 2nd ed.; John Wiley & Sons: Hoboken, NJ, USA, 1998.

applied sciences

MDPI

Article

PE-SLAM: A Modified Simultaneous Localization and Mapping System Based on Particle Swarm Optimization and Epipolar Constraints

Cuiming Li *, Zhengyu Shang *, Jinxin Wang, Wancai Niu and Ke Yang

School of Mechanical and Electrical Engineering, Lanzhou University of Technology, Lanzhou 730050, China
* Correspondence: li_goddess@163.com (C.L.); shang_02132024@163.com (Z.S.)

Featured Application: Simultaneous Localization and Mapping of autonomous cleaning and inspection robots operating in the photovoltaic power station scene.

Abstract: Due to various typical unstructured factors in the environment of photovoltaic power stations, such as high feature similarity, weak textures, and simple structures, the motion model of the ORB-SLAM2 algorithm performs poorly, leading to a decline in tracking accuracy. To address this issue, we propose PE-SLAM, which improves the ORB-SLAM2 algorithm's motion model by incorporating the particle swarm optimization algorithm combined with epipolar constraint to eliminate mismatches. First, a new mutation strategy is proposed to introduce perturbations to the pbest (personal best value) during the late convergence stage of the PSO algorithm, thereby preventing the PSO algorithm from falling into local optima. Then, the improved PSO algorithm is used to solve the fundamental matrix between two images based on the feature matching relationships obtained from the motion model. Finally, the epipolar constraint is applied using the computed fundamental matrix to eliminate incorrect matches produced by the motion model, thereby enhancing the tracking accuracy and robustness of the ORB-SLAM2 algorithm in unstructured photovoltaic power station scenarios. In feature matching experiments, compared to the ORB algorithm and the ORB+HAMMING algorithm, the ORB+PE-match algorithm achieved an average accuracy improvement of 19.5%, 14.0%, and 6.0% in unstructured environments, respectively, with better recall rates. In the trajectory experiments of the TUM dataset, PE-SLAM reduced the average absolute trajectory error compared to ORB-SLAM2 by 29.1% and the average relative pose error by 27.0%. In the photovoltaic power station scene mapping experiment, the dense point cloud map constructed has less overlap and is complete, reflecting that PE-SLAM has basically overcome the unstructured factors of the photovoltaic power station scene and is suitable for applications in this scene.

Keywords: ORB feature; image matching; multi-objective optimization; particle swarms; epipolar constraints; simultaneous localization and mapping (SLAM)

1. Introduction

Solar energy, as one of the main new energy sources, has become the backbone of the new energy industry [1]. Typically, large-scale photovoltaic power plants are built in geographically complex areas such as valleys and deserts, and utilizing mobile robots to carry out tasks such as cleaning and inspection in the photovoltaic power plants has obvious advantages [2–4]. The perception of their own status and the external environment by mobile robots is a prerequisite for their autonomous operation in the non-structured environment of photovoltaic power plants [5,6].

In applications such as autonomous driving, robot positioning, and map building, ORB-SLAM2 [7], as a visual SLAM (simultaneous localization and mapping) method based on image features, has become quite mature in scenes with rich textures and complex structures [8]. However, in special environments like photovoltaic power plants, there are unstructured factors: high feature similarity, weak scene textures, and simple structures. These unstructured factors impose strong interference on the ORB feature matching algorithm in the ORB-SLAM2 algorithm, leading to a decrease in tracking accuracy.

The images of the photovoltaic power plant scene are mainly composed of photovoltaic modules. As shown in Figure 1, the image of a photovoltaic module can be seen as a combination of multiple small images with similar texture and structural features. The appearance, shape, and color of photovoltaic modules are similar, lacking unique markings or sufficient distinguishing characteristics. The similar features lead to poor differentiation in feature descriptors, which results in a large number of mismatches during the feature matching stage, subsequently causing incorrect pose estimation. Inaccurate pose estimation can affect the tracking accuracy and mapping effectiveness of the SLAM system and may even result in tracking failures midway through. Figure 1 illustrates that the textures and structures of photovoltaic panels are extremely simple, with the extracted feature points concentrated at the connections of the photovoltaic panels. The limited number of feature points reduces the tracking stability and backend optimization capability of the SLAM system, leading to diminished system robustness. According to the analysis of a large number of feature matching experimental results, it was found that there is an inconsistency in the mismatches between the features in the images of the photovoltaic power plant scene, meaning they do not have the same constraints, whereas correct matches generally have the same constraints.

Figure 1. Photovoltaic power plant scene images.

ORB-SLAM2 uses the ORB [9] algorithm for image feature extraction and matching. In this context, this paper focuses on recent research on improving the ORB algorithm. Reference [10] proposed an improved ORB algorithm based on dynamic thresholds and enhanced quadtree methods to enhance the uniformity of feature point extraction. Reference [11] introduced an improved ORB algorithm combining adaptive histogram equalization techniques to improve feature point quality and matching efficiency under low or overexposed conditions. Reference [12] combined an improved ORB algorithm with the Lucas–Kanade algorithm to propose a new image feature matching algorithm, which enhances feature point uniformity and matching accuracy. Reference [13] proposed an improved ORB algorithm based on adaptive thresholds and local grayscale differences, improving the effectiveness of feature description. References [14,15] segmented and organized high-dimensional feature data by establishing kd trees and k-means trees, enabling faster nearest neighbor searches and accelerating feature matching speed. Reference [16] incorporated grid motion statistics into feature matching to eliminate incorrect matches, effectively improving feature matching accuracy. Reference [17] used the RANSAC algorithm to remove mismatches in coarse matching point sets, achieving high real-time performance and effectively enhancing feature matching accuracy. Reference [18] utilized the PROSAC algorithm to remove mismatches by sampling from an ever-expanding best set, reducing computational load compared to Reference [17], but the randomness of sampling decreased the algorithm's robustness. Reference [19] employed the particle swarm algorithm with feature point similarity as the fitness function for feature point matching, providing effective matching accuracy and robustness, albeit with poorer real-time performance. Reference [20] addressed mismatch removal in remote sensing images using the particle swarm algorithm combined with affine transformations, showing good real-time performance and matching accuracy but limited applicability across various scenes. Reference [21] introduces a hardware accelerator for ORB-SLAM that optimizes rBRIEF descriptor generation using a genetic algorithm, reducing energy consumption by 14,597× and 9609× on CPU and GPU, respectively. Reference [22] enhances ORB feature extraction in complex lighting by combining image enhancement with a truncated adaptive threshold, improving feature point detection and robustness. Reference [23] introduces AAM-ORB, which integrates an affine attention module into the ORB feature matching pipeline to improve feature matching accuracy and efficiency under scene changes, using grid-based motion statistics to enhance speed. Reference [24] introduces a vision-based spatial target recognition method for robotic arms, combining improved ORB feature extraction with an enhanced GMS-MLESAC algorithm to achieve precise, fast, and stable recognition of spatial targets for complex task execution. Reference [25] improves ORB feature matching by introducing a signature-based method that significantly reduces the number of features needing comparison, thereby speeding up the matching process while maintaining high precision.

The methods mentioned above enhance the accuracy or speed of the ORB algorithm. However, due to the challenges posed by unstructured factors in the photovoltaic power plant environment for feature matching and the real-time requirements of visual SLAM systems, some of these proposed improvements are not suitable for visual SLAM algorithms in unstructured environments. This paper proposes a method for removing ORB image feature mismatches based on an improved particle swarm algorithm combined with epipolar constraints (PE-match), focusing on finding the majority of correct matches with consistent constraint conditions among the coarse matching point pairs. Compared to matching methods based on random sampling and clustering, the particle swarm algorithm has better global search capability and, compared to genetic algorithms, has lower algorithm complexity. Epipolar constraints offer better robustness against changes in perspective compared to affine transformations, and the analytical solution provided by epipolar constraints ensures the real-time performance of the algorithm. PE-match addresses the sensitivity of epipolar constraints to mismatches by leveraging the optimization capability of the improved Particle Swarm Optimization Algorithm, effectively expanding the application environment of epipolar constraints. PE-SLAM employs PE-match to improve the motion model of ORB-SLAM2, enhancing the tracking accuracy of the motion model in unstructured environments of photovoltaic power plants. The improved motion model, as the main tracking module within ORB-SLAM2, further increases the localization and mapping accuracy of ORB-SLAM2 in photovoltaic power plant scenes.

2. PE-SLAM System Framework

PE-SLAM is an improvement of the ORB-SLAM2 framework, as shown in Figure 2, consisting of the following five modules: "TRACKING", "LOCAL MAPPING", "LOOP

CLOSING", "DENSE POINT CLOUD MAPPING", and "FULL BA". "TRACKING" focuses on real-time computation of the camera's motion trajectory by matching features in adjacent image frames. "LOCAL MAPPING" optimizes local poses and constructs local maps in the local scene. "LOOP CLOSING" corrects trajectory drift caused by error accumulation by determining if the camera has returned to a previously visited location. "DENSE POINT CLOUD MAPPING" is used to build dense point cloud maps. "FULL BA" further enhances the system's accuracy and stability through optimization algorithms.

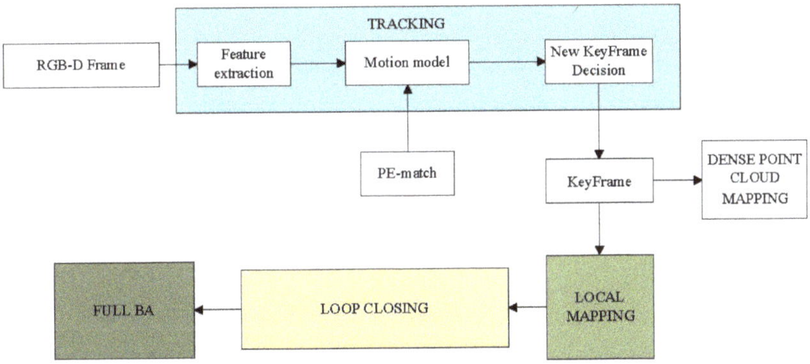

Figure 2. PE-SLAM system framework diagram.

PE-SLAM is based on the classic ORB-SLAM2 framework and combines particle swarm with the epipolar constraint feature matching algorithm (PE-match) to perform a mismatch removal operation on the feature matching completed by the motion model. At the same time, after generating keyframes in visual odometry, a dense point cloud map based on keyframes is established to compensate for the limitation of ORB-SLAM2 in only being able to generate sparse maps.

3. Improved Motion Model Based on PE-Match

The "TRACKING" module of ORB-SLAM2 can be divided into two main stages during the tracking process: pose estimation and pose optimization. The pose estimation stage employs three strategies: tracking based on reference keyframes, predictive tracking based on motion models, and relocalization. In the pose optimization stage, by establishing the matching relationship between map points of the current frame and its adjacent keyframes, more feature point matches are obtained to optimize and adjust the previous pose estimation.

The primary task of tracking in the pose estimation stage is handled by the motion model. As shown in Figure 3, its basic idea is to assume the camera is in uniform motion over a short period, using the pose and velocity estimated from the previous frame to predict the approximate positions of matching feature points between the current frame and the previous frame. A search area is then set around this position for feature descriptor matching, which improves matching speed while ensuring good accuracy. The specific process is as follows:

Step 1: Set the success criteria for feature matching: minimum distance $< 0.9\times$ next smallest distance.

Step 2: Initialize the bit position R, t of the current frame using the current motion velocity and the bit position of the previous frame.

Step 3: Using the bit position R and t of the current frame, project the 3D point of the previous frame to the current coordinate system, search for feature points that can be matched within the radius t of that 2D point, traverse the points that can be matched, calculate the descriptor distance, and record the minimum matching distance.

Step 4: Record the matching point feature directions for direction verification, eliminate the clusters of point pairs with a small number of the same directions in the direction histogram, and retain the clusters of the top three most numerous direction point pairs.

Step 5: If fewer than 20 matching point pairs are found, expand the search radius $t = 2 \times t$ and search again.

Step 6: Perform G2O map optimization of the bit pose of the current frame using matched point pairs.

Step 7: Based on the number of successful matches, determine whether the tracking is successful or not.

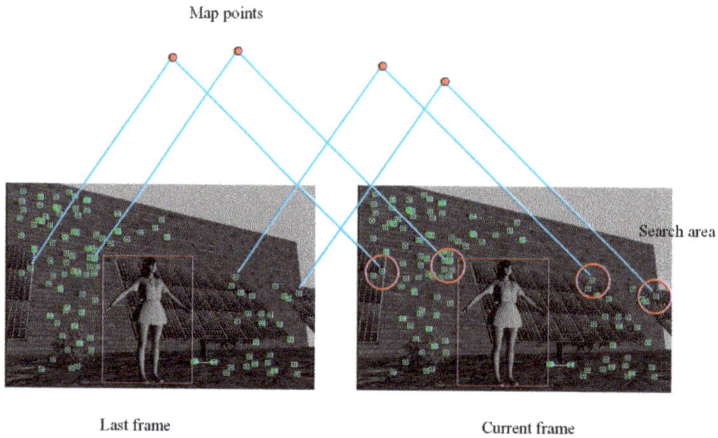

Figure 3. Schematic diagram of motion model.

The motion model in ORB-SLAM2 only utilizes orientation histograms to remove outliers from feature matching results, which leads to fast processing but mediocre removal of incorrect matches. Addressing this issue, this paper proposes a method that, after filtering with orientation histograms in Step 4, employs a particle swarm combined with epipolar constraints to further eliminate incorrect matches. This secondary filtering step can remove the majority of incorrect matches, thereby enhancing the accuracy of the SLAM system.

4. Feature Matching Algorithm Based on Particle Swarm Optimization and Epipolar Constraint (PE-Match)

The traditional ORB feature matching algorithm has the advantage of fast computation speed. However, in photovoltaic power plant images, there are non-structural factors such as high feature similarity and simple structures, leading to a high occurrence of incorrect matches with the ORB algorithm. As mentioned earlier, feature matching in photovoltaic power plant scenes exhibits global feature similarity, with incorrect matches lacking consistency and not possessing the same constraint properties, while correct matches generally exhibit the same constraint properties. Therefore, PE-match is designed to address the characteristics of incorrect matches in photovoltaic power plant scenes. It utilizes the particle swarm algorithm's information exchange and parallel search capabilities to search for constraints that match the majority of point pairs in the coarse matching point set and ultimately eliminates outliers based on these constraints.

The epipolar constraint leverages rotation and translation information in images and is minimally affected by large disparities between images. Its analytical solution guarantees real-time performance for algorithms and is suitable for image processing scenarios involving mobile robots. Therefore, this paper adopts the epipolar constraint as the required constraint for the algorithm.

The particle swarm optimization algorithm [26] has unique advantages in dealing with multi-objective optimization problems, but traditional particle swarm algorithms suffer from the drawback of getting trapped in local optima. This paper improves the mutation operations of pbest (personal best value) to enhance the global search capability of the particle swarm algorithm in the later stages of convergence.

4.1. Epipolar Constraint Principle

Epipolar constraints [27] reduce the distribution of possible matches of a point on another image from two dimensions to one dimension, providing constraints to weed out false matches. The epipolar constraint schematic is shown in Figure 4; I_1 and I_2 are the image planes of the camera at two moments, O_1 and O_2 are the camera centers corresponding to the two image planes, and according to the pinhole camera model, it can be known that p_1 and p_2 are the 2D pixel points mapped by the spatial point P on the two image planes. The ray O_1p_1 has a spatial point P corresponding to pixel p_1, and the projection of this ray on the image plane I_2 is the epipolar line L_2, the pixel point of P projected on the image plane I_2 must be on the epipolar line L_2, and p_2 matching p_1 must be on the epipolar line.

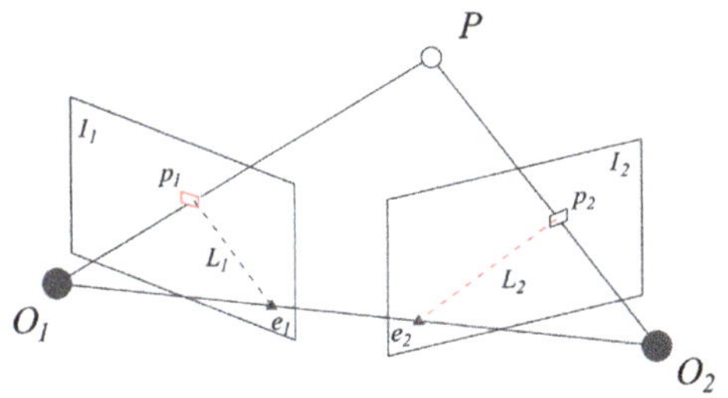

Figure 4. Schematic diagram of epipolar constraint.

A point p_1 in the image plane I_1 in Figure 4, whose corresponding epipolar line L_2 in the image plane I_2 can be expressed as follows:

$$\begin{cases} L_2 : ax + by + c = 0 \\ [a, b, c] = \boldsymbol{F}p_1 \end{cases} \tag{1}$$

In the formula, $\boldsymbol{F}$ is the fundamental matrix.

The distance from the pixel to be verified to L_2 is calculated using the point-to-straight line distance, and if the distance is within a certain threshold, then the pixel is identified as the correct match for p_1. As shown in Figure 5, if it is not within the threshold then this matched point pair is rejected.

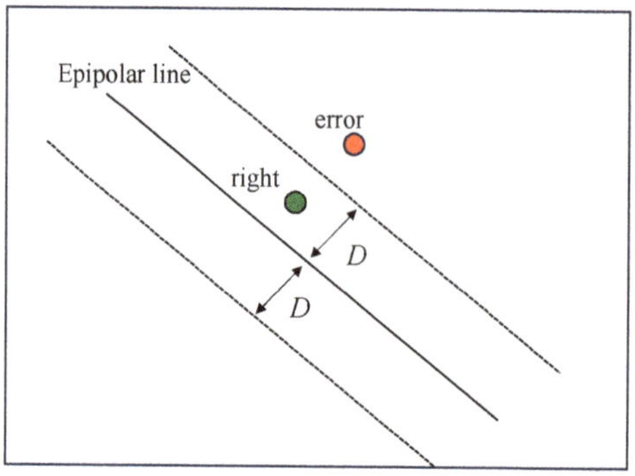

Figure 5. Schematic diagram of matching point pairs for pole constraint verification.

The fundamental matrix $\boldsymbol{F}$ is a mathematical representation of the correspondence between pairs of matching points and contains the camera's internal and external reference information. Given the fundamental matrix and the pixel coordinates of image feature points, we can calculate the epipolar line corresponding to the matching point on the other image.

Let the spatial position of point P be $P = [X, Y, Z]^T$, the pixel positions of the two pixels p_1 and p_2 are known from the pinhole camera model and represented by Equation (2) as follows:

$$\begin{cases} s_1 p_1 = KP \\ s_2 p_2 = K(\boldsymbol{R}P + \boldsymbol{t}) \end{cases} \tag{2}$$

In the formula, $\boldsymbol{K}$ is the camera internal reference, $\boldsymbol{R}$ and $\boldsymbol{t}$ are the camera motion information, and s_1 and s_2 are constants.

Equation (3) is derived from Equation (2) and written as follows:

$$p_2{}^T \boldsymbol{F} p_1 = 0 \tag{3}$$

In the formula: $\boldsymbol{F}$ is the fundamental matrix, $\boldsymbol{F} = \boldsymbol{K}^{-T}\boldsymbol{t}^{\wedge}\boldsymbol{R}\boldsymbol{K}^{-1}$, which contains the translation and rotation information of the camera.

Consider a pair of matching points whose normalized coordinates are $p_1 = (x_1, y_1, 1)^T$, $p_2 = (x_2, y_2, 1)^T$. Equation (4) is obtained from Equation (3) and normalized coordinates:

$$\begin{bmatrix} x_2 & y_2 & 1 \end{bmatrix} \begin{bmatrix} f_{11} & f_{12} & f_{13} \\ f_{21} & f_{22} & f_{23} \\ f_{31} & f_{32} & f_{33} \end{bmatrix} \begin{bmatrix} x_1 \\ y_1 \\ 1 \end{bmatrix} = 0 \tag{4}$$

Given m sets of matches: $p_1{}^i = (x_1^i, y_1^i, 1)^T$, $p_2{}^i = (x_2^i, y_2^i, 1)^T$, Equation (5) can be obtained:

$$\begin{bmatrix} x_2^1 x_1^1 & x_2^1 y_1^1 & x_2^1 & y_2^1 x_1^1 & y_2^1 y_1^1 & y_2^1 & x_1^1 & y_1^1 & 1 \\ \vdots & \vdots & \vdots & \vdots & \vdots & \vdots & \vdots & \vdots & \vdots \\ x_2^m x_1^m & x_2^m y_1^m & x_2^m & y_2^m x_1^m & y_2^m y_1^m & y_2^m & x_1^m & y_1^m & 1 \end{bmatrix} f = 0 \tag{5}$$

From Equation (4), the fundamental matrix $\boldsymbol{F}$ has nine unknowns. Since the rank of $\boldsymbol{F}$ is 2, it reduces the degree of freedom by one. Therefore, eight matching point pairs are needed to solve eight linear equations to derive the unique $\boldsymbol{F}$. This method is sensitive to incorrect matches and image noise [28], and the problem of solving the fundamental matrix can be seen as a multi-objective optimization problem for the set of matching point pairs.

4.2. Improved Mutation Strategy

One of the common issues with particle swarm optimization is the tendency to prematurely converge. As the number of iterations increases, the population diversity decreases, leading the particle swarm to easily get stuck in local optima. In contrast, the mutation strategy of the differential evolution algorithm [29] can introduce disturbances to particles during the search process, thus possessing strong global search capabilities. Therefore, integrating the mutation strategy can enhance the global convergence ability of the particle swarm optimization algorithm.

The traditional mutation strategy is to weight the difference between two random individuals in the population and add it to the third random individual to obtain the mutation individual, as shown in the following Equation (6):

$$\begin{cases} h_i = q_{r1} + f \\ f = W(q_{r2} - q_{r3}) \end{cases} \tag{6}$$

where q_{r1}, q_{r2}, q_{r3} are three particle individuals randomly selected in the population, with $r1 \neq r2 \neq r3$; h_i is the ith mutant individual; W is the scaling factor; and f is the perturbation. According to Equation (6), it can be seen that in the early stage of the search due to the randomness of the initial distribution of the particle swarm resulting in a large discrepancy between particles, the traditional mutation method applies a large perturbation to the particles. As the search proceeds, the particle swarm gradually converges, the difference between particles becomes smaller, and the perturbation becomes smaller.

The particle swarm optimization algorithm has the advantage of fast convergence in the early stage, but its drawback is that it easily gets stuck in local optima in the later stage. Directly applying traditional mutation strategies to the individual best values of the particle swarm can result in the individual best values being minimally disturbed in the late stage of convergence, making it difficult for the particle swarm to escape from local optima.

To address the aforementioned issue, this paper proposes a new mutation strategy, the core idea of which is to reduce disturbances to particles when the differences between particles are significant in the early convergence stage of the particle swarm, thereby accelerating the early convergence process. When the differences between particles are small in the later stages of convergence, the degree of disturbance to particles is increased to help the particle swarm escape from local optima. The new mutation strategy is shown in Equation (7), written as follows:

$$\begin{cases} h_i = q_{r1} + f(D) \\ f(D) = \frac{Dimension}{1+\exp^{\left(\frac{-1}{10\times D}\right)}} - \frac{Dimension}{2} \\ D = \frac{\sum \frac{q_{r2}-q_{r3}}{\max(q_{r2}-q_{r3})}}{n} \end{cases} \tag{7}$$

In the formula, $Dimension$ is the search range of particles, D is the difference degree of particles, and n is the number of perturbations. Assuming that the search range $Dimension = 100$, the result of perturbation $f(D)$ is shown in Figure 6, and the perturbation $f(D)$ of particles decreases as the degree of difference D of particles increases.

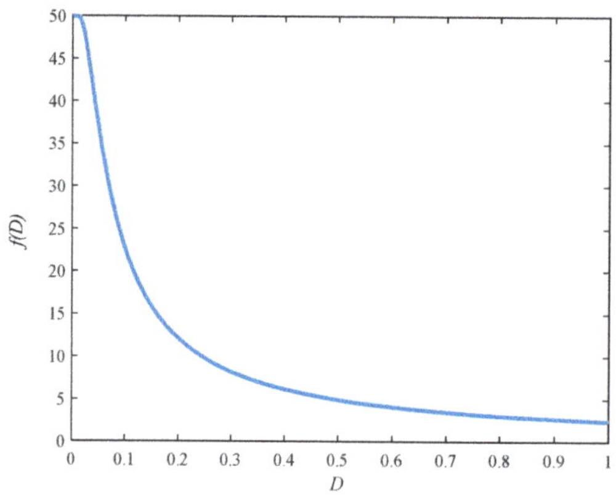

Figure 6. Improved mutation strategy for particle perturbation.

4.3. Specific Steps for PE-Match

The previous sections have elaborated on the epipolar constraint principle and the improved mutation strategy. The specific steps of the feature matching method based on improved particle swarm and epipolar constraint (PE-match) are as follows:

Step 1: Obtain a set of coarse matching point pairs through the traditional ORB algorithm.

Step 2: Initialize the particle population. Firstly, set the population number; after that, each particle reads randomly from the coarse matching without repeating eight pairs of matching point pairs of pixel coordinates and set the particle velocity to zero.

Step 3: Design a fitness function based on the characteristics of correct matches having the same constraint properties in a photovoltaic station scene, while incorrect matches do not possess the same constraint properties and calculate the fitness of each particle. The number of matching point pairs satisfying the epipolar constraints is used as the fundamental for evaluating the quality of the fundamental matrix for the particle's fitness assessment. The specific calculation steps are as follows:

(a) The fundamental matrix $\boldsymbol{F}$ is calculated from the coordinates of the eight pairs of pixels in the particle.

(b) The epipolar line is calculated from the fundamental matrix $\boldsymbol{F}$ and the pixel coordinates of the first point in the coarse matched point pair, and the distance d from this epipolar line is calculated using the second point in the matched point pair as in Equation (8). If the distance is less than a certain threshold, the matched pair is considered a correct match.

$$d = \left| \frac{ax + by + c}{\sqrt{a^2 + b^2}} \right| \tag{8}$$

(c) Using the set of $\boldsymbol{F}$ and coarse matching point pairs, we complete step (b) with j random pairs and count the number of correct matches as the fitness value. Since the fitness calculation takes up a lot of time in the particle swarm algorithm, setting the number of coarse matching point pairs participating in fitness calculation as j ensures that the algorithm does not increase in time consumption as the number of coarse matching point pairs increases. Usually, j is set to 50 or 100.

Step 4: Record and update $Gbest_t$ and $Pbest_i$. $Gbest_t$ stores the position of the particle with the highest fitness found by the entire population, while $Pbest_i$ stores the position of the particle with the highest fitness found by the individual particle itself.

Step 5: Update the velocity and position of the particles. The particle swarm optimization algorithm during the optimization process, the velocity and position of the particles are updated by the search experience of the whole swarm and the search experience of the individual particles as shown in Figure 7.

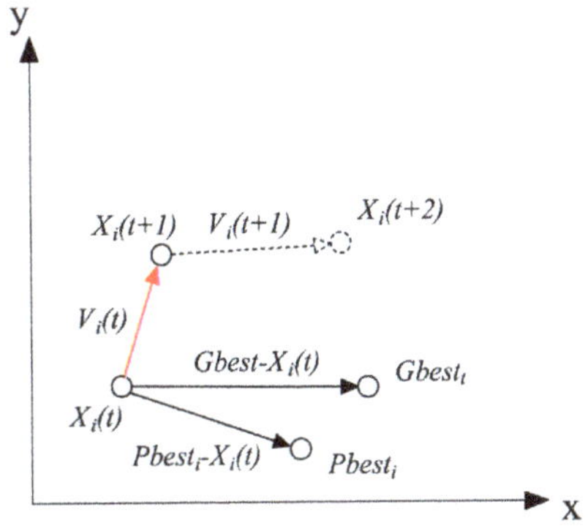

Figure 7. Particle update schematic.

The velocity and position update rules are as shown in Equation (9).

$$\begin{cases} V_i(t+1) = \omega V_i(t) + c_1 r_4 (Pbest_i - X_i(t)) \\ \qquad +c_2 r_5 (Gbest_t - X_i(t)) \\ X_i(t+1) = X_i(t) + V_i(t) \end{cases} \tag{9}$$

In the formula, $V_i(t)$ and $X_i(t)$ are the velocity and position of the ith particle at iteration t; r_4 and r_5 are two stochastic parameters uniformly distributed in [0, 1]; c_1 and c_2 are the learning weights for the individual particle extremes and the particle population extremes; and ω is the inertia weight.

Step 6: Perturb $Pbest_i$. In this paper, $Pbest_i$ is perturbed using improved mutation and crossover operations to prevent the algorithm from getting stuck in local optima. The mutation operation is represented by Formula (17).

$$\begin{cases} h_i = Pbest_i + \frac{Dimension}{1+\exp^{\left(\frac{-1}{10\times D}\right)}} - \frac{Dimension}{2} \\ D = \frac{\sum \frac{Pbest_{r2} - Pbest_{r3}}{\max(Pbest_{r2} - Pbest_{r3})}}{n} \end{cases} \tag{10}$$

In the formula, h_i is the mutated individual, i is the particle index, D is the degree of particle difference, $Dimension$ is the particle search range, n is the number of subtractions, and $r2$ $r3$ are random numbers and $r2 \neq r3$.

The crossover operation is expressed as Equation (11) and is performed on the $Pbest_i$ of the particle population. L is the crossover probability.

$$Pbest_i = \begin{cases} h_i, rand \leqslant L \\ Pbest_i, rand > L \end{cases} \tag{11}$$

Step 7: Verify that the $Gbest_t$ has not changed significantly over multiple iterations, and if it has not changed significantly or if the maximum number of iterations is reached then output the fundamental matrix with the highest fitness. If the output condition is not reached then go back to Step 3 and continue iterating.

Step 8: The false matches in the coarse matching are eliminated based on the fundamental matrix output from Step 7 according to the principle of epipolar constraints.

5. Results

5.1. Feature Matching Experiment

5.1.1. Description of the Data Set

For the feature matching experiment, the OxfordVGG [30] dataset was used, from which two sets of scene images with high feature similarity, simple structure, and other non-structural factors that are also present in photovoltaic power plants were selected, as shown in Figure 8. Scene 1 features a gradually blurry tree, while Scene 2 consists of a wall with changing perspectives, each set containing four images. The dataset provides an essential matrix for validating the matching results.

(1) (2) (3) (4)

(**a**) Scene 1

Figure 8. *Cont.*

(1) (2) (3) (4)

(**b**) Scene 2

Figure 8. Optical images of OxfordVGG dataset.

5.1.2. Experimental Environment Configuration and Parameter Setting

The computer hardware specifications used in this experiment are as follows: AMD Ryzen 7 5800H processor (USA), 16 GB of memory (USA), a 64-bit Windows 11 operating system, and Visual Studio 2017 as the programming software. The third-party library OpenCV 4.1.0 was used for ORB feature extraction and description.

To validate the advantage of the algorithm proposed in this paper (ORB+PEmatch), the matching performance is compared with the traditional ORB algorithm, ORB+HAMMING algorithm [31], and ORB+RANSAC algorithm [32] in the actual environment of a photovoltaic power plant. The traditional ORB algorithm provides the same set of coarse matching point pairs as for the other three algorithms. The ORB+HAMMING algorithm uses twice the minimum HAMMING distance as the threshold to eliminate mismatches. ORB+RANSAC employs the random sample consensus algorithm to compute the fundamental matrix, with an inlier distance threshold set to three. After obtaining the fundamental matrix, mismatches are eliminated using the epipolar constraint with a common threshold of three. The inlier distance threshold is the pixel distance from the matching points to the epipolar line. Due to the noise often present in images, setting the inlier distance threshold too small can filter out correct matches, while setting it too large may result in missing incorrect matches. Based on multiple experiments and considering commonly used values from past research, the experiment adopts an approximated optimal value of three as the inlier distance threshold.

Random sampling is contrasted with the particle swarm optimization sampling discussed herein. To ensure fairness in performance comparison, the PE-match's inner point distance threshold is set to three. Based on multiple experimental results, the near-optimal parameter settings for the particle swarm are determined: the population size is set to 20, which influences the search capability of the particle swarm algorithm, though an excessively high number can reduce the algorithm's real-time performance. The number of differences n is set to three, influencing the mutation strategy's sensitivity to differences among particles. The crossover probability L is set to 50%, determining the influence of the mutation strategy on *Pbest*. The learning weights for individual and global extrema c_1 and c_2 are both set to 0.5, while the inertia weight ω is also set to 0.5. The number of matching point pairs j involved in fitness calculation is set to 100; this value allows the particle swarm algorithm to maintain real-time performance without degrading despite an increase in feature points, though setting it too low can diminish the optimization capability. The iteration ends, and the optimal fundamental matrix is output if there are no significant changes in *Gbest* over 8 (It_max) iterations, a measure included to ensure the stability of the algorithm's real-time performance. Taking the 1–4 matching experiments of Scene 1 from the OxfordVGG dataset as an example, Figure 9 illustrates the influence of key parameters on the algorithm's accuracy and processing time. For different matching scenarios, this paper will make minor adjustments based on the aforementioned parameters to adapt to varying conditions. Because the outcomes of the particle swarm optimization are related to the number of iterations and the initial distribution of particles, subsequent experimental data related to this algorithm will be the averages of multiple trials.

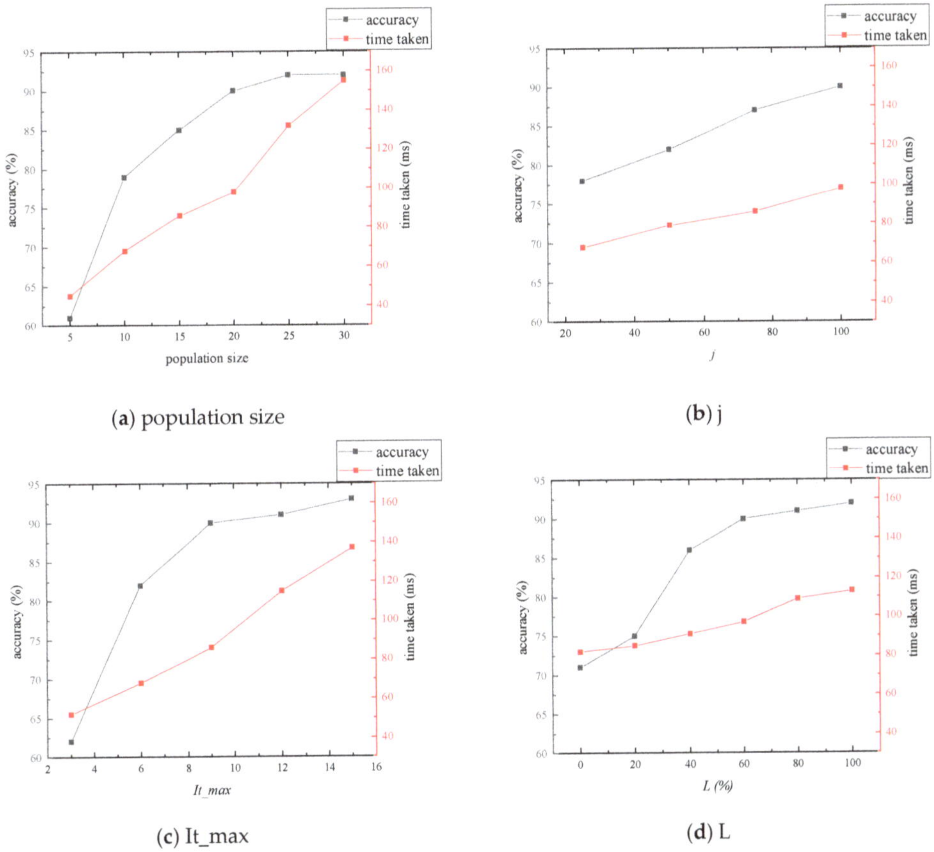

Figure 9. Influence of important parameters on accuracy and time consuming.

5.1.3. Experimental Results and Analysis

Figure 10 displays the fitness changes with the number of iterations t in the 1–3 multiple matching experiments of Scene 1 in the OxfordVGG dataset using the traditional particle swarm algorithm and the improved particle swarm algorithm after lifting the constraint on the number of coarse matches j involved in the computation. It can be observed that the traditional particle swarm algorithm is prone to falling into local optima, while the improved particle swarm algorithm inherits the fast convergence advantage of the traditional particle swarm algorithm in the early iterations and shows better capability to escape local optima in the later iterations.

The accuracy of each algorithm is shown in Figure 11, where accuracy refers to the ratio of correctly matched pairs to all matched pairs in the matching results of each algorithm. Image feature matching is performed using the dataset images, with the need to match three image pairs in each scenario, namely 1-2, 1-3, and 1-4, with images becoming more dissimilar as their numbers increase. The matching accuracy of the proposed algorithm is on average 19.5% higher than that of the ORB algorithm, 14.0% higher than that of the ORB+HAMMING algorithm, and 6.0% higher than that of the ORB+RANSAC algorithm.

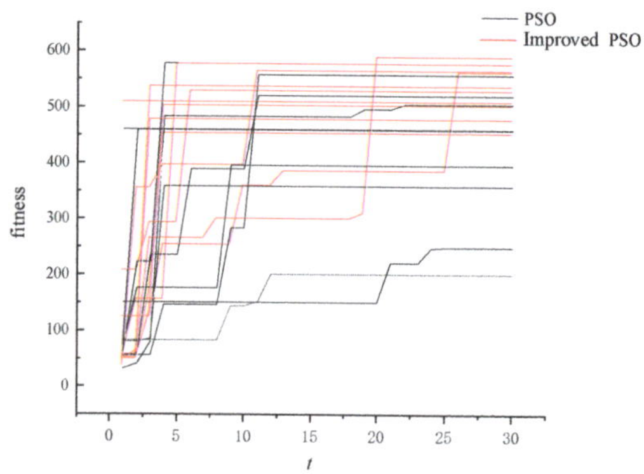

Figure 10. Line chart of fitness change.

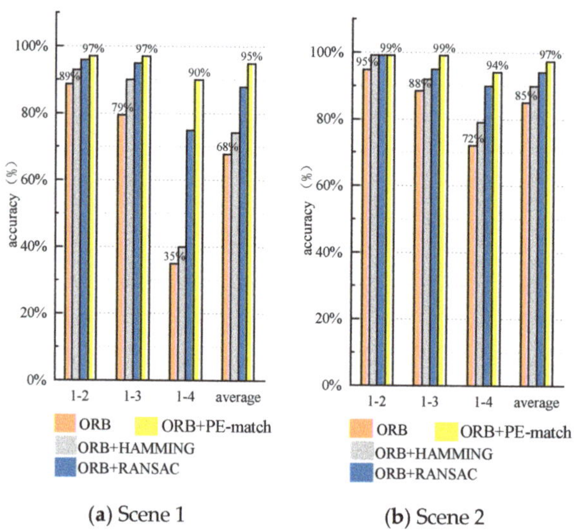

Figure 11. Matching accuracy based on the data.

Figure 12 presents the recall rate results of the three algorithms in the six matching experiments for the two scenarios. The recall rate is defined as the ratio of the number of correctly matched point pairs after eliminating misalignments to the number of correctly matched point pairs in the coarse matches (traditional ORB) before removing misalignments. From the graph, it can be observed that the ORB+HAMMING algorithm eliminates a large number of correctly matched point pairs in the coarse matches, resulting in sparse matching results. In contrast, both our algorithm and the ORB+RANSAC algorithm perform well in retaining correctly matched point pairs from the coarse matches, with our algorithm showing a significantly higher recall rate than the ORB+RANSAC algorithm. These results suggest that our algorithm can effectively preserve correctly matched point pairs while eliminating misalignments, thus improving the quality of matching.

The time consumption of each algorithm is shown in Table 1, which records the average time consumption results of multiple experiments in four scenarios for each algorithm. The ORB algorithm demonstrates excellent speed, with our algorithm taking an average of 2.39 times longer than the ORB algorithm.

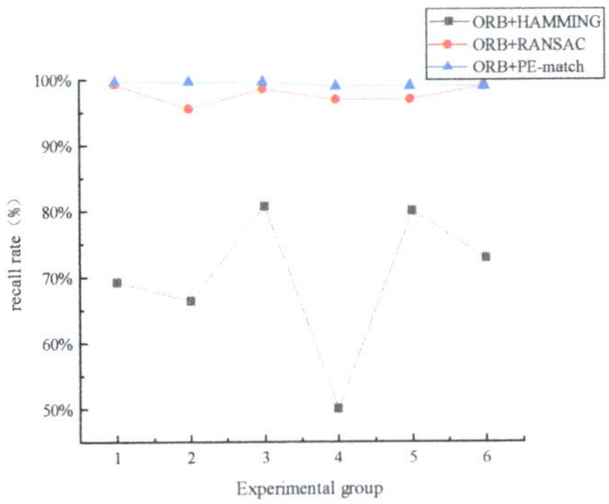

Figure 12. The recall rate of matching results based on the dataset.

Table 1. The time taken for each algorithm to run (unit: ms).

	ORB	ORB +HAMMING	ORB +RANSAC	ORB +PE-Match
Scene 1	62	62	63	151
Scene 2	55	55	56	132
average	59	59	60	142

5.2. SLAM Trajectory Experiment

5.2.1. Description of the Data Set

To validate the performance metrics of the improved algorithm, experiments were conducted using the widely used and authoritative TUM dataset [33] in the field of visual SLAM research. This dataset is provided by the Technical University of Munich in Germany and consists of a large number of data sequences, including RGB-D data captured by Microsoft Kinect sensors (USA) and ground truth pose data obtained from motion capture systems.

Based on research results regarding high feature similarity, weak texture, and simple structure in the field of photovoltaic power plant scenes, the algorithm was validated using the scenarios named structure_notexture_near, nostructure_texture_far, and structure_texture_far. The structure_notexture scenario refers to scenes with 3D contours but no texture features; the nostructure_texture scenario refers to scenes with texture features but no 3D contours; and the structure_texture scenario refers to scenes with both 3D contours and texture features. These three sequences can evaluate the algorithmic performance of

PE-SLAM and ORB-SLAM2 in unstructured scenes. The three sequences are shown in Figure 13.

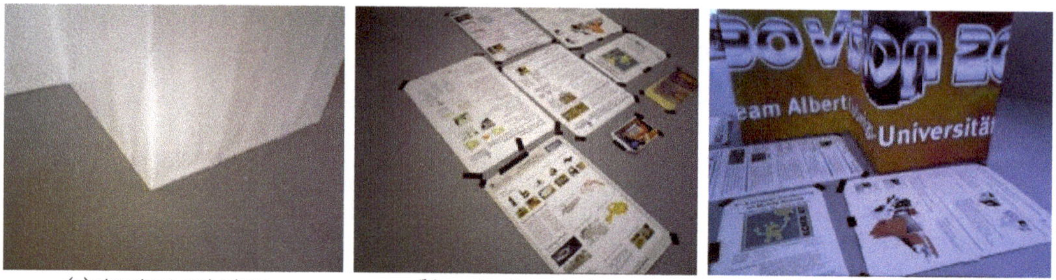

(**a**) structure_notexture (**b**) nostructure_texture (**c**) structure_texture

Figure 13. Unstructured scenarios for datasets.

5.2.2. Experimental Environment Configuration and Parameter Setting

The hardware configuration for the experiment is as follows: AMD Ryzen 7 5800H processor (USA), NVIDIA GeForce 3060 6 GB GPU, 16GB RAM, and 64-bit Ubuntu 18.04 operating system. The experimental parameters are set according to the TUM1.yaml file of ORB-SLAM2. The PE-match parameters are mostly consistent with the settings in Section 5.1.2, with the number of coarse matching point pairs involved in fitness calculation (denoted as "j") adjusted to 50. The iteration stops and outputs the optimal fundamental matrix when there is no significant change in $Gbest_t$ within 6 iterations. Such parameter adjustments are made with the consideration of reducing the time consumption of PE-match in the PE-SLAM system to free up time for backend optimization, meaning that PE-SLAM aims to improve the accuracy of the motion model while minimizing any negative impact on the effectiveness of the backend optimization.

The TUM dataset provides real camera motion trajectory information, which is used to evaluate the trajectory information obtained by algorithms against the ground truth trajectory. The evaluation metrics include absolute trajectory error (ATE) and relative pose error (RPE). ATE represents the direct difference between estimated and ground truth poses, reflecting the trajectory consistency at a global level. RPE measures the local accuracy of the trajectory within fixed time intervals, corresponding to trajectory drift.

The root mean square error (RMSE) is used to reflect ATE, with $trans(F_i)$ representing the translation part, written as follows:

$$RMSE(F_{1:n}) := \left(\frac{1}{n} \sum_{i=1}^{n} trans(F_i)^2 \right)^{\frac{1}{2}} \tag{12}$$

$$F_i := Q_i^{-1} S P_i \tag{13}$$

In the formula, Q_i represents the estimated pose of the i-th frame, P_i represents the real pose of the i-th frame, F_i represents the absolute trajectory error of the i-th frame, and S is the scale conversion matrix.

Given the known time interval $\triangle$ and the total number of image frames n, it is possible to calculate $m = n - \triangle$ RPEs, with the RMSE defined as follows:

$$RMSE(E_{1:n}, \triangle) := \left(\frac{1}{m} \sum_{i=1}^{m} trans(E_i)^2 \right)^{\frac{1}{2}} \tag{14}$$

$$E_i := \left(Q_i^{-1} Q_{i+\triangle} \right)^{-1} \left(P_i^{-1} P_{i+\triangle} \right) \tag{15}$$

In the formula: Q_i represents the estimated pose of the i-th frame, P_i represents the true pose of the i-th frame, and E_i represents the relative pose error of the i-th frame.

Due to the fact that various modules of ORB-SLAM2 and PE-SLAM, such as target detection, tracking, backend optimization, and mapping are independent threads, the most time-consuming among them is the tracking thread. The speed of the tracking thread determines the system processing speed. Additionally, since the backend optimization occurs between the tracking and the arrival of the next frame, its effectiveness is not only dependent on the tracking thread's performance but also on tracking time consumption. Moreover, PE-SLAM has made improvements to the motion model of ORB-SLAM2. Considering these factors, a tracking time evaluation algorithm is adopted to assess the real-time performance comprehensively.

5.2.3. Experimental Results and Analysis

Experimental verification of PE-SLAM and ORB-SLAM2 is conducted, with absolute trajectory errors shown in Table 2 and relative reduction rates shown in Table 3.

Table 2. Absolute trajectory errors (ATE, m) of PE-SLAM and ORB-SLAM2.

Experimental Sequences	ORB-SLAM2			PE-SLAM		
	REMS	MEAN	S.D.	REMS	MEAN	S.D.
structure_notexture	0.030	0.026	0.014	0.026	0.022	0.013
nostructure_texture	0.125	0.101	0.074	0.070	0.063	0.030
structure_texture	0.020	0.019	0.007	0.014	0.013	0.005

Table 3. Relative reduction in absolute trajectory error (%) of PE-SLAM compared to ORB-SLAM2.

Experimental Sequences	REMS	MEAN	S.D.
structure_notexture	13.3%	15.4%	7.1%
nostructure_texture	44.0%	37.6%	59.5%
structure_texture	30.0%	31.6%	28.6%

It can be seen that the motion model improvement based on PE-match has brought about an enhancement in system accuracy. In the case of nostructure_texture, where only texture is present without contours, PE-SLAM achieves a 44.0% reduction in error, indicating that the improved algorithm provides better system accuracy in situations where textures are similar and lack contour structures.

Table 4 shows the relative pose errors, while Table 5 presents the relative reduction rate of pose errors. It can be observed that the reduction effect of relative pose errors is consistent with that of absolute trajectory errors. In ORB-SLAM2, the strategy for backend optimization involves local trajectory optimization after tracking completion and before the arrival of the next frame. The tracking time needs to be minimized to ensure the effectiveness of backend trajectory optimization. The fact that the absolute trajectory errors are smaller than the relative pose errors in PE-SLAM indicates that the improved algorithm's tracking time meets the real-time requirements for system backend optimization. The average tracking time for each sequence is detailed in Table 6.

Table 4. Relative pose errors (RPEs, m) of PE-SLAM and ORB-SLAM2.

Experimental Sequences	ORB-SLAM2			PE-SLAM		
	REMS	MEAN	S.D.	REMS	MEAN	S.D.
structure_notexture	0.046	0.039	0.024	0.043	0.036	0.023
nostructure_texture	0.255	0.164	0.195	0.113	0.092	0.066
structure_texture	0.032	0.028	0.015	0.026	0.022	0.012

Table 5. Relative reduction in relative pose Error (%) of PE-SLAM compared to ORB-SLAM2.

Experimental Sequences	REMS	MEAN	S.D.
structure_notexture	6.5%	7.7%	4.2%
nostructure_texture	55.7%	44.0%	66.2%
structure_texture	18.8%	21.4%	20.0%

Table 6. Average tracking time (ms) of PE-SLAM and ORB-SLAM2.

Experimental Sequences	ORB-SLAM2	PE-SLAM
structure_notexture	10	35
nostructure_texture	21	68
structure_texture	25	70

As shown in Figures 14–16, based on the ground truth provided by the TUM dataset and the CameraTrajectory obtained from the SLAM algorithm, six trajectory error plots is generated. In the trajectory error plot, the actual trajectory is represented by a black line segment, while the estimated trajectory is depicted by a blue line segment. Red short lines indicate the distance difference between the predicted position and the actual position of the camera at the same time point. The length of the red short lines visually reflects the degree of error.

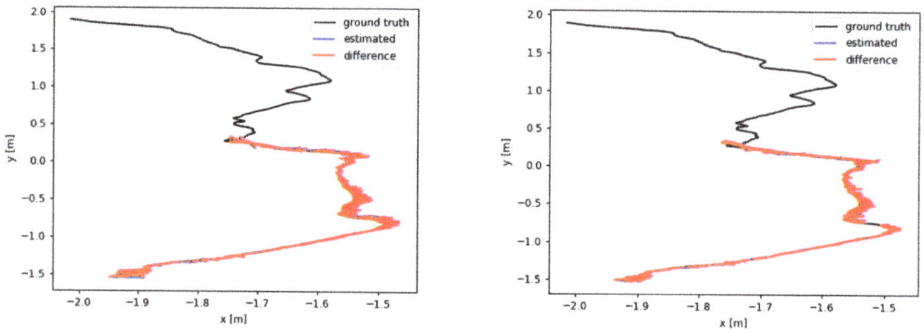

(**a**) ORB-SLAM2 (**b**) PE-SLAM

Figure 14. Trajectory error plot of the structure_notexture sequence.

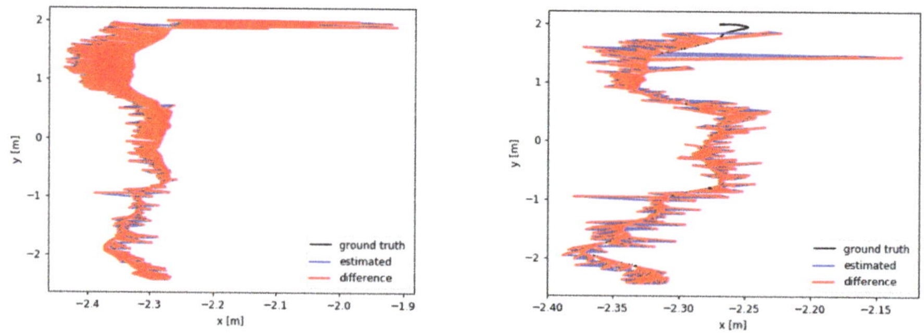

(**a**) ORB-SLAM2 (**b**) PE-SLAM

Figure 15. Trajectory error plot of the nostructure_texture sequence.

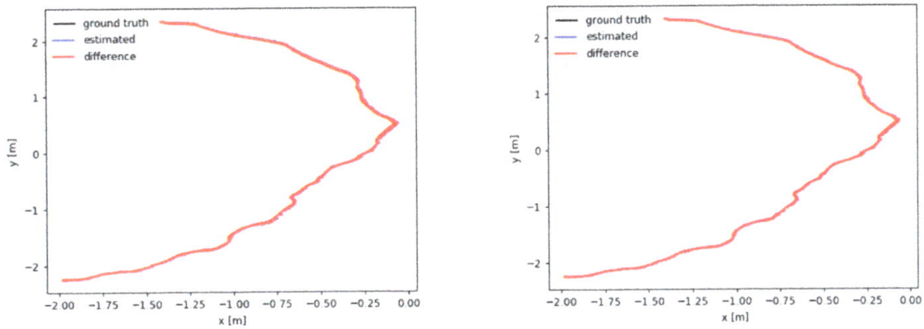

(**a**) ORB-SLAM2 (**b**) PE-SLAM

Figure 16. Trajectory error plot of the structure_texture sequence.

As shown in Figures 17–19, we generate six dense point cloud maps based on the three image sequences mentioned above: "structure_notexture", "nostructure_texture", and "structure_texture". In this map, ORB-SLAM2 represents the dense point cloud map established after integrating a dense mapping thread into the original algorithm, and PE-SLAM represents the dense point cloud map built from the improved tracking thread. The mapping results of each sequence are shown in the following image.

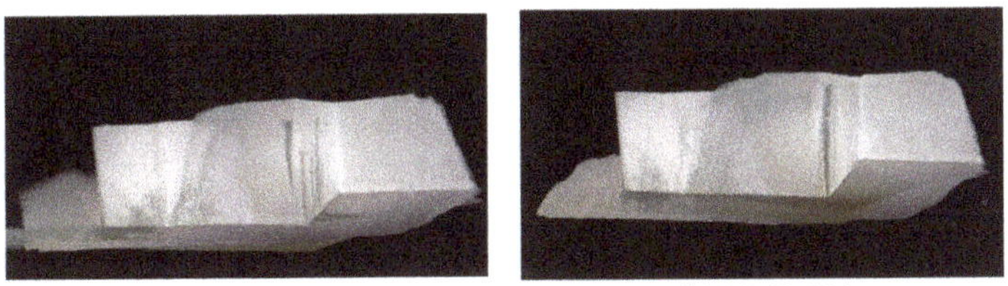

(**a**) ORB-SLAM2 (**b**) PE-SLAM

Figure 17. The 3D map generated from the structure_notexture sequence.

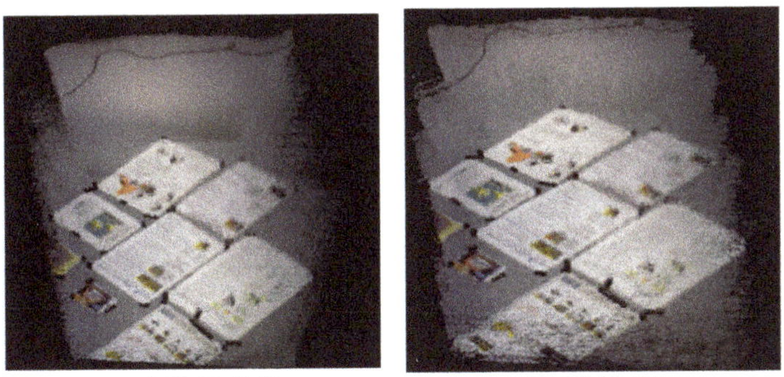

(**a**) ORB-SLAM2 (**b**) PE-SLAM

Figure 18. The 3D map generated from the nostructure_texture sequence.

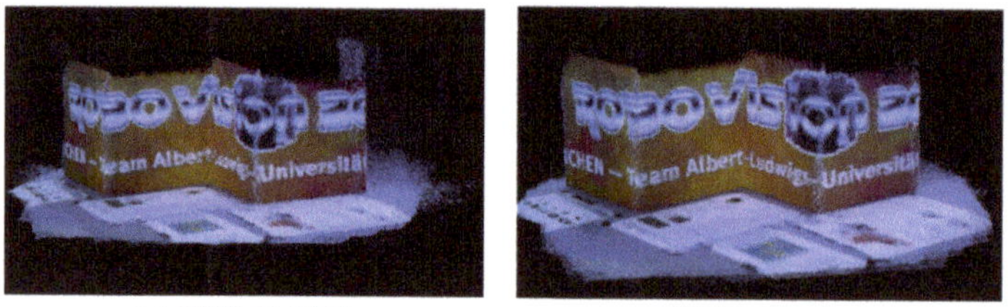

(**a**) ORB-SLAM2 (**b**) PE-SLAM

Figure 19. The 3D map generated from the structure_texture sequence.

5.3. Simulation Experiments of Photovoltaic Scenarios

5.3.1. Establishment of the Simulation Environment Based on Gazebo

Gazebo uses simple text formats (such as URDF, SDF, etc.) to describe and create simulation models, which include attributes such as geometric shapes, mass, inertia, kinematics, and dynamics. As shown in Figure 20, a robot model is built on Gazebo 9, with the blue block on top representing the camera appearance model. The Kinect camera simulation code provided by Gazebo is loaded onto the camera model, thereby realizing the robot's perception part. The robot is driven using a two-wheel drive, and the drive wheels are controlled using the differential drive model provided by Gazebo.

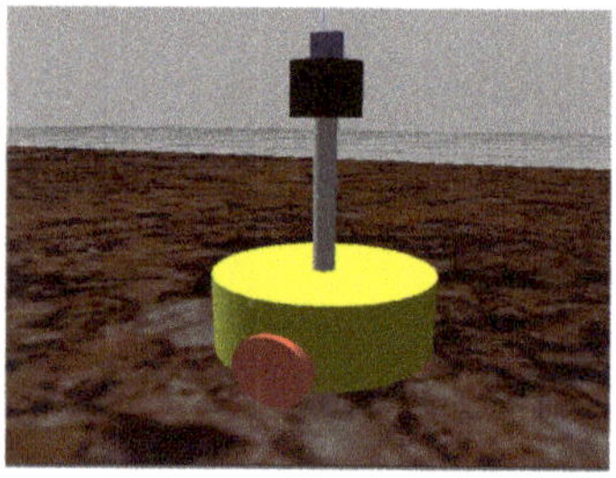

Figure 20. Robot model based on Gazebo.

The simulation of the photovoltaic scene is shown in Figure 21. It consists of a wall, photovoltaic modules, land, and shrub vegetation. To simulate the unstructured factors of a real photovoltaic power station, such as high feature similarity, weak textures, and simple structures, the wall is made of gray bricks with high feature similarity. The ground is made of the most common land found in photovoltaic power stations. A small number of green plants are represented by small, common shrubs and trees. These elements collectively create a localized simulation scene of a photovoltaic power station. The image sequence of this scene is named Si.

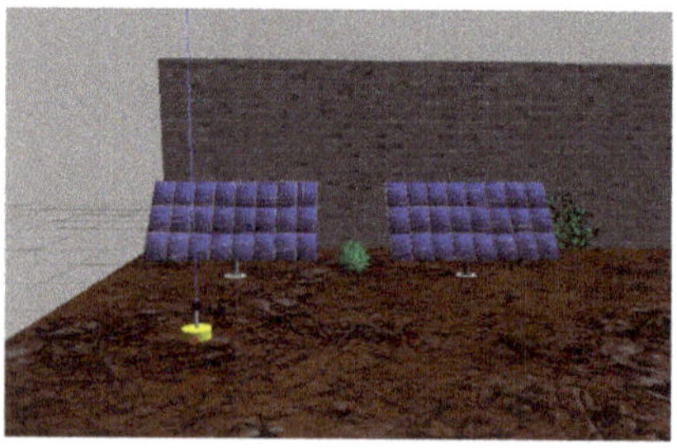

Figure 21. Photovoltaic power station simulation scene based on Gazebo.

5.3.2. Simulation Experiment Results and Analysis

The evaluation metrics for the simulation trajectory experiment are the same as those for the dataset experiment. By conducting simulation experiments on the aforementioned Si, PE-SLAM and ORB-SLAM2 were validated. The absolute trajectory error validation values are shown in Table 7, and the relative reduction rates are shown in Table 8.

Table 7. Absolute Trajectory Errors (ATE, m) of PE-SLAM and ORB-SLAM2.

Experimental Sequences	ORB-SLAM2			PE-SLAM		
	REMS	**MEAN**	**S.D.**	**REMS**	**MEAN**	**S.D.**
Si	0.026	0.024	0.009	0.023	0.021	0.009

Table 8. Relative Reduction in Absolute Trajectory Error (%) of PE-SLAM compared to ORB-SLAM2.

Experimental Sequences	REMS	MEAN	S.D.
Si	11.5%	12.5%	0%

As shown in Table 8, the trajectory errors of both ORB-SLAM2 and PE-SLAM in the Gazebo photovoltaic power station simulation scene are at the centimeter level, indicating that both algorithms achieve good localization accuracy in this scenario. Additionally, PE-SLAM demonstrates better system accuracy, reducing the absolute trajectory error by 11.5% compared to ORB-SLAM2.

Tables 9 and 10 present the results of the relative pose error in the photovoltaic scene simulation. The fact that the relative trajectory error is larger than the absolute trajectory error indicates that the back-end optimization of both algorithms was not affected by the tracking thread's processing time, as both performed local and global trajectory optimization. The average tracking time for each sequence is shown in Table 11.

Table 9. Relative pose errors (RPEs, m) of PE-SLAM and ORB-SLAM2.

Experimental Sequences	ORB-SLAM2			PE-SLAM		
	REMS	**MEAN**	**S.D.**	**REMS**	**MEAN**	**S.D.**
Si	0.095	0.066	0.067	0.090	0.064	0.063

Table 10. Relative reduction in relative pose error (%) of PE-SLAM compared to ORB-SLAM2.

Experimental Sequences	REMS	MEAN	S.D.
Si	5.2%	3.0%	5.9%

Table 11. Average tracking time (ms) of PE-SLAM and ORB-SLAM2.

Experimental Sequences	ORB-SLAM2	PE-SLAM
Si	23	49

As shown in Figure 22, the trajectory error plot s is drawn based on the ground truth of the robot's actual trajectory provided by the Gazebo simulation system and the camera trajectory (CameraTrajectory) derived from the algorithm. In the trajectory error graph, the ground truth trajectory is represented by black line segments, while the estimated trajectory is shown with blue line segments. Red short lines indicate the distance difference between the predicted and actual positions of the camera at the same time point, and the length of these red short lines visually reflects the degree of error.

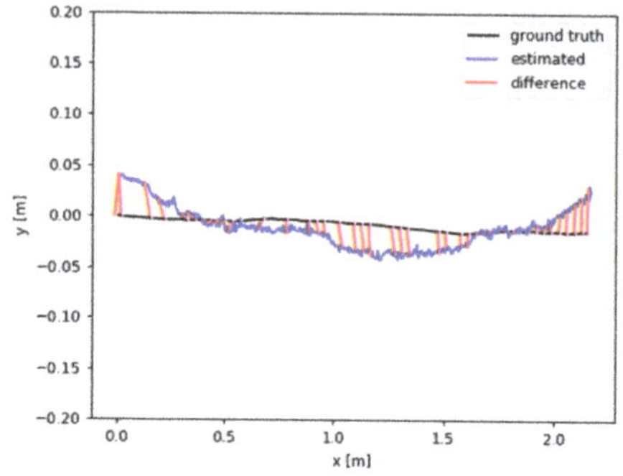

(a) ORB-SLAM2

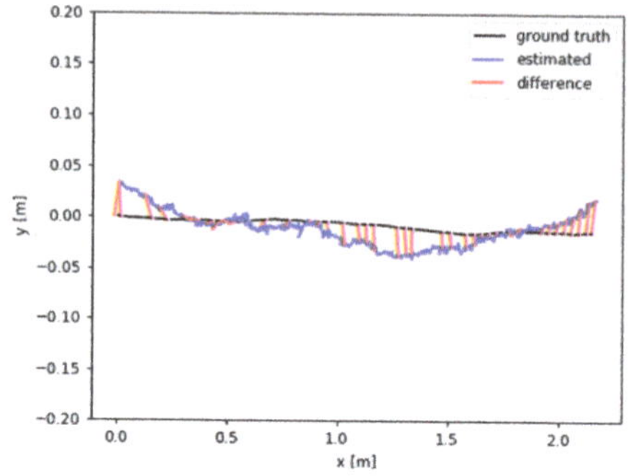

(b) PE-SLAM

Figure 22. Trajectory error plot of the Si.

As shown in Figure 23, two dense point cloud maps is generated from the Si image sequence mentioned earlier. ORB-SLAM2 represents the dense point cloud map created by adding a dense mapping thread to the original algorithm, while PE-SLAM represents the dense point cloud map built using the poses obtained from the improved tracking thread. The mapping results for each sequence are shown in the following figures.

(a) ORB-SLAM2

(b) PE-SLAM

Figure 23. The 3D map generated from the Si.

Due to the relatively uniform texture and structure of the photovoltaic power station simulation scene, which enhances the scene's unstructured factors, the mapping performance would be better in a real photovoltaic scene.

5.4. Experiments on Building Maps in Photovoltaic Scenarios

5.4.1. Experimental Scenario Composition

The mapping experiments used two image sequences captured by an RGB-D camera. One of the scenes is indoors, where controllable lighting conditions, such as changes in brightness and backlighting, are possible, as shown in Figure 24. Additionally, compared to outdoor scenes, indoor environments typically generate less noise with RGB-D cameras, leading to more accurate depth estimations. The outdoor scene, shown in Figure 25, uses a real photovoltaic power station scene. The figure illustrates that in a real photovoltaic power station, there are complex textures and structures on buildings and the ground, and the photovoltaic panels have stains with varying textures. These factors reduce the scene's unstructured characteristics. The indoor sequence is named In, and the outdoor sequence is named Real.

(a) bright (b) dark (c) backlighting

Figure 24. Partial images of indoor scenes.

(a) viewpoint 1 (b) viewpoint 2 (c) viewpoint 3

Figure 25. Partial images of outdoor scenes.

5.4.2. Mapping Results and Analyses

As shown in Figures 26 and 27, we generate four dense point cloud maps based on the two image sequences mentioned above, "In" and "Real". In this map, ORB-SLAM2 represents the dense point cloud map established after integrating a dense mapping thread into the original algorithm, and PE-SLAM represents the dense point cloud map built from the improved tracking thread. The mapping results of each sequence are shown in the following image.

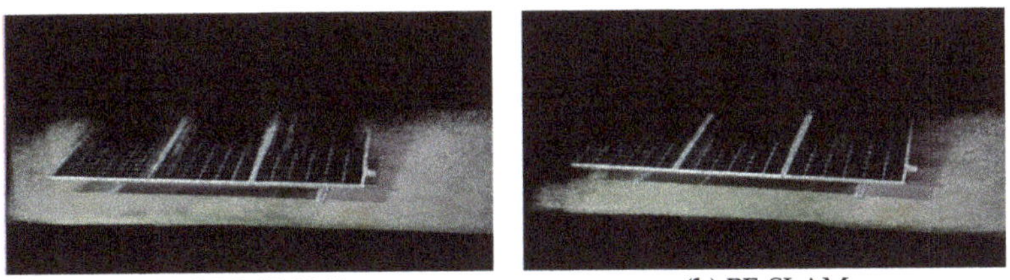

(a) ORB-SLAM2 (b) PE-SLAM

Figure 26. The 3D map generated from In sequence.

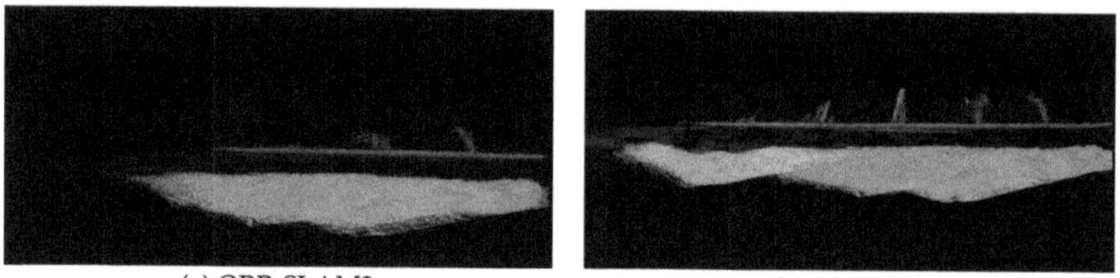

(a) ORB-SLAM2 (b) PE-SLAM

Figure 27. The 3D map generated from Real sequence.

The map established by PE-SLAM reduces the lateral and angular point cloud overlaps aligned with the camera motion direction compared to ORB-SLAM2. This reflects that the camera poses obtained by the improved algorithm are more accurate, allowing for better tracking performance in this scene. The established map shows a small amount of longitudinal overlaps inconsistent with the camera motion direction, with outdoor scenes exhibiting significantly more longitudinal overlaps than indoor scenes. This is mainly due to deviations in the camera's depth estimation, which become more pronounced under strong outdoor lighting conditions. As shown in Figure 19, the mapping area of PE-SLAM is larger than that of ORB-SLAM2, which is due to tracking failures midway with ORB-SLAM2, indicating that PE-SLAM has a better robustness.

6. Discussion

In the photovoltaic power station scenario, the feature similarity makes the ORB algorithm prone to mismatches during the feature matching stage. Extensive experiments have shown that these mismatches are random and lack consistent constraints, while correct matches share the same constraint conditions. Based on the characteristics of mismatches generated by ORB in the photovoltaic station scenario, an improved particle swarm optimization (PSO) algorithm is used to identify the constraint conditions that comply with the majority of matched point pairs. This helps in determining the correct matching constraints and subsequently eliminating mismatches. The epipolar constraint, which includes both intrinsic and extrinsic camera parameters, is suitable for both near and far points and has an analytical solution that meets the real-time requirements of SLAM algorithms. Therefore, this paper chooses the epipolar constraint as the constraint condition. Since the PSO algorithm is prone to local optima, a new mutation strategy is designed to perturb the Pbest of the particle swarm, enhancing the global search capability of the algorithm in the later stages. This improves the PSO algorithm's ability to address the sensitivity of epipolar constraints to mismatches. Based on the above theory, the paper designs PE-match to improve the motion model of ORB-SLAM2, thereby reducing the number of mismatches in photovoltaic power station scenarios and improving SLAM's localization accuracy and mapping performance.

Table 12 presents the characteristics and main experimental results of PE-SLAM in the six experimental scenarios mentioned earlier. The TUM dataset is more authoritative, and based on the experimental results, PE-SLAM has achieved better localization accuracy and mapping performance compared to ORB-SLAM2. In the structure_notexture and structure_texture scenarios, the ATE (REMS) is at the centimeter level, while in the nostructure_texture scenario, the ATE (REMS) is at the decimeter level, with the most significant improvement in this scenario, achieving an ATE (REMS) reduction rate of 44.0%. In the simulation experiments, the ATE (REMS) is only 0.023, indicating that PE-SLAM is suitable for the photovoltaic power station simulation scenarios in terms of localization accuracy. However, the mapping performance is moderate, which we believe is due to the high degree of texture similarity in the simulation scenarios. In real photovoltaic power station scenarios, the presence of background buildings, ground, and stained photovoltaic panels weakens the unstructured scene factors. Theoretically, this should result in better mapping performance, which is confirmed by the mapping experiments in the In scenario.

The mapping performance in the Real scenario is noticeably lower than in the In scenario, which is due to the significant influence of outdoor lighting on the RGB-D camera in outdoor environments. Overall, PE-SLAM has surpassed ORB-SLAM2 in both localization accuracy and mapping performance in the unstructured environments of photovoltaic power stations.

Table 12. Summary of experimental results.

Dataset.	Scene	Degree of Feature Similarity	Degree of Weak Texture	Degree of Structural Simplicity	Degree of Camera Noise	ATE (REMS) of PE-SLAM	ATE Reduction Rate	PE-SLAM Mapping Performance
TUM	structure_notexture	High	High	Moderate	Low	0.026	13.3%	High
	nostructure_texture	Moderate	Low	High	Low	0.070	44.0%	Moderate
	structure_texture	Low	Low	Moderate	Low	0.014	30.0%	High
Sim. Exp.	Si	High	High	Moderate	Low	0.023	11.5%	Moderate
Exp.	In	High	Moderate	Moderate	Moderate	-	-	High
	Real	High	Moderate	Moderate	High	-	-	Moderate

7. Conclusions

This study proposes a PE-SLAM algorithm based on ORB-SLAM2, which combines an improved particle swarm optimization algorithm with epipolar constraints to improve the accuracy and robustness of the motion model in a photovoltaic power station environment. Experimental results show that in unstructured scenes, ORB+PE-match has an average improvement in matching accuracy of 19.5%, 14.0%, and 6.0% compared to the traditional ORB algorithm, ORB+HAMMING algorithm, and ORB+RANSAC algorithm, with better recall rates. In trajectory experiments using the TUM dataset, PE-SLAM reduces the average absolute trajectory error and relative pose error by 29.1% and 27.0%, respectively, compared to the ORB-SLAM2 algorithm. Mapping experiment results demonstrate the successful application of PE-SLAM in a photovoltaic power station scenario, where the dense point cloud map accurately and completely reflects the scene, effectively overcoming challenges posed by unstructured factors. Therefore, the proposed PE-SLAM algorithm offers an effective solution to address the challenges that ORB-SLAM faces in a photovoltaic power station environment.

PE-SLAM can be applied to cleaning or inspection robots in photovoltaic power station scenarios. Without external information support (such as GPS), it provides the robot with self-localization and 3D point cloud information about the environment using only the robot's own vision sensors. The mapping experiments in photovoltaic power station scenarios show that camera noise has a certain impact on the mapping results. For practical application of the algorithm, we believe that using RGB-D cameras with better anti-glare capabilities can improve the performance of PE-SLAM. From the parameter selection experiments, it can be seen that some parameters significantly affect PE-SLAM's performance. Therefore, optimal parameter selection is necessary for different scenarios. Future research could consider incorporating adaptive parameter algorithms to make PE-SLAM simpler and more effective in various unstructured environments. Additionally, PE-SLAM currently cannot modify the already-generated 3D point cloud based on back-end optimization results. If future research can address this, we believe the mapping performance will improve.

Author Contributions: Conceptualization, C.L. and Z.S.; methodology, C.L. and Z.S.; software, Z.S.; validation, C.L., Z.S., J.W. and K.Y.; formal analysis, Z.S. and K.Y.; investigation, W.N. and Z.S.; resources, W.N.; data curation, Z.S.; writing—original draft preparation, C.L. and Z.S.; writing—review and editing, C.L.; visualization, Z.S.; supervision, C.L.; project administration, C.L.; funding acquisition, C.L. and W.N. All authors have read and agreed to the published version of the manuscript.

Funding: This research was funded by the National Natural Science Foundation of China, grant number 52265065.

Institutional Review Board Statement: Not applicable.

applied sciences

MDPI

Article

Neural-Network-Based Adaptive MPC Path Tracking Control for 4WID Vehicles Using Phase Plane Analysis

Yang Sun [1,2,3], Xuhuai Liu [1,2,3,*], Junxing Zhang [4], Bin Tian [5], Sen Liu [1,2,3], Wenqin Duan [1,2,3] and Zhicheng Zhang [1,2,3]

1 College of Mechanical and Equipment Engineering, Hebei University of Engineering, Handan 056038, China; sunyang28@hebeu.edu.cn (Y.S.); liusen012@gmail.com (S.L.); 15953853866@163.com (W.D.); 15203186361@163.com (Z.Z.)
2 Key Laboratory of Intelligent Industrial Equipment Technology of Hebei Province, Handan 056038, China
3 Handan Key Laboratory of Intelligent Vehicles, Handan 056038, China
4 Key Laboratory of Modern Manufacturing Technology, Ministry of Education, Guizhou University, Guiyang 550025, China; jx_zhanggz@163.com
5 Institute of Automation, Chinese Academy of Sciences, Beijing 100190, China; bin.tian@ia.ac.cn
* Correspondence: 15930393530@163.com

Abstract: To improve the adaptability of 4WID electric vehicles under various operating conditions, this study introduces a model predictive control approach utilizing a neural network for adaptive weight parameter prediction, which integrates four-wheel steering and longitudinal driving force control. To address the difficulty in adjusting the MPC weight parameters, the neural network undergoes offline training, and the Snake Optimization method is used to iteratively optimize the controller parameters under diverse driving conditions. To further enhance vehicle stability, the real-time stability state of the vehicle is assessed using the $\beta - \dot{\beta}$ phase plane method. The influence of vehicle speed and road adhesion on the instability boundary of the phase plane is comprehensively considered to design a stability controller based on different instability degree zones. This includes an integral sliding mode controller that accounts for both vehicle tracking capability and stability, as well as a PID controller, which calculates the additional yaw moment based on the degree of instability. Finally, an optimal distribution control algorithm coordinates the longitudinal driving torque and direct yaw moment while also considering the vehicle's understeering characteristics in determining the torque distribution for each wheel. The simulation results show that under various operating conditions, the proposed control strategy achieves smaller tracking errors and more concentrated phase trajectories compared to traditional controllers, thereby improving path tracking precision, vehicle stability, and adaptability to varying conditions.

Keywords: path tracking; model predictive control; stability control; neural network; phase plane

1. Introduction

As a key component of intelligent transportation systems, advancements in autonomous mobility have garnered significant attention, especially in academia and industry [1,2]. Path tracking control is a crucial technology for autonomous vehicles, aiming to reduce the discrepancy between the actual trajectory and the desired trajectory by considering the vehicle's real-time dynamics [3]. However, when the vehicle is traveling at a rapid pace or operating on a slippery road surface, it is prone to losing stability and deviating from the target trajectory [4]. Therefore, designing a path tracking control method that maintains high accuracy and strong robustness under complex driving conditions is of great significance for enhancing the safety and reliability of autonomous driving systems.

Existing path tracking techniques mainly include PID control, Pure Pursuit control [5], backstepping control [6], fuzzy logic control [7], LQR control [8], sliding mode approach [9], and Model Predictive Control [10,11]. Among these, Model Predictive Control has been extensively utilized in path tracking applications owing to its predictive nature and distinct advantages in handling nonlinear dynamics and multi-constraint problems [12]. However, it still faces some challenges. Many researchers have proposed improvements such as optimizing controller parameters, enhancing model accuracy [13], and adding constraints [14]. Wu et al. [15], by analyzing the impact of the prediction horizon and weighting factors on trajectory accuracy and vehicle yaw stability, plotted the corresponding relationships as three one-dimensional lookup tables, achieving adaptive adjustment of the controller parameters. Yang and Liu [16] proposed a rule-based adaptive prediction horizon MPC for lateral tracking control and fuzzy PID for forward motion control, which is adapted to changes in vehicle speed. Wang et al. [17] utilized a fuzzy logic-based approach to dynamically tune the lateral displacement weight coefficient in the MPC trajectory following controller, considering both the average deviation and its rate of change. Yang et al. [18] designed a trajectory tracking control strategy based on a T–S fuzzy variable-weight MPC controller. The lateral error and heading error are used as fuzzy inputs, while the weighting factors for lateral error, yaw rate error, velocity error, and control increment are treated as outputs. By analyzing the relationship between the outputs and the vehicle states, fuzzy rules are formulated to achieve adaptive adjustment of the controller parameters. Xue and Zheng [19] developed an adaptive model predictive controller with flexible sampling and prediction intervals, incorporating a rounding function and a contraction factor, where the adaptation is guided by the average curvature's second-order derivative. Although the aforementioned methods have improved the adaptability of path tracking controllers to complex driving environments to some extent, most of them rely on manually designed lookup tables or fuzzy logic rules. The parameter tuning process heavily depends on empirical knowledge or prior experience and lacks theoretical optimization foundations, making the controller performance is susceptible to subjective factors. In addition, these methods often suffer from insufficient tracking accuracy and degraded stability under extreme conditions such as high speeds, sharp turns, or low-adhesion road surfaces, making it difficult to meet the robustness and safety requirements of autonomous vehicles in complex driving scenarios. Wang et al. [20] developed a path-tracking strategy utilizing model predictive control with an adaptive prediction horizon, optimized using particle swarm optimization. The method theoretically possesses strong global optimization capabilities and can adapt to various road conditions and vehicle states. However, as an online optimization algorithm, particle swarm optimization typically requires a large number of iterations and multiple objective function evaluations, placing high demands on system computational resources and response time, which significantly increases the online computational burden of the controller. Moreover, due to the potential sensitivity to initial values, the stability and convergence of the optimization results cannot be reliably guaranteed under complex driving conditions.

The lateral controllers in the aforementioned studies only control the front-wheel turning angle. Advancements in autonomous driving technology have enabled the development of vehicles with independently controlled steering and drive systems, providing a platform for advanced control strategies like Active Front/Rear-Wheel Steering (AFS/ARS) and Direct Yaw Moment Control (DYC) [21,22].The direct yaw moment control strategy primarily adopts a traditional layered framework, where the higher-level controller obtains the wheel angles and generalized forces, while the lower level controller distributes the generalized forces to the four wheels [23]. Ren et al. [24] introduced a unified control approach that combines longitudinal force with front-axle turning angle MPC to fully leverage the advantages of the MPC controller in solving multi-constraint problems. Numerous studies have employed the phase plane analysis to examine the dynamic behavior of vehicles, aiming to resolve the coupling issue between vehicle tracking performance and stability [25]. Liu and Dai [26] proposed a coordinated control approach for AFS and DYC, where the instability degree obtained from the $\beta - \dot{\beta}$ phase plane analysis is used as the controller's intervention condition. A switching function is designed based on the edge density function of a normal distribution to smoothly transition between the weight coefficients of the two controllers. Zhu et al. [27] employed an adaptive fuzzy controller and the phase plane method to regulate the quantization and scale factors for calculating the required additional yaw moment. Wang and Song [28] introduced an error penalty into the optimal distribution control objective function and modified the weight matrix using the vehicle's steady-state coefficients obtained through $\beta - \dot{\beta}$ phase plane analysis, optimizing the torque distribution across the wheels during vehicle operation. This method considers the vehicle's dynamic response during driving to optimize the conditional constraints in the torque distribution strategy, thereby enhancing control performance under specific stability requirements. However, it does not incorporate the longitudinal dynamic response into the control framework, lacking coordinated control over the coupling between vehicle power performance and stability. In real-world scenarios, the vehicle's longitudinal driving capability and lateral stability are often mutually constrained. Ignoring such coordination under extreme conditions may lead to vehicle instability and compromise driving safety.

Based on the above analysis and discussion, this paper proposes an adaptive path tracking control algorithm for distributed drive electric vehicles, aiming to address several critical issues identified in relevant studies from 2023 to 2025, as summarized in Table 1. First, existing adaptive MPC approaches generally rely on either predefined rules or fully depend on optimization algorithms. The former suffers from strong subjectivity in controller design, while the latter incurs a high online computational burden, limiting their applicability under complex driving conditions. Second, many existing methods fail to adequately consider the vehicle's adaptability to varying environments, making it difficult for the system to handle highly dynamic and uncertain driving scenarios. Furthermore, under dynamic driving conditions, there exists a significant coupling between path tracking performance and vehicle stability. Current control strategies lack effective coordination mechanisms to address this coupling and often neglect the influence of longitudinal driving performance on vehicle stability, which ultimately compromises overall control performance. The main contributions of this paper are as follows:

(1) The Back Propagation (BP) neural network is utilized to model the intricate non-linear correlation between the MPC weight coefficients and trajectory tracking precision along with dynamic stability, reducing the reliance on manually designed lookup tables or fuzzy logic rules and minimizing the influence of subjective factors. To optimize the weights and biases of the neural network during backpropagation, the Snake Optimization method is employed for iterative tuning of neural network parameters. The neural network model is trained offline using extensive simulation data, and a fitness criterion is formulated by incorporating both trajectory accuracy and stability to reduce online computation costs. The Snake Optimization algorithm is employed to optimize the MPC weight parameters under different driving conditions.

(2) To overcome the coupling issue of vehicle tracking capability and stability, a stability control approach utilizing the phase plane method is developed. The dynamic state of the vehicle is analyzed in real-time using the phase plane, and different instability

degree-based zone controllers are designed. To ensure smooth switching between controllers, a joint control domain is introduced.

(3) Addressing the overlooked impact of the vehicle's longitudinal power performance on stability in the aforementioned studies, this paper ensures that dynamic performance and stability requirements are met by coordinating the longitudinal driving torque and additional yaw moment based on the instability degree and road friction coefficient. The optimal distribution algorithm is designed to determine the driving torques for each wheel, considering the vehicle's understeer characteristics.

Table 1. Literature review and analysis.

Ref.	Approach	Limitations
Wu [15]	By analyzing the impact of the prediction horizon and weighting factors on trajectory accuracy and vehicle yaw stability, corresponding relationships were plotted as three one-dimensional lookup tables to achieve adaptive adjustment of controller parameters.	Lookup tables can only express the relationship between individual control parameters and single factors, such as tracking accuracy or yaw stability. However, controller parameters are often influenced by the nonlinear coupling of multiple variables simultaneously. This may lead to decreased controller accuracy or oversimplification of system behavior.
Wang [20]	A path tracking strategy utilizing model predictive control with an adaptive prediction horizon was developed. This strategy was optimized using particle swarm optimization.	The particle swarm algorithm requires evaluating multiple particles and performing several updates in each iteration. If Particle Swarm Optimization (PSO) is used is used for parameter optimization in the online control phase, it will result in significant computational delays, making it difficult to meet real-time requirements.
Wang [28]	A comprehensive controller for wheel steering angles and additional yaw moment was developed, where real-time adjustment of the control matrix weights was achieved through the vehicle's steady-state parameters. The optimal torque distribution for each tire was ultimately achieved by minimizing the tire load ratio.	The adjustment of control weights, guided by vehicle steady-state parameters, serves as the optimization goal for torque distribution at the execution layer. However, the control strategy does not assign fixed weights to lateral and heading errors in the path tracking controller. In different speed or road conditions, fixed error weights cannot balance the conflict between path accuracy and vehicle stability.

2. Materials and Methods

This paper develops an integrated path tracking controller based on Model Predictive Control (MPC) for four-wheel independent drive (4WID) vehicles. The control system is based on a three-DOF vehicle dynamics model, which considers the longitudinal, lateral, and yaw motions of the vehicle. The MPC path tracking controller serves as the core module. To fully exploit the advantages of MPC in handling multi-constraint problems, this study selects the front and rear wheel steering angles and the longitudinal driving forces as control variables. The performance of the controller depends on the selection of the weighting matrices Q and R. Improper weighting may lead to reduced tracking accuracy or compromised vehicle stability. The reference model calculates the ideal sideslip angle and yaw rate of the vehicle's center of mass, which serve as target inputs for the stability control module under varying road conditions.

2.1. Vehicle Model and Ideal Reference Model

In this study, a three-DOF dynamic equation models the vehicle's motion characteristics during path tracking. The analysis assumes a small front-wheel steering angle and a constant longitudinal velocity along the x-axis. The model ignores the differences in steering angles between the inner and outer wheels (Ackermann steering geometry), as well as the effects of air resistance and suspension. The corresponding bicycle model is depicted in Figure 1. The lateral and forward movement of the vehicle, along with its yaw dynamics, are captured by Equation (1).

$$\begin{cases} F_y = \mathrm{m}(\dot{v}_y + \gamma v_x) = F_{yf} + F_{yr} \\ F_x = \mathrm{m}(\dot{v}_x - \gamma v_y) = F_{xf} + F_{xr} \\ I_z\dot{\gamma} = l_f F_{yf} - l_r F_{yr} \end{cases} \tag{1}$$

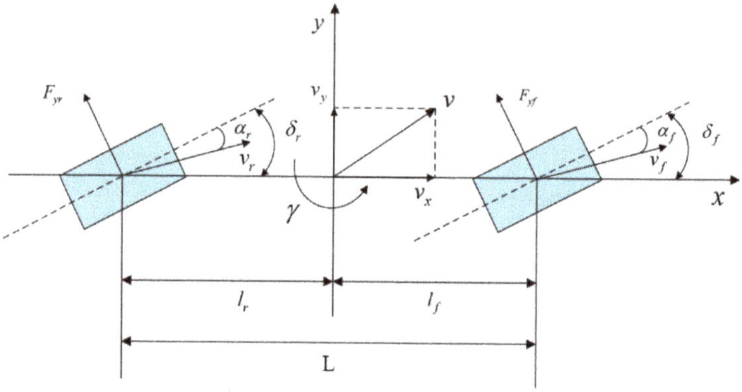

Figure 1. Three-DOF Bicycle Model.

In the equation, m represents the vehicle's total mass. v_x and v_y denote its forward and lateral velocities, respectively. The yaw angular velocity is denoted by γ. F_{xf}, F_{xr}, F_{yf} and F_{yr} correspond to the longitudinal and side forces acting on the front-axle and rear-axle tires. I_Z represents the moment of inertia; l_f and l_r indicate the distances from the center of mass to the front-axle and rear-axle. It should be noted that the three-DOF bicycle model used in this study offers a simplified representation of vehicle dynamics under medium- to high-speed driving conditions. Its accuracy decreases at low speeds and large steering angles, such as during parking maneuvers. Therefore, this study focuses on medium- to high-speed scenarios. Future work will extend the model to address low-speed and large steering angle conditions. When the tires function within the linear range, the side force is given by Equation (2).

$$\begin{cases} F_{yf} = k_1\alpha_f = k_1(\frac{v_y + l_f\gamma}{v_x} - \delta_f) \\ F_{yr} = k_2\alpha_r = k_2(\frac{v_y - l_r\gamma}{v_x} - \delta_r) \end{cases} \tag{2}$$

In the equation, k_1 and k_2 denote the equivalent cornering rigidity of the front-axle and rear-axle. The slip angles of the front-axle and rear-axle are denoted by α_f and α_r, while δ_f and δ_r correspond to their respective steering angles.

In the control framework adopted in this study, the effects of the suspension system and aerodynamic forces are neglected, and parameters such as tire cornering stiffness are simplified as constant values. This modeling approach has been widely adopted in numerous recent studies on lateral control, particularly under typical conditions such as medium- to high-speed driving and lane-keeping scenarios, where it shows strong applicability. When the tire operates within a small slip angle range (typically less than 6°), the variation in cornering stiffness is minimal, making it a reasonable simplification to treat it as a constant [29]. The cornering stiffness parameters were calibrated experimentally at 90 km/h, and have shown effective control performance across a wide range of medium- to high-speed driving conditions. However, it is acknowledged that under extreme dynamic scenarios involving large slip angles, this simplification may not fully capture the actual vehicle dynamics. In future work, we plan to incorporate online parameter estimation methods, as well as suspension and aerodynamic modeling, to improve model adaptability and robustness.

The vehicle's total mass and inertia properties, along with the front and rear axle cornering stiffness coefficients, were derived from CarSim 2020.0, a validated commercial vehicle dynamics simulation platform. These parameters correspond to a C-Class Hatchback passenger car model embedded in CarSim's database, which has been extensively adopted in both academic research and engineering practice due to its realism and representative nature.

The tire's rotational dynamics are expressed as:

$$I_w\dot{\omega} = T - R_w F \tag{3}$$

where I_w is the tire's rotational inertia, ω is the angular velocity of the tire. R_w is the tire radius, F and T respectively represent the longitudinal force and driving torque of the tire. Equation (1) can be derived and expressed as a differential equation in Equation (4).

$$\begin{cases} \dot{v}_x = \frac{mv_y R_w^2}{mR_w^2+I_w}\gamma + \frac{R_w}{mR_w^2+I_w}T \\ \dot{v}_y = \frac{k_1+k_2}{mv_x}v_y + (\frac{l_f k_1 - l_r k_2}{mv_x} - v_x)\gamma - \frac{k_1}{m}\delta_f - \frac{k_2}{m}\delta_r \\ \dot{\gamma} = \frac{l_f k_1 - l_r k_2}{I_z v_x}v_y + \frac{l_f{}^2 k_1 + l_r{}^2 k_2}{I_z v_x}\gamma - \frac{l_f k_1}{I_z}\delta_f + \frac{l_r k_2}{I_z}\delta_r \end{cases} \tag{4}$$

Taking road adhesion limitations into account, the maximum allowable angular velocity and body slip angle are derived from Equation (5).

$$\begin{cases} |\gamma_{\max}| = \frac{0.85\mu g}{v_x} \\ |\beta_{\max}| = 0.85\mu g(\frac{l_r}{v_x^2} + \frac{ml_f}{k_r L}) \end{cases} \tag{5}$$

The ideal reference model characterizes the vehicle's lateral dynamics under steady-state conditions. When the vehicle maintains a steady state, $\dot{\beta} = 0$ and $\dot{\gamma} = 0$. The required vehicle's ideal angular velocity and body slip angle for the system are computed as shown in Equation (6).

$$\begin{cases} \gamma_d = \min\left\{\left|\frac{v_x}{L(1+Kv_x^2)}\delta_f\right|, |\gamma_{\max}|\right\}\mathrm{sgn}(\delta_f) \\ \beta_d = \min\left\{\left|\frac{l_r/L+ml_f v_x^2/(k_r L^2)}{1+Kv_x^2}\delta_f\right|, |\beta_{\max}|\right\}\mathrm{sgn}(\delta_f) \end{cases} \tag{6}$$

where μ denotes the road friction coefficient, and g is the gravitational acceleration. The parameters L and K correspond to the wheelbase and vehicle stability factor, respectively. Additionally, γ_d signifies the ideal yaw rate, and β_d refers to the ideal center of mass slip angle.

2.2. MPC Path Tracking Model

During path tracking control, the following relationship exists:

$$\begin{cases} \dot{X} = v_x - v_y\varphi \\ \dot{Y} = v_x\varphi + v_y \\ \dot{\varphi} = \gamma \end{cases} \tag{7}$$

In the equation, φ represents the vehicle's heading angle, X and Y denote its forward and lateral positions in the global coordinate system.

Based on Equation (4), the system can be transformed into state-space representation. The state variables are taken as $x = \left[X, Y, \varphi, v_x, v_y, \gamma\right]^T$, the control parameters as $u = \left[\delta_f, \delta_r, T\right]^T$, and the output variables as: $y = [X, Y, \varphi]^T$. The resulting state-space equation is presented in Equation (8).

$$\begin{cases} \dot{x} = Ax + Bu \\ y = Cx + Du \end{cases} \tag{8}$$

where:

$$A = \begin{bmatrix} 0 & 0 & -v_y & 1 & 0 & 0 \\ 0 & 0 & v_x & 0 & 1 & 0 \\ 0 & 0 & 0 & 0 & 0 & 1 \\ 0 & 0 & 0 & 0 & 0 & \frac{mv_y R_w^2}{mR_w^2 + I_w} \\ 0 & 0 & 0 & 0 & \frac{k_1 + k_2}{mv_x} & \frac{(l_f k_1 - l_r k_2)}{mv_x} - v_x \\ 0 & 0 & 0 & 0 & \frac{(l_f k_1 - l_r k_2)}{I_z v_x} & \frac{(l_f{}^2 k_1 + l_r{}^2 k_2)}{I_z v_x} \end{bmatrix}, B = \begin{bmatrix} 0 & 0 & 0 \\ 0 & 0 & 0 \\ 0 & 0 & 0 \\ 0 & 0 & \frac{R_w}{mR_w^2 + I_w} \\ -\frac{k_1}{m} & -\frac{k_2}{m} & 0 \\ -\frac{l_f k_1}{I_z} & \frac{l_r k_2}{I_z} & 0 \end{bmatrix}, C = \begin{bmatrix} 0 & 0 & 1 \\ 0 & 1 & 0 \\ 1 & 0 & 0 \\ 0 & 0 & 0 \\ 0 & 0 & 0 \\ 0 & 0 & 0 \end{bmatrix}^T, D = \begin{bmatrix} 0 \\ 0 \\ 0 \end{bmatrix}$$

Assuming the current state is $\dot{x}_0 = f(x_0, u_0)$, performing the Taylor series expansion while retaining only the first-order terms yields:

$$\dot{x} = f(x_0, u_0) + A_t(x - x_0) + B_t(u - u_0) \tag{9}$$

By subtracting Equation (9) from the current state point, we obtain:

$$\dot{\widetilde{x}} = A_t(x - x_0) + B_t(u - u_0) \tag{10}$$

Applying the first-order difference approximation, Equation (10) is discretized using a fixed sampling interval $T_s = 0.01$, as shown in Equation (11).

$$\begin{cases} \widetilde{x}_{k+1} = A_d \widetilde{x}_k + B_d \widetilde{u}_k \\ y_k = C_d \widetilde{x}_k \end{cases} \tag{11}$$

In the equation, $A_d = A_t T_s + I_{Nx}$, $B_d = B_t T_s$ and $C_d = I_{Nx}$. In order to limit the control inputs, the state and control variables are augmented into a new set of state variables, denoted as $\xi_k = \begin{bmatrix} \widetilde{x}_k \\ \widetilde{u}_{k-1} \end{bmatrix}$, and the variation in control inputs at each iteration relative to the preceding one is represented by $\Delta\widetilde{u}_k$, which gives Equation (12).

$$\begin{cases} \zeta_{k+1} = A_1 \xi_k + B_1 \Delta\widetilde{u}_k \\ \eta_k = C_1 \xi_k \end{cases} \tag{12}$$

where: $A_1 = \begin{bmatrix} A_d & B_d \\ 0_{Nu \times Nx} & I_{Nu \times Nu} \end{bmatrix}$, $B_1 = \begin{bmatrix} B_d \\ I_{Nu \times Nu} \end{bmatrix}$, $C_1 = \begin{bmatrix} C_d \\ 0_{Nx \times Nu} \end{bmatrix}^T$. N_x represents the number of state variables. N_u denotes the number of control inputs. Let k denote the current time step. To predict the states and outputs for the next N_p steps, the control input is represented as Equation (13).

$$\Delta U = [\Delta\widetilde{u}(k), \Delta\widetilde{u}(k+1), \ldots, \Delta\widetilde{u}(k + Nc - 1)]^T \tag{13}$$

where: $\Delta\tilde{u}(i) = \left[\Delta\delta_f(i), \Delta\delta_r(i), \Delta T(i)\right]^T, i = k \ldots k + Nc - 1$. From the above equation, the controller predicts the output Y over the next N_p steps as illustrated in Equation (14).

$$Y = \Psi\zeta_k + \Theta\Delta U \tag{14}$$

where:

$$Y = \begin{bmatrix} \eta_{k+1} \\ \eta_{k+2} \\ \cdots \\ \eta_{k+Nc} \\ \cdots \\ \eta_{k+Np} \end{bmatrix}; \Psi = \begin{bmatrix} C_1A_1 \\ C_1A_1^2 \\ \cdots \\ C_1A_1^{Nc} \\ \cdots \\ C_1A_1^{Np} \end{bmatrix}; \Theta = \begin{bmatrix} C_1B_1 & 0 & 0 & 0 \\ C_1A_1B_1 & C_1B_1 & 0 & 0 \\ \vdots & \vdots & \ddots & \vdots \\ C_1A_1^{Nc}B_1 & C_1A_1^{Nc-1}B_1 & \cdots & C_1A_1B_1 \\ \vdots & \vdots & \ddots & \vdots \\ C_1A_1^{Np-1}B_1 & C_1A_1^{Np-2}B_1 & \cdots & C_1A_1^{Np-Nc-1}B_1 \end{bmatrix}$$

To ensure smoothness and stability in the tracking process, the optimization objective is to minimize state deviations and control efforts. This objective is formulated as a quadratic programming problem, as shown in Equation (15).

$$J = (Y - Y_{ref})^T Q(Y - Y_{ref}) + \Delta U^T R\Delta U + \rho\varepsilon^2 \tag{15}$$

where: $Q_i = \begin{bmatrix} q_1 & 0 \\ 0 & q_2 \end{bmatrix}$, $R_i = diag(r, r, r)$, $i = 1, 2, 3 \ldots Np$. Q_i and R_i are the prediction and control matrices for the MPC, respectively. While q_1, q_2 and r represent the weights for lateral deviation, yaw rate error, and control input variation, respectively. Model Predictive Control needs to solve the optimal control sequence over the prediction horizon while satisfying the objective function and various constraints. The constraints on the four-wheel steering angles, as well as the longitudinal driving force and its rate of change, are expressed as shown in Equation (16).

$$\begin{cases} \tilde{u}_{\min} \le \tilde{u}(t+k) \le \tilde{u}_{\max}, k = 0, 1, \ldots, N_C - 1 \\ \Delta\tilde{u}_{\min} \le \Delta\tilde{u}(t+k) \le \Delta\tilde{u}_{\max}, k = 0, 1, \ldots, N_C - 1 \end{cases} \tag{16}$$

Considering the limitations of the steering and drive motors, the front wheel steering angle is constrained within the range of [−0.44, 0.44] radians, with a rate limit of [−0.02, 0.02] radians. The rear wheel steering angle is constrained within [−0.44, 0.44] radians, with a rate limit of [−0.005, 0.005] radians. The longitudinal driving torque is constrained to [−600, 600] Nm, with a rate limit of [−50, 50] Nm. These constraints are based on the technical specifications of typical wheel-side motors and steering actuators, and are compatible with the discretization interval of the MPC used in this study [30]. By solving the constrained quadratic optimization problem, the optimal front and rear steering angles and longitudinal driving torque can be obtained. Only the first element of the optimal control sequence is applied at each sampling instant, and the optimization process is repeated at every time step.

2.3. Trajectory Tracking Control System Design for 4WID Vehicles

Figure 2 illustrates the framework of the trajectory tracking strategy for a distributed drive vehicle. The planning layer provides road information and vehicle state parameters as inputs to the MPC. Heidfeld et al. [31] use the Unscented Kalman Filter (UKF) method, combining a nonlinear double-track model with a brush tire model, to achieve real-time estimation of vehicle state parameters including vehicle speed, yaw rate, and road friction coefficient. This approach improves the accuracy and robustness of the vehicle system control. Li et al. [32] propose a method for estimating the vehicle's center of mass side-slip angle based on the non-linear Dugoff tire model and the non-singular terminal sliding mode observer (NS-TSMO). This method improves the accuracy of side-slip angle estimation for vehicles driving under varying road adhesion conditions. The signal acquisition method is not the focus of this study. It is assumed that sensor data is accurately obtained, and the vehicle state parameters as well as the road adhesion coefficient are precisely obtained through estimation methods. To enable adaptive adjustment of the controller parameters, a neural network is used to offline-train the nonlinear relationship between the controller parameters and vehicle states under varying driving conditions, such as vehicle speed (v) and road adhesion coefficient (μ). The Snake Optimization Algorithm is utilized to iteratively optimize the controller parameters, which are then allocated to the controller to calculate the front and rear wheel steering angles and the longitudinal driving torques. The stability controller uses phase-plane analysis and an ideal reference model to track the desired yaw rate and sideslip angle of the vehicle's center of mass, thereby calculating the additional yaw moment. The optimal torque distribution controller coordinates the allocation of longitudinal driving torque and the additional yaw moment, and considers the vehicle's understeer characteristics to determine the driving torque for each wheel, ultimately achieving yaw moment control to enhance vehicle stability.

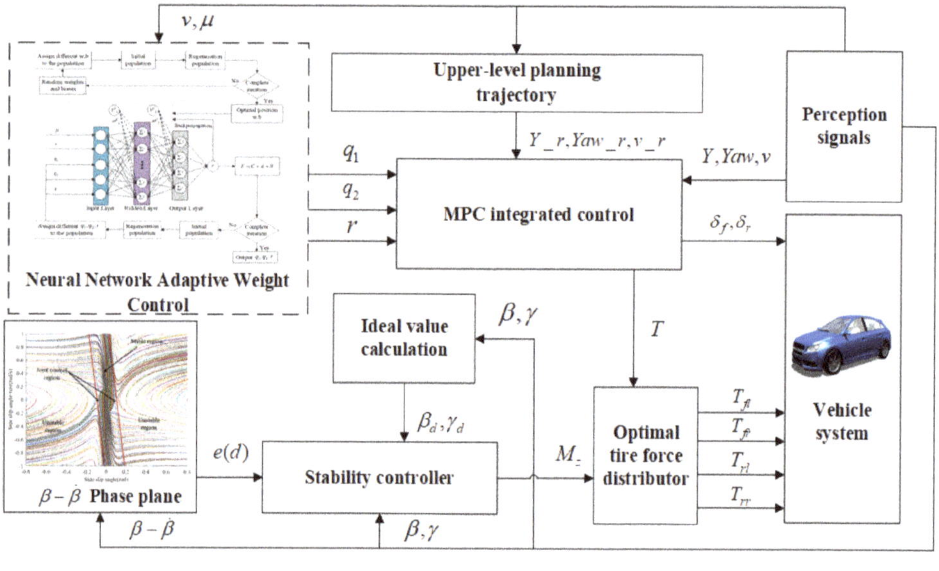

Figure 2. Framework of a Trajectory Tracking Control System for 4WID Vehicles.

3. Adaptive Adjustment Strategy for MPC Controller Parameters

Under varying driving conditions, such as changing road adhesion coefficients and vehicle speeds, there exists a complex nonlinear relationship between the parameters of the MPC controller and the accuracy and stability of path tracking. Specifically, the weight coefficients for lateral tracking error and heading error are denoted as q_1 and q_2, respectively; the weight coefficient for control increment is denoted as r. When vehicle stability is low, increasing the weight on lateral tracking error to improve accuracy may cause instability, which results in larger lateral errors. Therefore, adaptive adjustment of controller parameters under varying driving conditions is essential to maintain both tracking accuracy and stability. Leveraging the universal approximation capability of three-layer neural networks, this paper employs such a network to model the nonlinear mapping between MPC weight parameters and path tracking accuracy and stability under varying driving conditions. A fitness function is designed, and the Snake Optimization Algorithm is utilized to iteratively optimize the MPC weight parameters.

To avoid local optima in gradient descent-based error back-propagation, the Snake Optimization Algorithm is utilized for global search. This helps improve prediction accuracy and generalization by optimizing the neural network's weights and biases. Vehicle speed, road adhesion coefficient, and MPC weight coefficients serve as the neural network's inputs, while the maximum lateral error, average lateral error, maximum centroid sideslip angle, and maximum yaw rate are defined as the outputs, as shown in Figure 3.

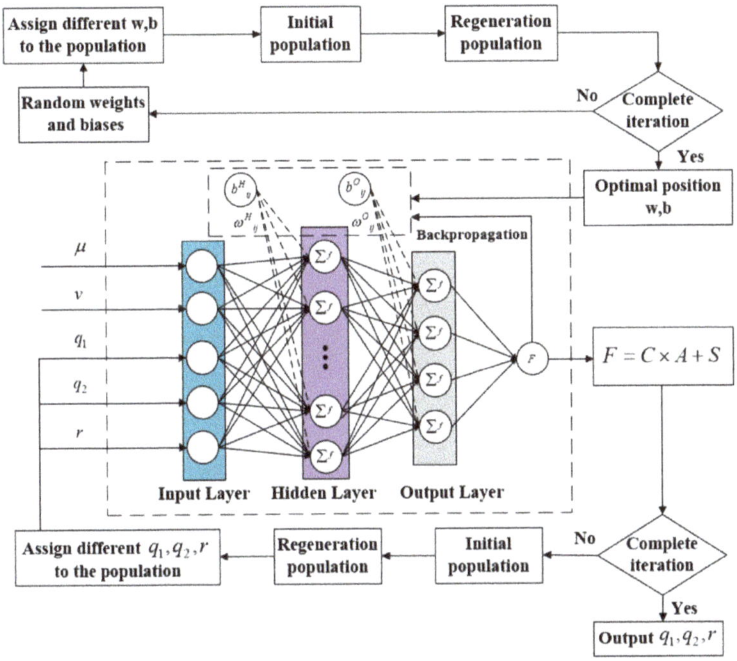

Figure 3. Nodes Neural Network Adaptive Weight Control Strategy.

To determine the number of hidden layer nodes in the neural network, an empirical formula $N_h = \sqrt{N_i \times N_o} + \alpha$ was used to narrow down the search range, where N_h denotes the number of hidden nodes, N_i the number of input nodes, N_o the number of output nodes, α is a tuning parameter. A sensitivity analysis was conducted with the number of hidden nodes varying from 6 to 14. As shown in Figure 4, the RMSE values on both the training and testing datasets initially decrease and then increase as the number of hidden nodes increases. When the number of hidden nodes is 11, the test RMSE reaches its minimum. Consequently, 11 hidden nodes are adopted in this study as the final network structure to achieve a trade-off between prediction accuracy and computational efficiency. This configuration not only ensures prediction accuracy but also effectively prevents overfitting. The iteration limit is set to 1000, with a target error of 1×10^{-6} and a learning rate of 0.001.

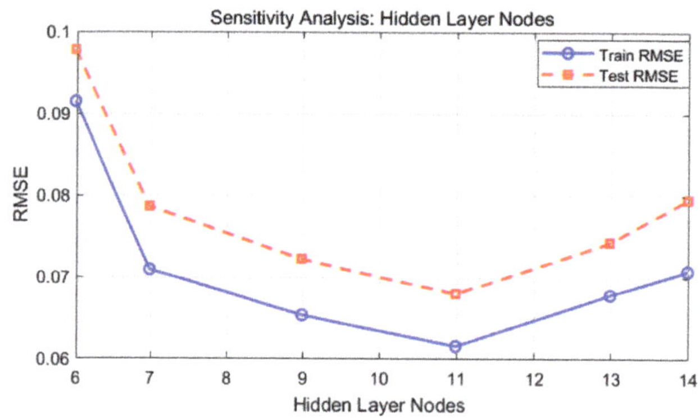

Figure 4. Sensitivity Analysis on the Number of Hidden Layer Nodes.

To evaluate the effectiveness of the proposed Snake Optimization (SO) algorithm for optimizing neural network weights and biases, a comparative study was conducted with two well-established methods: Genetic Algorithm (GA) and Particle Swarm Optimization (PSO). To ensure a fair comparison, all three algorithms were implemented under identical experimental settings. Specifically, the number of hidden layer neurons was fixed at 11, the learning rate was set to 0.001, the maximum number of training iterations was limited to 1000, and the target error threshold was defined as 1×10^{-6}. The convergence behavior of each algorithm during training is illustrated in Figure 5. The results indicate that the SO algorithm surpasses both GA and PSO in convergence efficiency and final prediction accuracy. In particular, when the convergence criteria were met, GA required 22 iterations to reach a final fitness value of 6.86×10^{-5}, while PSO also converged in 22 iterations with a final fitness of 5.04×10^{-5}. In contrast, SO achieved convergence in just 6 iterations, reaching a lower final fitness value of 3.75×10^{-5}. These outcomes demonstrate that SO not only converges more rapidly but also yields a more accurate solution, indicating superior parameter optimization capability and higher overall efficiency.

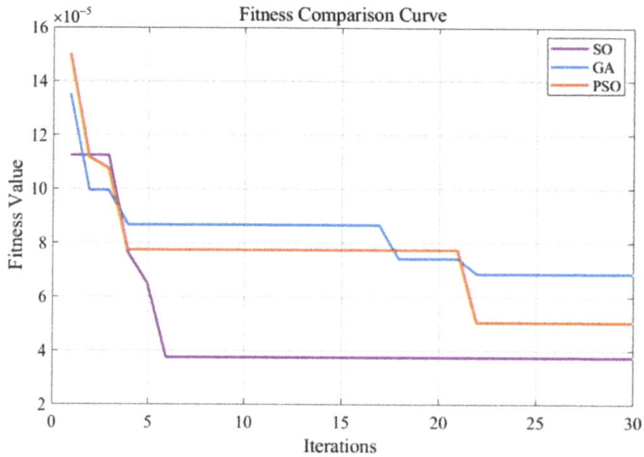

Figure 5. Comparison of Fitness Convergence Curves for Different Optimization Algorithms.

To enhance the overall performance and training stability of the neural network model, a systematic sensitivity analysis was conducted on several critical hyperparameters. These included the activation function, learning rate, regularization coefficient, batch size, and early stopping. The detailed results of this analysis are illustrated in Figure 6. The test root mean square error (Test RMSE) was used as the evaluation metric, with lower values indicating better generalization. Regarding the selection of activation functions, three commonly employed types were evaluated: hyperbolic tangent sigmoid (tansig), logarithmic sigmoid (logsig), and linear (purelin). The results demonstrated that the tansig function achieved the lowest Test RMSE of 0.06, suggesting its superior ability to capture non-linear relationships in the data for this regression task. The learning rate was explored over a range of values: 0.001, 0.01, 0.05, 0.1, 0.15, and 0.2. Among these, the network achieved optimal performance when the learning rate was set to 0.001, with a corresponding Test RMSE of 0.048. To mitigate overfitting and regulate model complexity, the regularization parameter was examined at values of 0.001, 0.05, 0.1, and 0.2. The best result was observed when the parameter was set to 0.001, yielding a Test RMSE of 0.075. The impact of batch size on training dynamics was also analyzed by testing values of 1, 5, 10, 20, 30, 50, and 100. A batch size of 30 resulted in the lowest Test RMSE of 0.054, indicating a favorable balance between convergence speed and stability. Furthermore, the early stopping strategy was evaluated by comparing training outcomes under enabled and disabled settings. Interestingly, disabling early stopping led to improved generalization, achieving a Test RMSE of 0.061. Based on these experimental findings, the final hyperparameter configuration adopted in this study included: the tansig activation function, a learning rate of 0.001, regularization parameter of 0.001, batch size of 30, and training without early stopping. This configuration provides an effective trade-off between generalization and training stability.

3.1. BP Neural Network Training

To enable the MPC controller to adapt to different driving conditions, the road surface is categorized into three types: dry, wet, and icy, with friction coefficients of 0.8, 0.5, and 0.3, respectively. The weight ranges for lateral, yaw angle, and control error are set from 1 to 100. The minimum and maximum speeds are set at 54 km/h and 120 km/h, respectively. A simulation platform is established to conduct MPC integrated control system simulations under various driving conditions, ultimately generating 2376 sample data points.

The forecasted outcomes for the training dataset input and output are presented in Figure 7a,c, while the predicted results for the test set input and output are shown in Figure 7b,d. In the figure, RMSE denotes the Root Mean Squared Error. The Snake algorithm optimized neural network parameters converged after the ninth iteration, with a fitness of 0.000128, as shown in Figure 8. The dataset was divided such that 70% was used for training and the remaining 30% was kept aside for testing. During model training, a subset of the training data was automatically selected for validation to assess learning performance and reduce the risk of overfitting. The test set was solely employed in the final evaluation stage to ensure unbiased performance assessment. The regression functions and correlation are presented in Table 2. Regarding the training, validation, and testing datasets, the regression slopes remain at 1, with correlation coefficients exceeding 0.98, validating the reliability of the predictive model. Thus, the neural network effectively captures the relationship between MPC weight parameters and vehicle state.

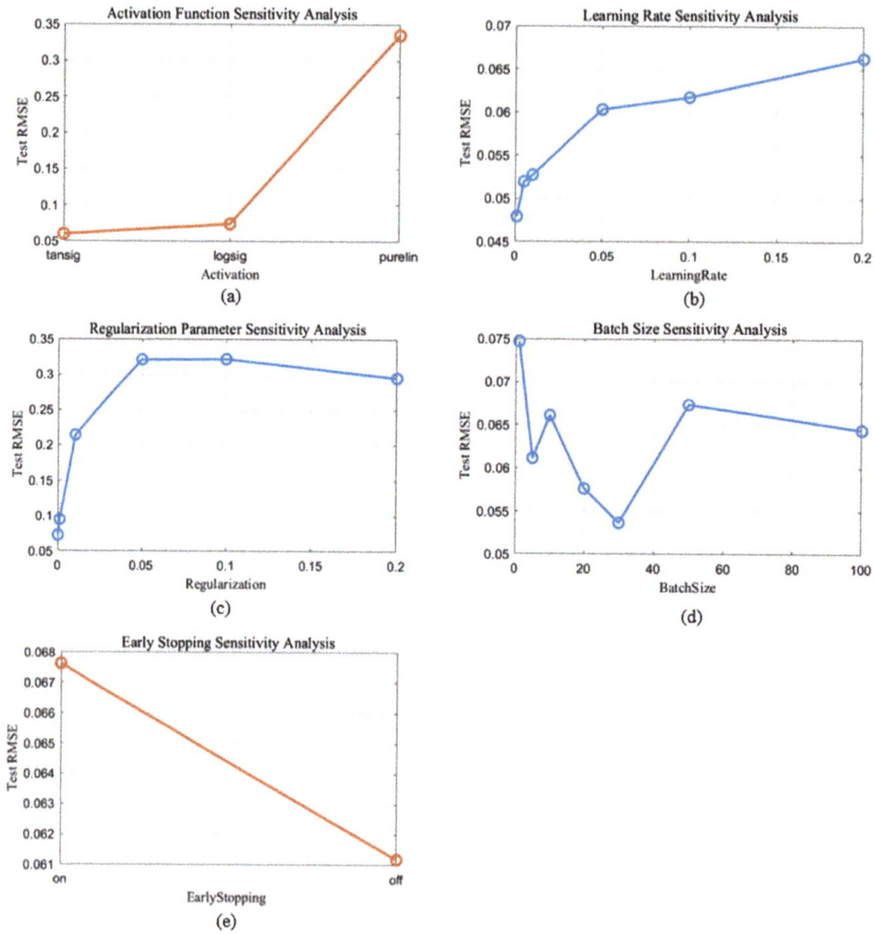

Figure 6. Sensitivity Analysis of Hyperparameters: (**a**) Activation Function; (**b**) Learning Rate; (**c**) Regularization Parameter; (**d**) Batch Size; (**e**) Early Stopping.

Table 2. Results of Regression Models and Correlation Analysis.

	Regression Models	Correlation
Training Set	Prediction = target + 0.0045	0.98703
Validation Set	Prediction = target + 0.0015	0.99076
Test Set	Prediction = target + 0.0036	0.98797
All	Prediction = target + 0.0039	0.98775

To address the potential risk of overfitting and more thoroughly evaluate the generalization capability of the proposed neural network model, this study employed the K-fold cross-validation technique. Specifically, the entire dataset was randomly divided into five equal subsets (K = 5). In each iteration, one subset was used as the validation set, while the remaining four were used for training. This procedure was repeated five times, ensuring that each subset served as the validation set exactly once. The final performance metrics were obtained by averaging the results across all folds, thereby improving the robustness of the evaluation and reducing dependence on any single data split.

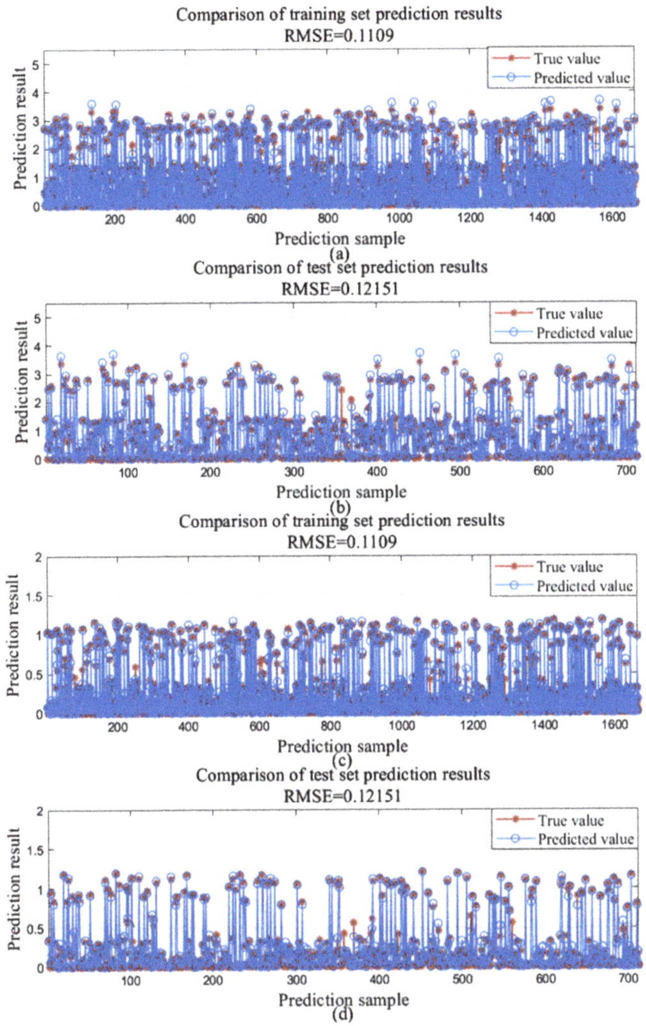

Figure 7. Prediction Performance of Training Set and Test Set: (**a**) Training Set Input; (**b**) Test Set Input; (**c**) Training Set Output; (**d**) Test Set Output.

Model performance was assessed using the coefficient of determination (R^2), which quantifies the proportion of variance in the observed data that is explained by the model's predictions. The coefficient of determination is calculated as $R^2 = 1 - \frac{\sum_{i=1}^{n}(y_i - \hat{y}_i)^2}{\sum_{i=1}^{n}(y_i - \overline{y})^2}$, where y_i is the observed value, $\hat{y}_i$ is the predicted value, and $\overline{y}$ is the mean of the observed values. The results demonstrate that the model achieved an average coefficient of determination of 0.9879 on the training folds and 0.9863 on the validation folds, reflecting strong training stability and robust convergence without evident signs of overfitting.

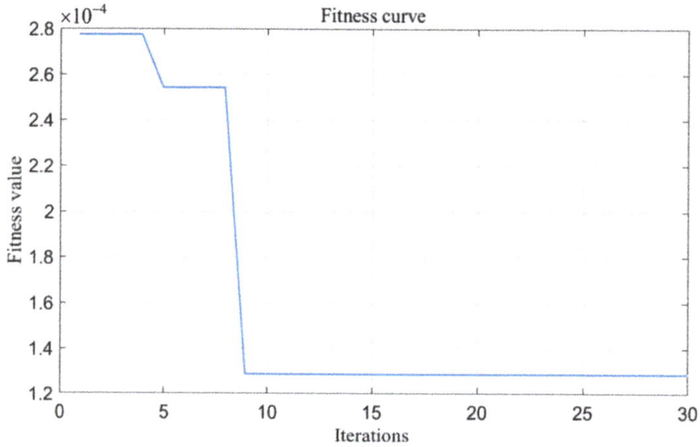

Figure 8. Snake Algorithm Optimization Fitness Curve.

3.2. Snake Optimization Algorithm

The Snake Optimization (SO) method is an optimization technique based on the foraging and reproductive behaviors of snakes. The reason for choosing SO in this study lies in its unique ability to dynamically balance global exploration and local exploitation, which is particularly advantageous for handling high-dimensional and complex problems. Through its two-phase design, SO avoids the premature convergence commonly seen in Genetic Algorithms (GA) and the tendency of Particle Swarm Optimization (PSO) to become trapped in local optima in multimodal problems. In the phase of food scarcity, SO performs extensive global exploration, increasing the likelihood of escaping local optima; when food is sufficient and the temperature is high, the algorithm shifts to local search around the target region, accelerating convergence and improving solution accuracy. Therefore, SO is well suited for solving optimization problems characterized by large search spaces and multiple local extrema, as in this study. The population size for the neural network parameter optimization using the SO algorithm is set to $n = 30$, and the iteration count is set to $T = 30$. In the first phase, when food is insufficient (i.e., food quality $Q(t) = 0.5e^{\frac{t-T}{T}} < 0.25$), the snake swarm explores and searches for food at random positions, updating their positions as described in Equation (17).

$$\begin{cases} x_{i,g}(t+1) = x_{i,g}(t) \pm c_1 \times A_g(x_{\min} + (x_{\max} - x_{\min}) \times rand) \\ A_g = \exp(-\frac{f_{rand,g}}{f_{i,g}}) \end{cases} \tag{17}$$

where $x_{i,g}(t)$ represents the $i-th$ element of the position vector of an individual in the $t-th$ generation, where $g = f, m$ denotes either a female or male individual, and $rand$ represents a randomly generated number within the closed interval [0, 1]. $x_{\min}, x_{\max}$ define the lower and upper limits of the boundary values. A_g represents the food-seeking ability of individuals in the population, while $f_{rand,g}, f_{i,g}$ represent the fitness of a randomly chosen individual and the $i-th$ individual, respectively.

In the second phase, when food is abundant but the temperature is too high (i.e., $Q > 0.25$ and $T(t) = e^{-\frac{t}{T}} > 0.6$), the snake swarm explores around the food and enters a local search phase, as illustrated in Equation (18).

$$\begin{cases} x_{i,g}(t+1) = x_{best} \pm x_Q \\ x_Q = c_2 \times Temp \times rand(x_{best} - x_{i,j}(t)) \end{cases} \tag{18}$$

In the equation, x_{best} indicates the best individual's position in the $t-th$ generation. If $Rand < 0.6$, the population engage in battle mode, and individuals fight to obtain the best mate of the opposite sex, as described in Equation (19).

$$\begin{cases} x_{i,m}(t+1) = x_{i,m}(t+1) \pm x_{Fm} \\ x_{i,f}(t+1) = x_{i,f}(t+1) \pm x_{Ff} \\ x_{Fm} = c_2 \times F_m \times rand \times (x_{best,f} - x_{i,m}(t)) \\ x_{Ff} = c_2 \times F_f \times rand \times (x_{best,m} - x_{i,f}(t)) \end{cases} \tag{19}$$

When $Rand > 0.6$, the population enter the mating mode, and their movement equation is given as follows:

$$\begin{cases} x_{i,m}(t+1) = x_{i,m}(t) \pm x_{Mm} \\ x_{i,f}(t+1) = x_{i,f}(t) \pm x_{Mf} \\ x_{Mm} = c_2 \times M_m \times rand \times (Q \times x_{i,f} - x_{i,m}(t)) \\ x_{Mf} = c_2 \times M_f \times rand \times (Q \times x_{i,m} - x_{i,f}(t)) \end{cases} \tag{20}$$

In the equation:

$$\begin{cases} F_m = \exp(-\frac{f_{best,f}}{f_{i,m}}) \\ F_f = \exp(-\frac{f_{best,m}}{f_{i,f}}) \end{cases} ; \begin{cases} M_m = \exp(-\frac{f_{i,f}}{f_{i,m}}) \\ M_f = \exp(-\frac{f_{i,m}}{f_{i,f}}) \end{cases}$$

where $f_{best,m}$, $f_{best,f}$ signify the optimal fitness levels in the male and female groups, while $f_{i,m}$ and $f_{i,f}$ represent the fitness scores of the $i-th$ individual within the male and female populations. In mating mode, there is a 50% probability of generating a new individual, which will replace the least fit individual in the population.

$$x_{worst,g}(t+1) = x_{\min} + rand \times (x_{\max} - x_{\min}) \tag{21}$$

Based on the results generated by the neural network model, the maximum and average lateral errors are normalized and represented as A, while the maximum centroid slip angle and maximum yaw rate are normalized and represented as S. The fitness function $F = C \times A + S$ is designed, where C is a constant. The Snake Algorithm is applied to iteratively optimize the controller parameters to ensure tracking accuracy and stability. The population size for controller parameter optimization is set to 30, with a maximum iteration count of 30, and position limits are set between -1 and 1. The constants c_1, c_2 are set to 0.5 and 2, respectively.

4. Stability Control Strategy

4.1. Phase Plane Control Domain Division

Based on Equation (4), using the centroid side slip angle and yaw rate as stability control objectives, the nonlinear vehicle model with direct yaw moment incorporated is derived as follows:

$$\begin{cases} \dot{\beta} = \frac{k_1+k_2}{mv_x}\beta + (\frac{l_f k_1 - l_r k_2}{m{v_x}^2} - 1)\gamma - \frac{k_1}{mv_x}\delta_f - \frac{k_2}{mv_x}\delta_r \\ \dot{\gamma} = \frac{l_f k_1 - l_r k_2}{I_z}\beta + \frac{{l_f}^2 k_1 + {l_r}^2 k_2}{I_z v_x}\gamma - \frac{l_f k_1}{I_z}\delta_f + \frac{l_r k_2}{I_z}\delta_r + \frac{1}{I_z}M_1 \end{cases} \tag{22}$$

At elevated speeds, the vehicle experiences increased lateral acceleration, and the tires primarily function under non-linear conditions. In this research, the Magic Tire Model combined with the phase plane method is utilized to investigate the vehicle's nonlinearity under extreme conditions. The model is expressed as shown in Equation (23).

$$F_{yi} = F_{zi} D_y \sin(C_y a\tan(B_y \alpha_i + E_y(B_y \alpha_i - a\tan(B_y \alpha_i)))) \tag{23}$$

The side force of the wheel is denoted as F_{yi}, while the vertical load is denoted by F_{zi}. α_i is the tire's sideslip angle, $i = f, r$. D_y, C_y, B_y and E_y are the fitting coefficients.

When the vehicle is in motion, the normal forces exerted on the front and rear tires are formulated in Equation (24).

$$\begin{cases} F_{zf} = mgl_r/(l_f + l_r) \\ F_{zr} = mgl_f/(l_f + l_r) \end{cases} \tag{24}$$

The phase plane method does not require solving nonlinear equations. Instead, it visually describes the system's stability through phase trajectories. The state points within the stable region, determined by the double-line approach, converge to the origin, while the points in the unstable region diverge. When the steering angle is minor, the intercept of the stability threshold is significantly influenced by the road friction coefficient, while the slope is affected by both the travelling speed and the road friction coefficient [33]. Therefore, in this study, the vehicle speed is incremented by 9 km/h, spanning from 54 km/h to 126 km/h, and the road adhesion value is assigned to fixed intervals of 0.1, ranging from 0.3 to 1, resulting in 72 sample sets. The boundary parameters, including the slope B_1 and the intercept x_0, are fitted using the least squares method, as shown in Figure 9. The parameters B_1 and B_2, obtained through least squares fitting, yielded coefficients of determination R^2 of 0.9575 and 0.9542, respectively. The RMSE of the fitting results for B_1 and B_2 are 0.6056 and 0.000802, respectively. The results indicate that the constructed model effectively captures the data characteristics and fits the observed data with high fidelity. The relative errors, calculated based on the magnitudes of the two datasets, are 4.04% and 1.04%, respectively. Some data is shown in Table 3. The expressions for the boundary parameters B_1 and B_2 are given by Equation (25).

$$\begin{cases} \left|\dot{\beta} + B_1\beta\right| \le B_2 \\ B_1 = 0.0006v^2 - 26.135\mu^2 + 0.0007v^2\mu + 0.1198v\mu^2 \\ \quad -0.2878v\mu - 0.1367v + 42.2786\mu + 8.9941 \\ B_2 = x_0 B_1 = (0.094\mu^2 - 0.014\mu + 0.053)B_1 \end{cases} \tag{25}$$

Table 3. Fitting parameters for a road surface friction coefficient of 0.8 at different speeds.

Vehicle Speed (m/s)	B_1	B_2
15.0	12.54	1.26
17.5	11.75	1.18
20.0	10.71	1.07
22.5	9.84	0.98
25.0	9.39	0.94
27.5	8.57	0.86
30.0	8.30	0.83
32.5	7.73	0.77
35	7.37	0.74

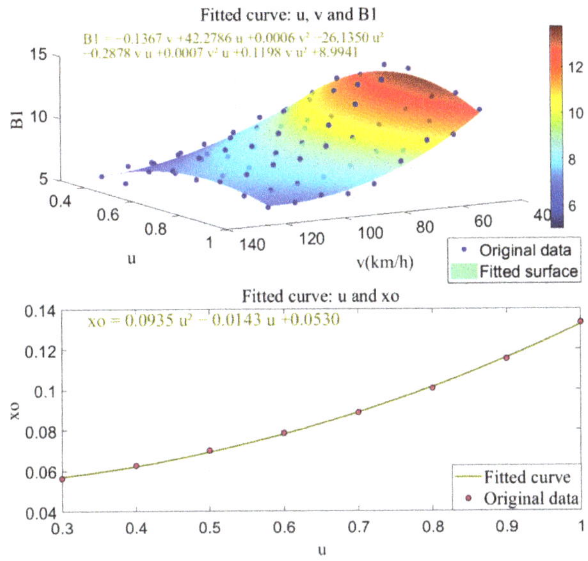

Figure 9. Boundary Parameter Fitting.

To accelerate the convergence process, the stable region is further divided into stable region and joint control domain. The region division formula is expressed by Equation (26).

$$\begin{cases} \left|\frac{B_1}{B_2}\beta + \frac{1}{B_2}\dot{\beta}\right| \le \mu & (stable\ region\ control\ domain) \\ \mu \le \left|\frac{B_1}{B_2}\beta + \frac{1}{B_2}\dot{\beta}\right| \le 1 & (joint\ control\ domain) \\ \left|\frac{B_1}{B_2}\beta + \frac{1}{B_2}\dot{\beta}\right| \ge 1 & (unstable\ region\ control\ domain) \end{cases} \quad (26)$$

Given the vehicle's state and using Equations (21)–(25), the system's phase trajectory is plotted to obtain the $\beta - \dot{\beta}$ phase plane diagram. Figure 10 displays the phase trajectory when the steering angle is 0 degrees, with a speed of 72 km/h and a road friction value of 0.8.

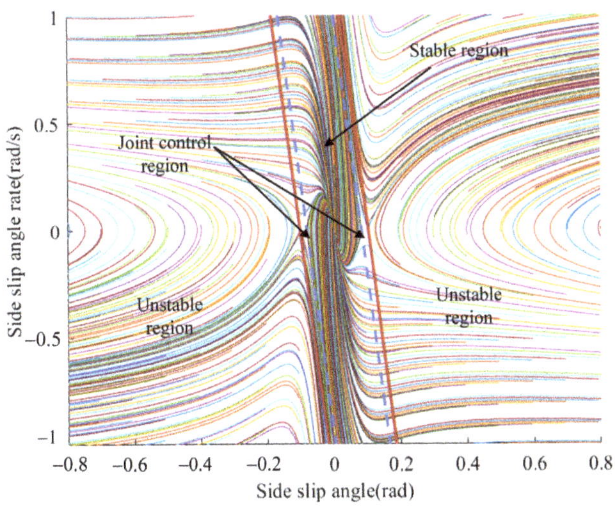

Figure 10. Division of phase plane control domains.

In the $\beta - \dot{\beta}$ phase plane diagram, the separation from the state point to the stability boundary indicates the vehicle's instability level. Inside the stable region, the instability degree is zero, as shown in Equation (27).

$$e(d) = \begin{cases} \frac{\left|B_1\beta+\dot{\beta}-B_2\right|}{\sqrt{1+B_1{}^2}}, (B_1\beta + \dot{\beta} > B_2) \\ 0, (\left|B_1\beta + \dot{\beta}\right| \leq B_2) \\ \frac{\left|B_1\beta+\dot{\beta}+B_2\right|}{\sqrt{1+B_1{}^2}}, (B_1\beta + \dot{\beta} < -B_2) \end{cases} \tag{27}$$

4.2. Sliding Mode Controller Design

The primary strength of the Sliding Mode Controller (SMC) lies in its robustness to vehicle nonlinearities and external disturbances [34]. The control variable is defined as the weighted sum of the deviations of the actual side slip angle and yaw rate from their respective reference values, as shown in Equation (28).

$$e = Pe_\beta + (1 - P)e_\gamma \tag{28}$$

To enhance the precision of the approximation process, an integral element is added to the sliding surface.

$$S = e + c\int_0^t edt = Pe_\beta + (1 - P)e_\gamma + c\int_0^t Pe_\beta + (1 - P)e_\gamma dt \tag{29}$$

The sliding mode approximation adopts an exponential convergence rate.

$$\dot{S} = -\varepsilon \mathrm{sgn}(S) - KS = P\dot{e}_\beta + (1 - P)\dot{e}_\gamma + cPe_\beta + c(1 - P)e_\gamma \tag{30}$$

In the sliding mode controller, the parameter c is a key factor in regulating the system's response characteristics, primarily influencing both the convergence rate and the amplitude of chattering. While a larger value of c can accelerate convergence to the steady state, it may also induce high-frequency oscillations, which can degrade control accuracy and potentially affect the robustness of the controller. To balance dynamic performance and system stability, the parameter c was gradually increased while comprehensively considering the impact of system oscillations. After multiple rounds of simulation and experimental comparisons, the parameter c was ultimately determined to be set at 0.5. By combining Equation (22) with Equations (28)–(30), the yaw moment M_1 is formulated as follows:

$$M_1 = -I_z\left[\frac{l_f k_1 - l_r k_2}{I_z}\beta + \frac{l_f{}^2 k_1 + l_r{}^2 k_2}{I_z v_x}\gamma - \frac{l_f k_1}{I_z}\delta_f + \frac{l_r k_2}{I_z}\delta_r - \dot{\gamma}_d + \frac{P}{1-P}(\dot{\beta} - \dot{\beta}_d) + c(\gamma - \gamma_d) + \frac{cP}{1-P}(\beta - \beta_d) + \frac{\varepsilon sign(S) + kS}{1-P}\right] \tag{31}$$

In the above equation, ε represents the thickness of the sliding mode controller boundary layer, while K denotes the convergence rate exponent. Both ε and K are positive numbers. P is defined as:

$$P = \begin{cases} 1 - \frac{2h}{H} & 0 < \mathrm{h} < 0.4\mathrm{H} \\ 0.2 & 0.4 < \mathrm{h} < 0.5\mathrm{H} \end{cases} \tag{32}$$

where h denotes the separation between the phase plane state point and the nearest stability limit, and H represents the spacing between stable boundaries. The continuous function is employed to replace the sign function in order to suppress system chattering, as shown in Equation (33).

$$sat(S) = \begin{cases} 1 & S > \Delta \\ \lambda S & |S| \leq \Delta,\ \Delta = \frac{1}{\lambda} \\ -1 & S < -\Delta \end{cases} \tag{33}$$

Based on Lyapunov stability theory, the system is asymptotically stable if the positive-definite scalar function $V(x) = S^2/2$ has a negative-definite time derivative. The Lyapunov function is defined as:

$$\dot{V} = \dot{S} \times S = -\varepsilon \text{sgn}(S) \times S - KS^2 = -\varepsilon|S| - KS^2 \tag{34}$$

When $S > 0$ or $S < 0$, $\dot{V} = -\varepsilon S - KS^2 < 0$; When $S = 0$, $\dot{V} = 0$. Therefore, $\dot{V} \leq 0$ always holds, which proves that the system is asymptotically stable.

Once the vehicle enters the instability region, the control objective at this point is to minimize the instability degree $e(d)$. Using $e(d)$ as the input to the controller, the PID controller is utilized to calculate the needed yaw moment within the instability region, with the output being the yaw moment M_2.

$$M_2 = k_p e(d) + k_i \int e(d)dt + k_d \frac{de(d)}{dt} \tag{35}$$

PID parameter tuning was conducted under high-instability and low-adhesion driving conditions. Specifically, a road friction coefficient of μ = 0.3 and a vehicle speed of 60 km/h were selected as the test conditions. The proportional gain was gradually increased to achieve a rapid response to the onset of instability, until the system approached the instability boundary. This ensured quick system response while maintaining an adequate gain margin. The integral term was introduced to eliminate residual yaw rate errors, and the derivative term was tuned to suppress oscillations near the stability boundary. Based on multiple simulation iterations, the final controller parameters were set to k_p = 8000, k_i = 90, and k_d = 50. Finally, with the set values fluctuating by ±25%, the controller maintained stable performance., indicating robustness to parameter variations. Incorporating the phase plane region division, a weighting coefficient $\lambda = \frac{x-1}{\mu-1}$ is introduced in the joint control domain, where μ represents the road adhesion coefficient and $x = \left|\frac{1}{c}\dot{\beta} + \frac{k}{c}\beta\right|$. The total additional yaw moment is calculated as follows:

$$M_z = \lambda M_1 + (1 - \lambda)M_2 \tag{36}$$

According to vehicle stability theory, when the centroid slip angle is small, the vehicle's stability is primarily affected by the yaw rate. In contrast, when the slip angle becomes large, stability mainly depends on the slip angle itself. Therefore, the integrated error-weighted distribution method, combined with coordinated control regions and designed using the phase-plane approach, can adapt the torque distribution under various driving conditions.

4.3. *Optimal Distribution of Longitudinal Driving Torque and Additional Yaw Moment*

The optimal distribution is a torque allocation strategy that considers three key factors: tire utilization, motor drive torque, and road friction [35]. This strategy aims to meet the required total longitudinal driving force F_x. Simultaneously, the additional yaw moment M_z, assigned by the upper-level controller, is allocated to each wheel. Through this allocation, direct yaw moment control is achieved. The method can be expressed as shown in Equation (37).

$$\begin{cases} F_{xq1} = F_x = \frac{T}{R_w} \\ F_{xm1} = \frac{2M_z}{B} \end{cases} \tag{37}$$

where F_{xq1} denotes the control effect of the longitudinal driving force, and F_{xm1} represents the control effect of the stability controller. B denotes track width. The coupling between the longitudinal and lateral forces can be further illustrated by the friction ellipse model, as shown in Figure 11. According to the friction ellipse theory, the lateral force of a tire decreases as the longitudinal force increases. The upper limit of the lateral force is inversely proportional to the limit of the longitudinal force, thus indicating that there is a coupling between the longitudinal and lateral forces of the tire. To coordinate the longitudinal control with the stability control, w_1 and w_2 are introduced. The adjusted F_{xq1} and F_{xm1} are expressed as:

$$\begin{cases} F_{xq2} = w_1 F_{xq1} \\ F_{xm2} = w_2 F_{xm1} \end{cases} \tag{38}$$

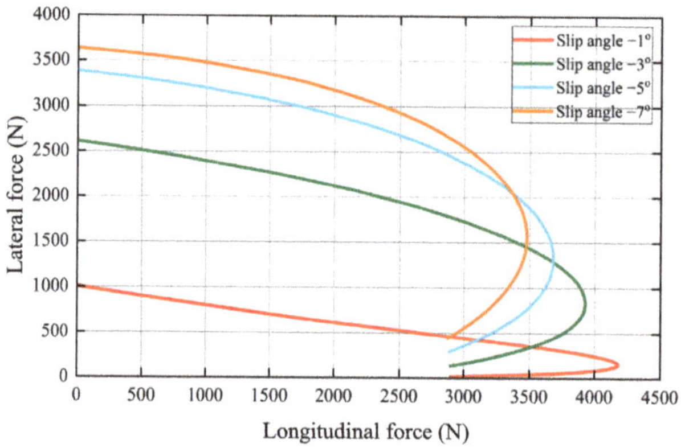

Figure 11. Friction Ellipse of Tire Forces.

Introducing the weighting coefficient η transforms the optimization problem for solving w_1 and w_2 into a single-objective optimization problem, as shown in Equation (39).

$$\begin{aligned} &\max f(w_1, w_2) = \eta w_1 + (1-\eta) w_2 \\ &st \begin{cases} 0 \le w_1, w_2 \le 1 \\ -2(A_{fl} + A_{rl}) \le F_{xq2} + F_{xm2} \le 2(A_{fl} + A_{rl}) \\ -2(A_{fr} + A_{rr}) \le F_{xq2} - F_{xm2} \le -2(A_{fr} + A_{rr}) \end{cases} \end{aligned} \tag{39}$$

where $A_i = \sqrt{(\mu F_{zi})^2 + F_{zi}{}^2}, (i = fl, fr, rl, rr)$, and F_{zi} signifies the vertical force on the tire. The larger the value of $\eta (0 < \eta < 1)$, the greater the contribution of the longitudinal controller, and conversely, the greater the contribution of the stability controller. Considering the vehicle's instability and road adhesion coefficient, parameter η is designed using the sigmoid function, as expressed by:

$$\eta = \frac{1}{1 + e^{10e(d) - 2\mu}} \tag{40}$$

The parameter η is formulated as a sigmoid-type exponential function to represent the nonlinear and gradual variation in the coupling between the longitudinal and lateral forces of the tire. When the phase-plane instability metric $e(d)$ is low, indicating near-stable tire conditions, and the road friction coefficient μ is relatively high, η approaches 1. This suggests that the longitudinal force has a limited influence on the lateral force, meaning the coupling between these forces is weak. Conversely, when $e(d)$ increases, indicating greater instability, or μ decreases, indicating smoother road surfaces, η decreases. This shows that the longitudinal force has a stronger suppressive effect on the lateral force. This behavior reflects the dynamic coupling characteristics of tire forces under varying operating conditions.

Finally, considering the restrictions of peak motor torque and road adhesion, the driving torque delivered to each tire is determined. The cost function and constraints for optimal wheel torque allocation are formulated using a quadratic programming algorithm as follows:

$$\min J = \sum_{i=fl,fr,rl,rr} \frac{\zeta_i {F_{xi}}^2}{(\mu F_{zi})^2}$$
$$\begin{cases} F_{xi} < \mu F_{zi} \\ \dfrac{T_{\min}}{r_w} \leq F_{xi} \leq \dfrac{T_{\max}}{r_w} \\ F_{fl} + F_{fr} + F_{rl} + F_{rr} = F_{xq2} \\ F_{xfr} - F_{xfl} + F_{xrr} - F_{xrl} = F_{xm2} \end{cases} \quad (41)$$

To further enhance vehicle stability and account for the vehicle's understeer characteristics, the front wheel weight is set to 1, while the rear wheel weight is derived from phase-plane analysis, as shown in Equation (42).

$$\zeta_{3,4} = \begin{cases} 1 + |\beta| \frac{B_1}{B_2} & |\beta| \leq \frac{B_2}{B_1} \\ 2 & |\beta| > \frac{B_2}{B_1} \end{cases} \quad (42)$$

5. Results and Discussion

In this part, to assess the performance of the introduced control approach, a CarSim/Simulink joint simulation platform is established. Double line change path and slalom path are set as reference trajectories. Controller A denotes the proposed control system developed in this study. For comparison, three traditional controllers, Controller B (a traditional fixed-parameter model predictive controller), Controller C (a linear quadratic regulator, LQR), and Controller D (a sliding-mode rear-wheel active steering controller based on a single-point look-ahead driver model), are also implemented. The basic vehicle parameters employed in this study are presented in Table 4. The values of the control weight parameters under various driving conditions are summarized in Table 5.

Table 4. Basic vehicle parameters.

Parameters	Units	Value
Vehicle mass	kg	1412
Distance from the front axle to the mass center	m	1.015
Distance from the rear axle to the mass center	m	1.895
Track width	m	1.675
Vehicle yaw moment of inertia	$kg \cdot m^2$	1536.7
Effective rolling radius	m	0.325
Front tire stiffness cornering	N/rad	134,035
Rear tire stiffness cornering	N/rad	77,802
Tire rotational inertia	$kg \cdot m^2$	1.5

Table 5. Controller parameters.

Vehicle Speed (km/h)	Road Friction Coefficient	q_1	q_2	R
120	0.8	1	100	$diag(43, 43, 43)$
72	0.5	45	100	$diag(45, 45, 45)$
65	0.3	1	100	$diag(41, 41, 41)$

The lane-change maneuver at the high-speed dry road conditions and low-speed slippery road conditions, as well as the sharp turn maneuver under low-speed icy and snowy road conditions, were set up to validate the control strategy introduced in this paper. These three scenarios were used to verify that the introduced control strategy can ensure high precision and stability in tracking the target trajectory under different driving conditions. Figure 12a–c show the fitness curve variations during the iterative calculation process of the Snake algorithm after assigning values to the offline-trained model under the given driving conditions. The fitness curves in Figure 12a,b converge on the 10th and 11th iterations, with fitness values of 2.40628 and 0.01125, respectively. The fitness curve in Figure 12c converges on the 7th iteration, with a fitness value of 0.22572.

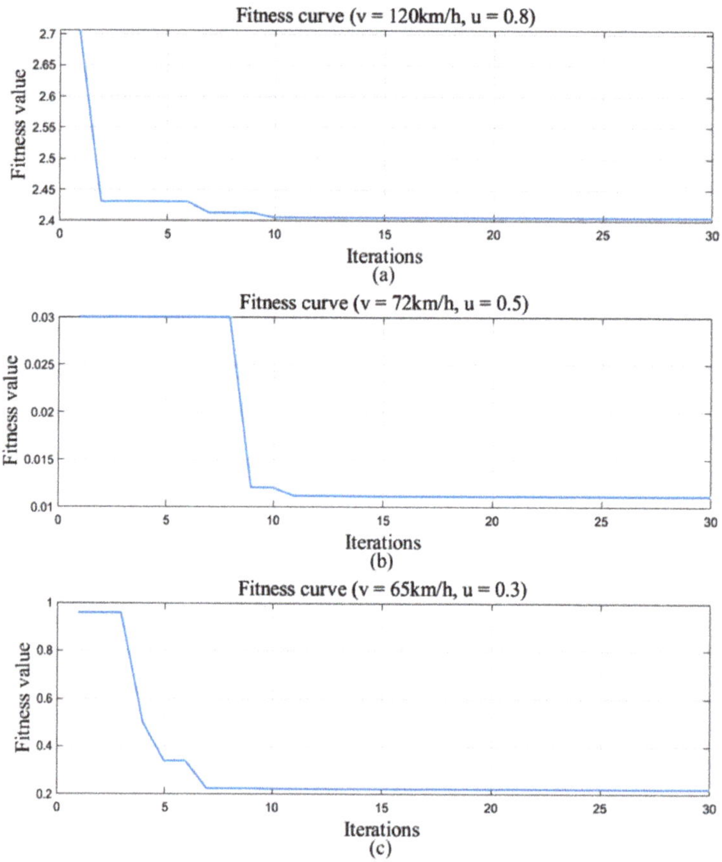

Figure 12. Snake algorithm fitness curve under different driving conditions: (**a**) 120 km/h, adhesion 0.8; (**b**) 72 km/h, adhesion 0.5; (**c**) 65 km/h, adhesion 0.3.

5.1. High-Speed Double Lane Change Maneuver

To evaluate the performance of the proposed path tracking controller under high-speed conditions, a lane-change and overtaking scenario on a dry road surface is simulated. The road friction coefficient is set to 0.8, and the vehicle maintains a constant speed of 120 km/h during the maneuver. The reference path is a typical double lane-change trajectory, with a total longitudinal length of 240 m and a lateral displacement of 3.6 m, forming a curve composed of two connected lateral shifts. The corresponding simulation results are shown in Figure 13, while the statistical results are presented in Table 6.

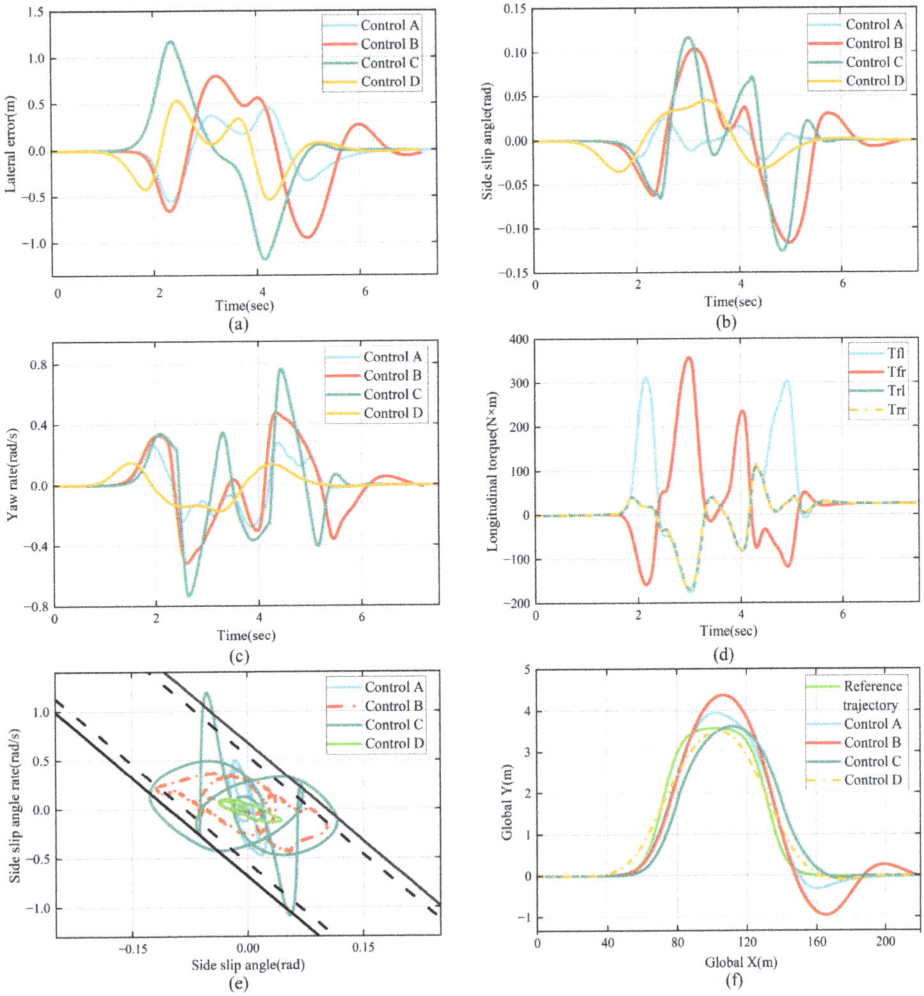

Figure 13. Simulation results for high-speed double lane change maneuver: (**a**) Lateral Error; (**b**) Side Slip Angle; (**c**) Yaw Rate; (**d**) Longitudinal Driving Torque of the Wheels; (**e**) Phase Plane Diagram; (**f**) Tracking Performance.

Table 6. Statistical Results of High-Speed Double Lane Change Maneuver.

	Maximum Tracking Error (m)	Mean Tracking Error (m)	Maximum CG Sideslip Angle (rad)	Mean CG Sideslip Angle (rad)	Maximum Yaw Rate (rad/s)	Mean Yaw Rate (rad/s)
Control A	0.54	0.142	0.024	0.001	0.28	0.08
Control B	0.95	0.304	0.125	0.032	0.51	0.16
Control C	1.18	0.253	0.131	0.031	0.77	0.17
Control D	0.56	0.153	0.052	0.013	0.17	0.04

From Figure 13a, the lateral errors of Controller A and Controller D appear nearly identical, with a peak lateral deviation of approximately 0.54 m, which is significantly lower than those of Controller B and Controller C. The maximum lateral error of Controller C reaches 1.18 m, showing the worst tracking performance. Compared to the fixed-weight

MPC controller, Controller A reduces the maximum lateral error by approximately 40%, demonstrating significant optimization. Figure 13b shows that although Controller A does not exhibit a significant advantage over Controller D in tracking performance, its maximum centroid side-slip angle, at around 0.024 rad, is the smallest among the four approaches. This demonstrates that the introduced method significantly enhances the vehicle's lane-changing stability under high-speed conditions.

Figure 13c shows that the peak yaw rate of Controller A reaches 0.28 rad/s, with an average yaw rate of 0.08 rad/s, slightly higher than that of Controller D, but much lower than those of Controller B and Controller C. From the phase plane plot in Figure 13e, it is apparent that the phase trajectories of Controller A and Controller D are within the stable region and more concentrated compared to the other controllers. The stability of controller C is the worst, and certain state points exceed the stability boundary. The phase plane plot visually demonstrates that the introduced control method greatly enhances the vehicle stability of the MPC controller under high-speed maneuver. Figure 13d presents the drive torque outputs of each wheel, where the optimal distribution strategy is designed to coordinate the total driving force and extra yaw moment, achieving direct yaw moment control to sustain vehicle body stability. Figure 13f shows the path tracking performance of each controller, with Controllers A and D aligning more closely with the target trajectory, while the other controllers exhibit larger lateral deviations.

5.2. Slippery Road Double Lane Change Maneuver

The low-adhesion double-lane-change maneuver serves to simulate vehicle overtaking maneuvers on a wet, slippery road in rainy conditions. The road friction coefficient is assigned a value of 0.5, and the vehicle travels at 72 km/h. Simulation results are presented in Figure 14, while the statistical results are summarized in Table 7.

From Figure 14a, it can be observed that Controllers A and B show better tracking accuracy relative to the other controllers, indicating that the MPC controller demonstrates better tracking capability under low-adhesion conditions. Among these con-trollers, Controller C demonstrates the poorest tracking performance, with a maximum lateral error of 0.69 m. Controller B's maximum lateral error is 0.08 m, while Controller A exhibits an exceptionally small peak lateral error of 0.01 m. Analysis reveals that the following accuracy of model predictive control-based controllers is largely determined by the choice of weight parameters. Proper tuning of these parameters can substantially enhance the precision of vehicle trajectory following. As a result, the introduced control approach greatly enhances the tracking accuracy of the MPC controller on low-adhesion road surfaces. As illustrated in Figure 14b, the centroid side-slip angle of Controller A is the smallest among all controllers, with a maximum value of 0.008 rad, much lower than that of the other controllers. Compared to Controller B, the centroid slip angle of Controller A is reduced by 65%.

From Figure 14c, it can be found that the yaw rate of each controller is quite similar, with a maximum yaw rate around 0.2 rad/s. Although the yaw rate of Controller D is relatively lower than that of Controller A, its lateral error is higher. Figure 14e shows the phase plane trajectories for different control strategies. It is visually evident that the vehicle state points of Controllers A and C are more concentrated compared to the other controllers. However, Controller C exhibits the largest lateral error, failing to maintain the tracking accuracy of the vehicle when operating on low-adhesion surfaces. Figure 14f illustrates the following effectiveness of each controller. Compared to the other controllers, the displacement curve of Controller A almost coincides with the target trajectory, indicating that under slippery road conditions, the proposed control approach is able to maintain high tracking accuracy while ensuring the vehicle's stability.

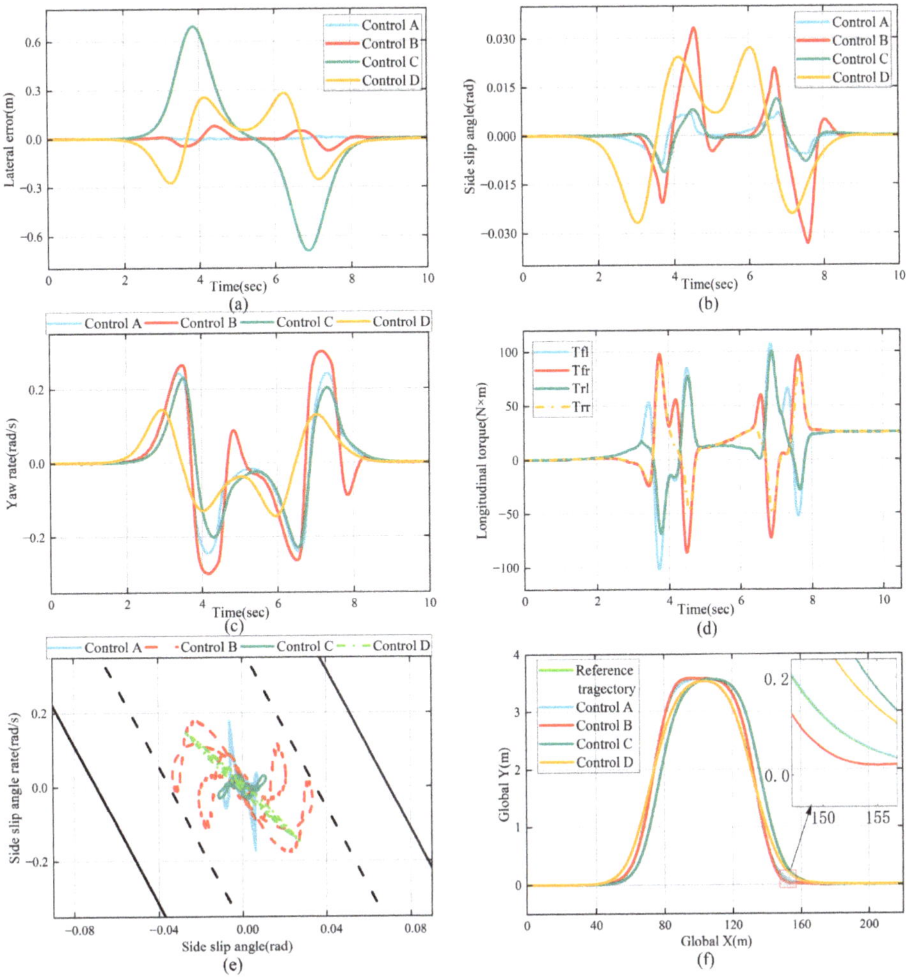

Figure 14. Simulation results for low adhesion double lane change maneuver: (**a**) Lateral Error; (**b**) Side Slip Angle; (**c**) Yaw Rate; (**d**) Longitudinal Driving Torque of the Wheels; (**e**) Phase Plane Di-agram; (**f**) Tracking Performance.

Table 7. Statistical Results of Medium-Speed Double Lane Change Maneuver.

	Maximum Tracking Error (m)	Mean Tracking Error (m)	Maximum CG Sideslip Angle (rad)	Mean CG Sideslip Angle (rad)	Maximum Yaw Rate (rad/s)	Mean Yaw Rate (rad/s)
Control A	0.01	0.002	0.008	0.001	0.25	0.05
Control B	0.08	0.012	0.032	0.003	0.30	0.05
Control C	0.69	0.144	0.014	0.002	0.23	0.05
Control D	0.28	0.053	0.032	0.012	0.16	0.03

5.3. Slalom Maneuver

To assess the performance and robustness of the proposed path tracking controller under low-adhesion conditions, a slalom road profile is used to simulate vehicle behavior on icy or snowy surfaces. The reference trajectory has a total longitudinal length of 370 m and a lateral displacement amplitude of 2 m, resulting in a smooth and continuous sequence of alternating turns. The road surface is assigned a friction coefficient of 0.3 to reflect low-μ conditions, and the vehicle travels at a constant speed of 65 km/h throughout the maneuver. This scenario is designed to evaluate the controller's ability to maintain stability and tracking accuracy under limited tire-road friction. The corresponding simulation results are presented in Figure 15, while the statistical results are summarized in Table 8.

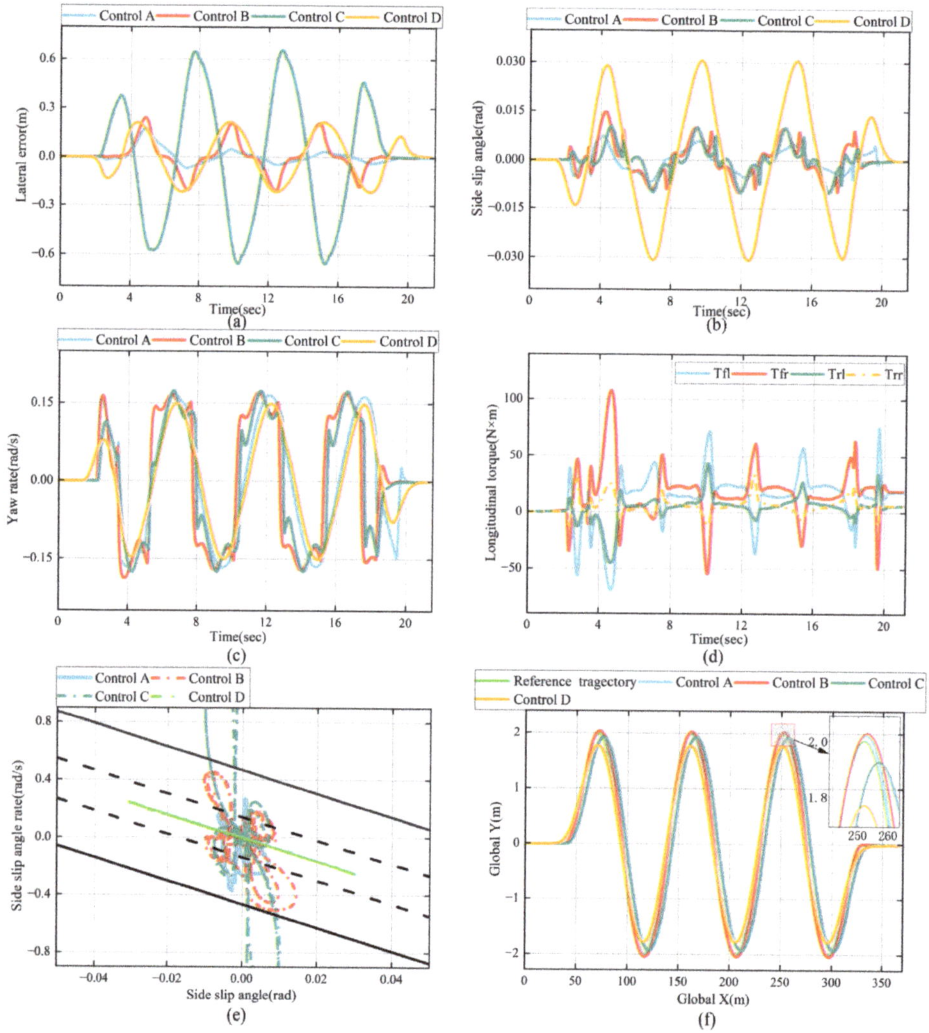

Figure 15. Simulation results for slalom maneuver: (**a**) Lateral Error; (**b**) Side Slip Angle; (**c**) Yaw Rate; (**d**) Longitudinal Driving Torque of the Wheels; (**e**) Phase Plane Diagram; (**f**) Tracking Performance.

Due to the vehicle driving on an icy and snowy road surface, the lateral force provided by the road decreases sharply. As shown in Figure 15a, at 4 s, the vehicle, based on Controller A, makes a sharp left turn. At this point, the decrease in tire lateral force causes the lateral offset to continuously increase until reaching a maximum value of 0.15 m. Afterwards, the stability controller based on optimal torque distribution stabilizes the vehicle, with the maximum lateral error gradually decreasing and stabilizing at a maximum value of 0.04 m. The average lateral error of Controller A is 0.025 m, much lower than that of the other controllers. The highest lateral deviation for Controller B and Controller D is similar, both around 0.25 m, while Controller C exhibits the largest maximum lateral error of 0.66 m. From Figure 15b, it is evident that Controller A not only exhibits better tracking accuracy than the other controllers but also has the smallest peak centroid side-slip angle, which is 0.007 rad, lower than that of the other controllers. Although Controller D achieves higher tracking accuracy than Controller C, its centroid slip angle is the largest, reflecting the coupling relationship between stability and tracking accuracy during operation.

From Figure 15c, It is observed that the yaw rate for all four controllers is nearly identical, peaking at approximately 0.15 rad/s. The phase-plane diagram in Figure 15e visually demonstrates the impact of different controllers on vehicle stability while driving on an icy road surface. Only the state points of Controller A and Controller D are located farther from the instability boundary, while the state points of the other controllers are close to or beyond the stability boundary. Additionally, the state points of Controller A are more concentrated compared to those of Controller D. From Figure 15f, it can be found that Controller A's trajectory is closer to the target trajectory than that of the other controllers. Due to the inability of MPC controllers based on fixed weights to adapt well to driving environments under different vehicle speeds and road conditions, they tend to exhibit larger lateral deviations and center of mass side slip angles than controller A. This clearly shows that the control approach introduced in this study significantly improves the vehicle's tracking capability and stability on icy and snowy roads, while also validating the adaptability of the neural network-based adaptive weight MPC controller to varying vehicle speeds and road adhesion conditions.

Table 8. Statistical Results of Slalom Maneuver.

	Maximum Tracking Error (m)	Mean Tracking Error (m)	Maximum CG Sideslip Angle (rad)	Mean CG Sideslip Angle (rad)	Maximum Yaw Rate (rad/s)	Mean Yaw Rate (rad/s)
Control A	0.15	0.025	0.007	0.001	0.17	0.08
Control B	0.24	0.062	0.012	0.002	0.19	0.11
Control C	0.66	0.291	0.011	0.002	0.17	0.09
Control D	0.22	0.113	0.034	0.011	0.15	0.07

6. Conclusions

This paper addresses the issue where distributed-driven vehicles cannot effectively adapt to variations in speed road friction coefficient. An adaptive weight MPC-based integrated control strategy, utilizing a neural network, is proposed. The strategy coordinates path tracking accuracy and stability control by employing phase plane analysis and an optimal torque distribution algorithm. A CarSim/Simulink co-simulation platform is established, with double lane change and slalom paths used as reference trajectories to assess the control strategy's performance under different vehicle speeds and road friction coefficient. Simulation results demonstrate that the proposed control strategy significantly improves the adaptability of the fixed-weight MPC controller to various driving environments while achieving high tracking accuracy and stability. By utilizing an offline-trained neural network model and the Snake Optimization algorithm to iteratively adjust the controller weights, the adaptive adjustment strategy reduces computational cost and enhances the real-time performance of the controller. However, this paper does not take into account the influence of prediction horizon parameters, road curvature, and slope variation on vehicle tracking accuracy and stability. Although the proposed control strategy exhibited some optimization in the high-speed double lane change test, the observed improvements were relatively marginal. Upon analysis, this may be attributed to the choice of prediction horizon. A longer prediction horizon can improve control accuracy, especially in complex road environments, but it also leads to increased computational burden. Therefore, in practical applications, a trade-off between control accuracy and computational cost is essential to ensure the real-time performance and stability of the control system. The current control strategy is primarily composed of a low-complexity controller model and an online parameter optimization algorithm, which theoretically offers low computational complexity and is thus potentially suitable for deployment in resource-constrained onboard systems. However, due to current experimental limitations, the runtime performance of critical modules, such as QP solving, neural network inference, and stability control, has not yet been evaluated on embedded platforms. To comprehensively validate the feasibility of the proposed method for practical deployment, future research will introduce hardware-in-the-loop (HIL) testing and conduct systematic runtime evaluations of the controller modules on representative embedded platforms. In addition, this study will further explore the impact of different prediction horizon settings on system real-time performance by balancing control accuracy and computational resource consumption. The effects of varying road curvature and slope on path tracking performance will also be systematically evaluated. The ultimate goal is to develop an MPC path tracking control strategy that is better suited to real-world driving conditions, thereby further enhancing the vehicle's tracking performance and overall stability.

Author Contributions: Conceptualization, Y.S.; Methodology, Y.S.; Software, S.L., W.D. and Z.Z.; Validation, X.L.; Investigation, X.L., J.Z. and B.T.; Resources, J.Z. and B.T.; Data Curation, X.L., S.L., W.D. and Z.Z.; Writing—Original Draft, X.L.; Writing—Review and Editing, Y.S. and X.L.; Visualization, X.L. and S.L.; Supervision, Y.S. All authors have read and agreed to the published version of the manuscript.

Funding: This research is supported by the National Natural Science Foundation of China (Grant No. 52465059) and the Natural Science Foundation of Hebei Province (Grant No. F 2021402011).

Data Availability Statement: All data, simulation models, parameter configurations, and scripts used in this study are publicly available at the following GitHub repository: https://github.com/Eric-svg617/mpc. This repository ensures transparency and supports reproducibility and further research extensions.

Conflicts of Interest: The authors confirm that there are no conflicts of interest associated with this article.

References

1. Xia, X.; Xiong, L.; Huang, Y.; Lu, Y.; Gao, L.; Xu, N.; Yu, Z. Estimation on IMU yaw misalignment by fusing information of automotive onboard sensors. *Mech. Syst. Signal Process.* **2022**, *162*, 107993. [CrossRef]
2. Hu, C.; Gao, H.; Guo, J.; Taghavifar, H.; Qin, Y.; Na, J.; Wei, C. RISE-based integrated motion control of autonomous ground vehicles with asymptotic prescribed performance. *IEEE Trans. Syst. Man Cybern. Syst.* **2019**, *51*, 5336–5348.
3. Zuo, Z.; Yang, X.; Li, Z.; Wang, Y.; Han, Q.; Wang, L.; Luo, X. MPC-based cooperative control strategy of path planning and trajectory tracking for intelligent vehicles. *IEEE Trans. Intell. Veh.* **2020**, *6*, 513–522. [CrossRef]
4. Zhang, J.; Shi, Z.; Yang, X.; Zhao, J. Robust adaptive control for continuous wheel slip rate tracking of vehicle with state observer. *Meas. Control* **2020**, *53*, 1331–1341.
5. Wang, W.-J.; Hsu, T.-M.; Wu, T.-S. The improved pure pursuit algorithm for autonomous driving advanced system. In Proceedings of the 2017 IEEE 10th International Workshop on Computational Intelligence and Applications (IWCIA), Hiroshima, Japan, 11–12 November 2017; pp. 33–38.
6. Norouzi, A.; Masoumi, M.; Barari, A.; Farrokhpour Sani, S. Lateral control of an autonomous vehicle using integrated backstepping and sliding mode controller. *Proc. Inst. Mech. Eng. Part K J. Multi-Body Dyn.* **2019**, *233*, 141–151. [CrossRef]
7. Hwang, C.-L.; Yang, C.-C.; Hung, J.Y. Path tracking of an autonomous ground vehicle with different payloads by hierarchical improved fuzzy dynamic sliding-mode control. *IEEE Trans. Fuzzy Syst.* **2017**, *26*, 899–914. [CrossRef]

8. Ye, Q.; Huang, R.; Zhang, Y.; Wang, R.; Chadli, M.; Chen, L. *Trajectory Tracking for Autonomous Multi-Axle Vehicles Based on Linear Quadratic Regulator*; SAE Technical Paper 2024-01-5023; SAE International: Warrendale, PA, USA, 2024.
9. Sun, Z.; Zou, J.; He, D.; Zhu, W. Path-tracking control for autonomous vehicles using double-hidden-layer output feedback neural network fast nonsingular terminal sliding mode. *Neural Comput. Appl.* **2022**, *34*, 5135–5150.
10. Pang, H.; Liu, N.; Hu, C.; Xu, Z. A practical trajectory tracking control of autonomous vehicles using linear time-varying MPC method. *Proc. Inst. Mech. Eng. Part D J. Automob. Eng.* **2022**, *236*, 709–723.
11. Lin, F.; Wang, S.; Zhao, Y.; Cai, Y. Research on autonomous vehicle path tracking control considering roll stability. *Proc. Inst. Mech. Eng. Part D J. Automob. Eng.* **2021**, *235*, 199–210.
12. Yang, L.; Yue, M.; Wang, J.; Hou, W. RMPC-based directional stability control for electric vehicles subject to tire blowout on curved expressway. *J. Dyn. Syst. Meas. Control* **2019**, *141*, 041009.
13. Cheng, S.; Li, L.; Chen, X.; Wu, J. Model-predictive-control-based path tracking controller of autonomous vehicle considering parametric uncertainties and velocity-varying. *IEEE Trans. Ind. Electron.* **2020**, *68*, 8698–8707. [CrossRef]
14. Jing, C.; Shu, H.; Shu, R.; Song, Y. Integrated control of electric vehicles based on active front steering and model predictive control. *Control Eng. Pract.* **2022**, *121*, 105066. [CrossRef]
15. Wu, D.; Guan, Y.; Xia, X.; Du, C.; Yan, F.; Li, Y.; Hua, M.; Liu, W. Coordinated control of path tracking and yaw stability for distributed drive electric vehicle based on AMPC and DYC. *Proc. Inst. Mech. Eng. Part D J. Automob. Eng.* **2024**, *239*, 1745–1762. [CrossRef]
16. Yang, C.; Liu, J. Trajectory tracking control of intelligent driving vehicles based on MPC and Fuzzy PID. *Math. Probl. Eng.* **2023**, *2023*, 2464254. [CrossRef]
17. Wang, J.; Teng, F.; Li, J.; Zang, L.; Fan, T.; Zhang, J.; Wang, X. Intelligent vehicle lane change trajectory control algorithm based on weight coefficient adaptive adjustment. *Adv. Mech. Eng.* **2021**, *13*, 16878140211003393. [CrossRef]
18. Yang, Z.; Li, S.; Wang, Z. Trajectory tracking control of distributed driving intelligent vehicles based on adaptive variable parameter MPC. *J. Mech. Eng.* **2024**, *60*, 363–377.
19. Xue, W.; Zheng, L. Active collision avoidance system design based on model predictive control with varying sampling time. *Automot. Innov.* **2020**, *3*, 62–72. [CrossRef]
20. Wang, H.; Wang, Q.; Chen, W.; Zhao, L.; Tan, D. Path tracking based on model predictive control with variable predictive horizon. *Trans. Inst. Meas.* **2021**, *43*, 2676–2688. [CrossRef]
21. Liang, Y.; Li, Y.; Yu, Y.; Zheng, L. Integrated lateral control for 4WID/4WIS vehicle in high-speed condition considering the magnitude of steering. *Veh. Syst. Dyn.* **2020**, *58*, 1711–1735. [CrossRef]
22. Li, Y.; Cai, Y.; Sun, X.; Wang, H.; Jia, Y.; He, Y.; Chen, L.; Chao, Y. Trajectory tracking of four-wheel driving and steering autonomous vehicle under extreme obstacle avoidance condition. *Veh. Syst. Dyn.* **2024**, *62*, 601–622. [CrossRef]
23. Liang, Y.; Li, Y.; Khajepour, A.; Zheng, L. Holistic adaptive multi-model predictive control for the path following of 4WID autonomous vehicles. *IEEE Trans. Veh. Technol.* **2020**, *70*, 69–81. [CrossRef]
24. Ren, Y.; Zheng, L.; Khajepour, A. Integrated model predictive and torque vectoring control for path tracking of 4-wheel-driven autonomous vehicles. *IET Intell. Transp. Syst.* **2019**, *13*, 98–107. [CrossRef]
25. Cui, Q.; Ding, R.; Wei, C.; Zhou, B. Path-tracking and lateral stabilisation for autonomous vehicles by using the steering angle envelope. *Veh. Syst. Dyn.* **2021**, *59*, 1672–1696.
26. Liu, J.; Dai, A. Distributed-Drive Vehicle Lateral-Stability Coordinated Control Based on Phase-Plane Stability Region. *World Electr. Veh. J.* **2024**, *15*, 202. [CrossRef]
27. Zhu, S.; Xu, Y.; Li, L.; Ren, Y.; Kuang, C.; Chen, H.; Gao, J. Adaptive Fuzzy Control of an Electronic Differential Based on the Stability Criterion of the Phase Plane Method. *World Electr. Veh. J.* **2024**, *15*, 243. [CrossRef]
28. Wang, G.; Song, Q. The Control of Handling Stability for Four-Wheel Steering Distributed Drive Electric Vehicles Based on a Phase Plane Analysis. *Machines* **2024**, *12*, 478. [CrossRef]
29. Zhang, L.; Chen, H.; Huang, Y.; Guo, H.; Sun, H.; Ding, H.; Wang, N. Model predictive control for integrated longitudinal and lateral stability of electric vehicles with in-wheel motors. *IET Control Theory Appl.* **2020**, *14*, 2741–2751.
30. Wang, Q.; Zhao, Y.; Lin, F.; Zhang, C.; Deng, H. Integrated control for distributed in-wheel motor drive electric vehicle based on states estimation and nonlinear MPC. *Proc. Inst. Mech. Eng. Part D J. Automob. Eng.* **2022**, *236*, 893–906. [CrossRef]
31. Heidfeld, H.; Schünemann, M.; Kasper, R. UKF-based State and tire slip estimation for a 4WD electric vehicle. *Veh. Syst. Dyn.* **2020**, *58*, 1479–1496.
32. Li, B.; Du, H.; Li, W.; Zhang, B. Non-linear tyre model–based non-singular terminal sliding mode observer for vehicle velocity and side-slip angle estimation. *Proc. Inst. Mech. Eng. Part D J. Automob. Eng.* **2019**, *233*, 38–54. [CrossRef]
33. Tang, H.; Bei, S.; Li, B.; Sun, X.; Huang, C.; Tian, J.; Hu, H. Mechanism analysis and control of lateral instability of 4WID vehicle based on phase plane analysis considering front wheel angle. *Actuators* **2023**, *12*, 121. [CrossRef]

34. Asiabar, A.N.; Kazemi, R. A direct yaw moment controller for a four in-wheel motor drive electric vehicle using adaptive sliding mode control. *Proc. Inst. Mech. Eng. Part K J. Multi-Body Dyn.* **2019**, *233*, 549–567. [CrossRef]
35. Wang, H.; Han, J.; Zhang, H. Lateral stability analysis of 4WID electric vehicle based on sliding mode control and optimal distribution torque strategy. *Actuators* **2022**, *11*, 244. [CrossRef]

MDPI AG
Grosspeteranlage 5
4052 Basel
Switzerland
Tel.: +41 61 683 77 34

Applied Sciences Editorial Office
E-mail: applsci@mdpi.com
www.mdpi.com/journal/applsci

www.ingramcontent.com/pod-product-compliance
Lightning Source LLC
Chambersburg PA
CBHW061507210926
51850CB00008B/1842
9783725875573